BORON IN SOILS AND PLANTS: REVIEWS

Developments in Plant and Soil Sciences

VOLUME 77

The titles published in this series are listed at the end of this volume.

Boron in Soils and Plants: Reviews

Invited review papers for Boron [97], the International Symposium on 'Boron in Soils and Plants', held at Chiang Mai, Thailand, 7–11 September 1997

Edited by

B. DELL, P. H. BROWN AND R. W. BELL

Reprinted from *Plant and Soil*, Volume 193, Nos. 1–2 (1997)

SPRINGER SCIENCE+BUSINESS MEDIA, B.V.

A C.I.P. Catalogue record for this book is available from
the Library of Congress.

ISBN 978-94-010-6352-4 ISBN 978-94-011-5580-9 (eBook)
DOI 10.1007/978-94-011-5580-9

Printed on acid-free paper

Contents

Foreword vii–viii

Preface ix-x

1. The chemistry of boron and its speciation in plants
   P.P. Power and W.G. Woods 1
2. Techniques for boron determination and their application to the analysis of plant and soil samples
   R.N. Sah and P.H. Brown 15
3. Reactions of boron with soils
   S. Goldberg 35
4. Absorption of boron by plant roots
   H. Hu and P.H. Brown 49
5. Boron in plant cell walls
   T. Matoh 59
6. Boron deficiency-induced impairments of cellular functions in plants
   I. Cakmak and V. Römheld 71
7. Boron mobility in plants
   P.H. Brown and B.J. Shelp 85
8. Physiological response of plants to low boron
   B. Dell and L. Huang 103
9. The occurrence and correction of boron deficiency
   V.M. Shorrocks 121
10. Diagnosis and prediction of boron deficiency for plant production
    R.W. Bell 149
11. Genotypic variation in plant response to low boron and implications for plant breeding
    B. Rerkasem and S. Jamjod 169
12. Boron toxicity
    R.O. Nable, G.S. Bañuelos and J.G. Paull 181
13. Boron in human and animal nutrition
    F.H. Nielsen 199

Author index 209

Subject index 211–215

Common name citation index 217–219

Foreword

This symposium on 'Boron in Soils and Plants' completes a quartet of reviews of the behaviour in soils and plants of four trace elements, copper, manganese, zinc and boron, selected for their importance in agriculture.

The series had its genesis in a suggestion from Professor Robin Graham of the Waite Agricultural Research Institute that it would be appropriate in 1981 to celebrate the Golden Jubilee of the publication in 1931 of the first definitive evidence for the essentiality of copper in plants. The previous decade had seen a resurgence of interest in copper deficiency and toxicity in agriculture and an expansion of our understanding of the behaviour of copper in soils and plants. We therefore decided to promote a review of our understanding of the behaviour of copper in soils and plants by inviting appropriate authors to prepare reviews of 14 topics for publication in a book and presentation at a Golden Jubilee International Symposium on 'Copper in Soils and Plants'. Posters of current research were also displayed and published. Murdoch University, Perth, Western Australia was chosen as the venue because of its then current research on copper, the importance of copper in Western Australian agriculture, and the presence in Perth of many international nutritionists due to the fortuitous scheduling in 1981 of the 'Fourth International Symposium on Trace Element Metabolism in Man and Animals'.

The success of the copper symposium led us to consider the value of similar reviews for manganese and zinc, two trace elements of outstanding importance to agriculture, particularly in Australia. As a result, symposia reviewing the behaviour of each of these nutrients and resulting in the publication of two books were organised along similar lines. The International Symposium on 'Manganese in Soils and Plants' was held at the Waite Agricultural Institute of the University of Adelaide in 1988, just 60 years after the publication from the same Institute of the seminal research of Samuels and Piper which established, for the first time, a link between a field disease ('roadside take-all' or 'grey speck' in oats) and a trace element deficiency (manganese). In 1993, the International Symposium on 'Zinc in Soils and Plants' was held at the University of Western Australia, where the zinc nutrition of plants and animals has been actively researched for many years. This meeting followed the XII International Plant Nutrition Colloquium at the same locale.

Originally, boron had not been considered as a subject for review in this series, partly because it was not as important in Australian agriculture as copper, manganese and zinc, and partly because few new research findings were being published. However, by 1993 research activity on boron had re-awakened, and Australian researchers had become involved in studies of boron toxicity in Australian crops and boron deficiency in Asian countries where it is a major problem. With this new information, it seemed timely for an International Symposium to review 'Boron in Soils and Plants' along similar lines to the previous three in the series. The Multiple Cropping Centre, University of Chiang Mai, Thailand was chosen as a venue because of the large volume of outstanding work on boron deficiency being done there and the importance of boron deficiency in the region. The Symposium is a satellite conference to and immediately precedes the XIII International Plant Nutrition Colloquium in Tokyo.

While Australian scientists have been primarily responsible for the organisation of the first three Symposia and, together with Thai scientists, for the fourth, the success of all the Symposia has depended heavily on the reviews contributed by international authors. We are pleased to record our heartfelt thanks for their generous and enthusiastic support.

April 1997 *Emeritus Professor Jack F Loneragan*

Plant and Soil **193:** ix–x, 1997.

Preface

This symposium on 'Boron in Soils and Plants' is fourth in a series of symposia focussing on essential micronutrients. This volume is particularly timely as it coincides with a period of significant progress in B research. Of the many recent advances reported in this volume, perhaps the most significant are the identification of the physical and chemical role of B in the cell wall, the characterisation of the genetic basis for differences in B accumulation and tolerance to B excesses and deficiencies, and the identification and characterisation of the mechanisms of phloem B transport. Each of these advances has contributed not only to our understanding of the fundamental behaviour of B in plants, but also has a direct impact on B management in agricultural systems.

A companion volume of contributed papers to 'Boron in Soils and Plants' reports significant new findings on B functions in cell walls, B uptake and translocation. Contributed papers also indicate significant current interest in documenting the incidence of B deficiency in Asia, and developing procedures for the prediction of B deficiency and for correcting it in horticultural and field crops.

Throughout the history of B research, it has been observed that B deficiency markedly affects cell wall structure. It was not until very recently, however, that physical properties of cell walls under B deficiency have been directly measured. The identification of physical changes in cell wall properties, in combination with the identification and characterisation of B containing cell wall complexes, clearly suggests that B plays a structural role in plant cell walls. The precise function of these cell wall B complexes is unknown, but evidence suggests that B is critical to the formation and physical characteristics of plant cell walls. This most likely represents a primary function of B in plants.

The identification of a primary role for B in the growing cell wall is at once exciting and problematic. Clearly, if B plays a critical role in cell wall formation, then a deficiency of B would impact on meristematic activity and cell growth. The changes in carbon allocation and cellular metabolism that would probably occur as a consequence of this growth inhibition, would have significant secondary effects on cellular metabolism and hence would make determination of any additional roles of B very difficult.

The uptake of B from soils and the movement of B within plants has a profound influence upon the occurrence and correction of B deficiency and toxicity. Defining the mechanism of B uptake has been a matter of controversy for over 30 years. Evidence reviewed in this volume suggests that B uptake is a non-metabolic process, with the rate of uptake determined by external solution B concentration, internal B complexation and plant water use. In apparent contradiction with these findings, it is frequently observed that diverse species, growing under identical environmental conditions in which external B concentration and water use are similar, can vary dramatically in B uptake and accumulation. This cannot be explained by current theories of B uptake. Whereas the uptake of B from soils has been the subject of many studies and considerable controversy, the transport of B within the plant has received much less attention. Boron has long been considered to be phloem immobile, and hence research on this topic has been very limited. Recent evidence demonstrates, however, that this is not generally true for all plants, and that B is in fact readily transported in the phloem of species that are rich in polyols. Boron is thus unique among the essential elements, in that its phloem mobility varies dramatically among species.

The advances described above and in the following pages are both fundamentally significant and of considerable practical relevance. In many areas, however, our knowledge of the function of B is clearly inadequate. Additional efforts to determine the role of B in flowering and seed set, to fully define the functions of B in plants, and to establish the mechanism of differential B uptake between species and cultivars, are of primary importance. New analytical tools that allow for the localisation of B within cells, the identification of B containing compounds, and the monitoring of B movement within plants and soils, will greatly aid this task. Genetic variability in sensitivity to B deficiencies and excesses provide an approach to B research that has not yet been fully utilised, while new insights into B transport and complexation might provide additional insight into the fundamental role of B in plants.

The next decade of B research offers great promises. This volume provides a foundation on which to build.

We are grateful for the financial support provided by the XIII International Plant Nutrition Colloquium, Borax International, Thai Soil and Fertiliser Society, The Australian Agency for International Development, and the Australian Centre for International Agricultural Research.

We acknowledge the help we received from the following scientists who acted as referees: Associate Professor FPC Blamey, Dr R J Brotherton, Associate Professor I Cakmak, Professor Dr H E Goldbach, Professor R D Graham, Dr U C Gupta, Associate Professor E J Hanson, Dr Z Rengel, Dr V M Shorrocks and Professor R M Welch; and the late Professor H Marschner, Emeritus Professor J F Loneragan, Associate Professor B Rerkasem and Professor A D Robson for their guidance through the planning stage for the book.

Finally, we thank Kay Howard who provided assistance in the final stages of editing.

Plant and Soil **193:** 1–13, 1997.

Chapter 1

The chemistry of boron and its speciation in plants

Philip P. Power[1] and William G. Woods[2]
[1]*Chemistry Department, University of California, Davis CA 95616, USA and* [2]*Office of Environmental Health and Safety, University of California, Riverside, CA 92521, USA**

Abstract

The chemistry and usage of B, as well as its speciation in plants, are reviewed. In the context of biology, the chemistry of the surprisingly rare element B is dominated by B-oxygen compounds. The occurrence, distribution and chemistry of these compounds are briefly described. In addition, the chemistry involved in the interaction of B–O compounds with naturally occurring molecules, particularly polyhydroxy compounds, is summarised. The essentiality of B in plants has been known for 75 years, but the biomolecule(s) with which it interacts to cause its essential function(s) remain largely unknown, although many have been suggested. Recent research on the complexation, isolation and partial characterisation of B complexes of polysaccharides, diols, and hydroxyacids is outlined. The particular importance of B interactions with cell wall components, membranes, enzymes, sugars, and polyols is discussed.

Chemistry

With the possible exception of carbon, B has the most interesting and diverse chemistry of any element (Greenwood and Earnshaw, 1984). Much of the interest in B chemistry arises from the so-called electron deficiency of many of its compounds. This is due to the fact that B possesses only three valence electrons which, when incorporated in three bond pairs, leave a p-orbital unfilled in the valence shell. This often results in association of B compounds through three-centre, two-electron bonds. This type of bonding is particularly common in the hydride derivatives (boranes). These compounds have played a very large role in the development of modern chemistry in areas as disparate as the development of handling techniques for highly air sensitive compounds, the molecular orbital theory of chemical bonding and organic synthetic methodology. Two Nobel awards, in 1976 to W N Lipscomb and in 1979 to H C Brown (shared with G Wittig), have underlined the importance of B compounds to bonding theory and organic chemistry. In addition, Alfred Stock is a name recognisable to many chemists for his pioneering studies of B hydrides (Stock, 1933) and the development of vacuum line techniques for air sensitive compounds.

Like other areas of chemistry, there has been an almost exponential increase in publication rate of papers and patents on the chemistry and applications of B since the end of World War II. However, many aspects of B chemistry have received a further impetus owing to the possible military applications of certain B compounds. These include the possible use of B hydrides as fuels in aviation and rocketry or the use of refractory binary metal or metalloid derivatives of B which have applications in protective armour or electronics. These significant research efforts however, have not had a major impact on industrial B chemistry, the vast majority of whose production is concerned with B–O compounds (Thompson, 1980) such as boric acid, inorganic borate salts and related species.

Currently, the principal industrial uses of B compounds are in the production of fibreglass, heat-resistant borosilicate glass (e.g. Pyrex) and of detergents. They are also of major importance in the manufacture of enamels, synthetic herbicides and fertilisers; smaller quantities are used in metallurgy and in nuclear shielding. There is a growing interest in the use of thermally sensitive B derivatives as precursors for the synthesis of elemental B itself or refractory B

* E-mail: William.Woods@ucr.edu

compounds such as boron nitride (Paine and Narula, 1990).

Atomic and physical properties

Boron is a highly refractory metalloid (Emsley, 1991) which is difficult to prepare in a state of high purity owing to its very high melting point (MP $\approx$ 3400 K) and its corrosive properties as a liquid. Its covalent radius, estimated from trivalent compounds involving B–B bonds (Power, 1992), is 0.085 nm, significantly larger than that of neighbouring carbon (0.77 pm). Its electronegativity (2.04 Pauling, 2.01 Allred–Rochow) is less than that of either hydrogen (2.2) or carbon (2.55). The stable isotopes of B are of mass 10 and 11 which are present in an approximate 20:80% ratio to afford an atomic wt of 10.81. The atomic mass is given to only four significant figures, since B minerals from the two major sources of B (the USA and Turkey) have slightly different isotopic ratios. The ^{11}B isotope (nuclear spin I = −3/2) is relatively easy to detect by NMR (Emsley, 1991) since its relative sensitivity is 0.17 (H = 1). NMR provides a powerful diagnostic tool in elucidating the structures of B compounds. However, there can be problems with broadening of the resonances due to the nuclear quadrupole moment if the chemical environment is of low symmetry. NMR studies using the ^{10}B isotope (I = 3) are not normally used owing to its lower sensitivity and higher quadrupole moment. The high thermal neutron capture cross section of ^{10}B (3837 barns cf. 0.005 barns for ^{11}B) has led to its use in nuclear shielding and in neutron capture therapy (Hawthorne, 1993) for malignant tumours.

The most important oxidation state is B^{3+}. The lower oxidation state of +1, which is common in heavier elements of the B group (e.g. Tl^{+}), is of essentially no importance in industrial B chemistry or in the chemistry of naturally occurring B compounds. The small, highly polarising B^{3+}cation does not exist under chemically significant conditions. This can be rationalised (Cotton and Wilkinson, 1989) on the basis that the ionisation energies for the valence electrons in B are too great to be compensated for by lattice energies of ionic compounds it might form or by the hydration of B^{3+} ion in solution. Thus, although B(III) is present in most compounds, its bonding has considerable covalent character. The lower oxidation states +1, 0 or <0 occur only in compounds such as the higher boranes (e.g. B_5H_9), the subvalent halides (Massey, 1983) like B_4Cl_4, the metal borides such as Ti_2B (Wells, 1984) or in a few compounds containing multiple B–B bonds (Power, 1992). Boron normally has a coordination number of either 3 or 4 in naturally occurring compounds. Higher coordination numbers (i.e. 5 or 6) are observed only under laboratory conditions in the so-called electron deficient compounds such as the icosahedral $B_{12}H_{12}^{2-}$ anion.

Boron minerals and distribution

In spite of its importance B, along with its neighbours Li and Be, are not abundant elements since they are bypassed in the normal chain of thermonuclear reactions in stars (Reeves, 1974). It is probable that these elements are produced by fragmentation reactions resulting from the collisions of cosmic ray particles which produce lighter nuclei from heavy ones. However, more recent work has proposed that B may also be produced during explosions of massive stars (Crosswell, 1992). In spite of its low natural abundance, B is widely distributed (Morgan, 1980) in both the lithosphere and hydrosphere. In rocks its concentration averages about 10–20 mg B kg^{-1}. In sea water it can range from ca. 1–10 mg B kg^{-1}, while its concentration in river water is about 1/350 that of sea water. Soils may be divided into two types: those with low B content (<10 mg B kg^{-1}) or high B content (10–100 mg B kg^{-1}). Most soils have a low B content; high B content soils are those associated with recent volcanism. Usually only soluble B is available to plants and this is usually about 10% of the total soil B content. The B level in animal tissue averages ~0.3 mg B kg^{-1} on a dry wt basis (Barr et al., 1993) with a range of 0.03–0.4 mg B kg^{-1} in blood and as high as 1.2 mg B kg^{-1} in areas with a high B content in the water supply. Boron is found in erythrocytes, blood serum and plasma and at about $10^{-5} - 10^{-6}M$ (0.01-0.1 mg B kg^{-1}) and in plants at 10–100 mg kg^{-1} dry wt.

Boron is obtained commercially from concentrated mineral deposits that are associated with former geologic volcanic activity. The steam volatility of boric acid accounts for these rare deposits, which formed after the carrying water evaporated. The primary minerals that crystallise are normally ulexite $NaCa[B_5O_6(OH)_6]\cdot 5H_2O$, along with borax (tincal) $Na_2[B_4O_5(OH)_4]\cdot 8H_2O$; natural boric acid (sassolite) can form near hot springs. Leaching and weathering afford the less soluble colemanite $Ca[B_3O_4(OH)_3]\cdot 2H_2O$, while high pressure and temperature produce kernite $Na_2[B_4O_5(OH)_4]\cdot 2H_2O$, a less hydrated form of borax. A summary of the vari-

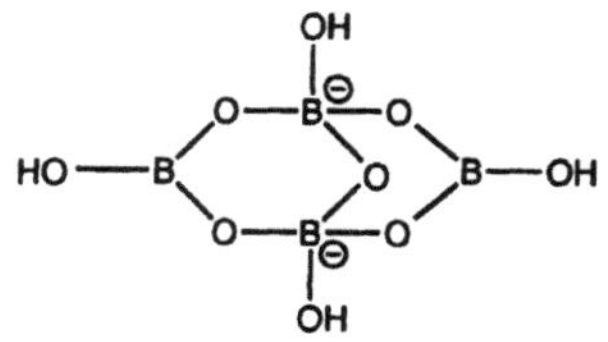

Figure 1. Schematic drawing of the bicyclic $[B_4O_5(OH)_4]^{2-}$ anion in borax.

ous B minerals (Erd, 1980) as well as their geochemistry (Morgan, 1980) has been published.

Naturally occurring B is found exclusively bound to oxygen as borates, less often as boric acid or, much more rarely, to fluorine as in the BF_4^- ion. The coordination number of the B centres in minerals is either three or four or a combination thereof. The borates can be regarded as hydrated B oxide salts most of which incorporate a boroxine B_3O_3 structural motif in which the ratio of tetrahedral BO_4 to the total B content is the same as the ratio of the charge on the anion to total number of B atoms. For example, in tincal, $Na_2[B_4O_5(OH)_4]{\cdot}8H_2O$, there are two BO_4 moieties and the remaining two Bs are trigonally coordinated BO_3 units in a bicyclic structure (Figure 1). Colemanite contains the polymeric anion $[B_3O_4(OH)_3]_n^{2n-}$. The stability of the six-atom B_3O_3 boroxine ring depends on the presence of either one or two BO_4 units.

The formation of such rings may be most easily understood in terms of the interaction of boric acid $B(OH)_3$ in aqueous solution. Essentially $B(OH)_3$ behaves as a very weak Lewis acid ($K_a = 6{\times}10^{-10}$, pK_a 9.1) according to the equilibrium in Equation (1).

$$B(OH)_3 + H_2O \leftrightharpoons B(OH)_4^- + H^+ \qquad (1)$$

At low concentrations ($\leq$0.02 *M*), only the mononuclear species $B(OH)_3$ and $B(OH)_4^-$ are present; these are the species expected in plant fluids. At higher concentrations and with increasing pH, polynuclear ions such as $B_2O(OH)_6^{2-}$ or those incorporating B_3O_3 rings such as $B_3O_3(OH)_4^-$, $B_4O_5(OH)_4^{2-}$ (the borax anion) and $B_5O_6(OH)_4^-$ are formed. Thus, the interaction between $B(OH)_3$ and $B(OH)_4^-$ may afford $[B_3O_3(OH)_5]^{2-}$ as described by Equation (2).

$$B(OH)_3 + 2B(OH)_4^- \leftrightharpoons [B_3O_3(OH)_5]^{2-} + 3H_2O \qquad (2)$$

Further increases in pH result in higher nuclearity borates, but above pH10 $B(OH)_4^-$ is produced exclusively.

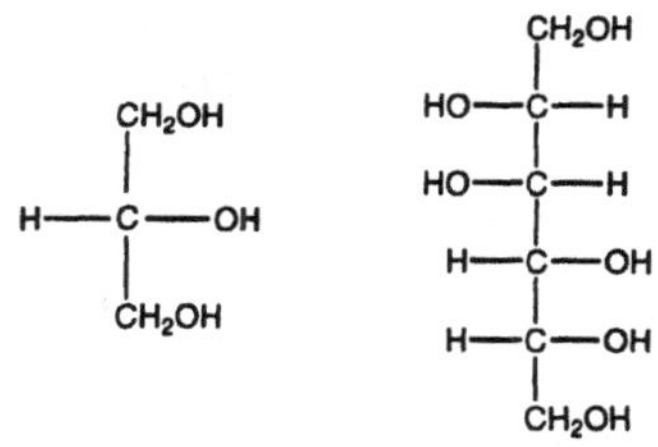

Figure 2. Structural formulae of glycerol and mannitol, used to enhance the acidity of boric acid to allow its titration.

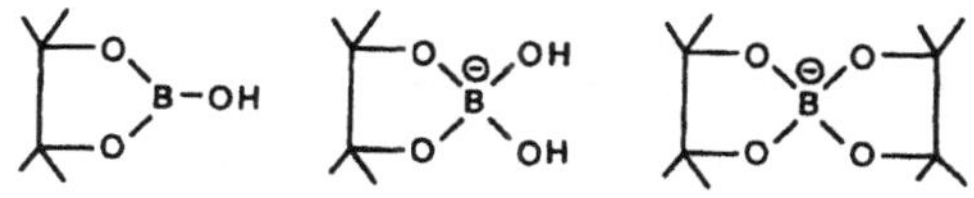

Figure 3. Schematic drawings (l. to r.) of the neutral *cis*-diol monoborate ester, the monoborate complex and the bis(diol) borate complex.

Borate esters and borate complexes

Boric acid reacts with alcohols in accordance with Equation (3).

$$B(OH)_3 + 3ROH \leftrightharpoons B(OR)_3 + 3H_2O \qquad (3)$$

The reaction is an equilibrium which lies well to the left. The reaction may be driven to completion by removal of the water component in accordance with Le Chatelier's principle. This can be done by the addition of a hydrocarbon, e.g. benzene or toluene, to form an azeotrope (either binary or ternary) which may be removed by distillation. A wide variety of boric acid esters may be synthesised in this way. However, the more bulky tertiary alcohols require different synthetic methods such as the reaction of the alcohol with $NaBH_4$ or transesterification via reaction with a B ester of a more volatile alcohol. The displaced lighter alcohol may thus be distilled away which serves to drive the equilibrium toward the desired product. Treatment of an alkali metal borate with methanol in the presence

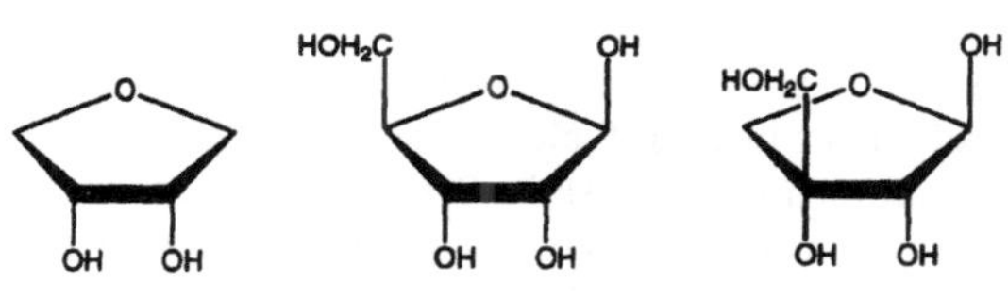

Figure 4. Schematic drawings (l. to r.) of erythritan, ribose and apiose, *cis*-dihydroxytetrahydrofurans which form particularly stable borate complexes.

Table 1. Equilibrium constants[a] for boric acid complexes with polyols (Sanderson, 1980)[b].

	k_1	k_2
1,2-ethanediol	2.15	1.15
1,3-propanediol	1.27	0.11
glycerol	16.0	41.2
catechol	7.8×10^3	1.42×10^4
D-mannitol	1.1×10^2	1.37×10^5
D-glucose	1.50×10^3	7.60×10^2
D-sorbitol		4.44×10^5
D-ribose		1.57×10^7

[a] k_1 and k_2 are as defined by Equations (5) and (6).
[b] The pK_a of erythritan (3,4-*cis*-dihydroxytetrahydrofuran, Figure 4) is 4.8 (Mazurek and Perlin, 1963).

of a strong mineral acid also affords borate esters as illustrated by Equation (4). Ignition of the volatile ester gives a green flame, used as a field test for the presence of B.

$$Na_2B_4O_7 + H_2SO_4 + 12MeOH \rightarrow 4B(OMe)_3 + Na_2SO_4 + 7H_2O \quad (4)$$

It has been known for a long time that the addition of polyols such as glycerol or mannitol (Figure 2) increases the acidity of boric acid or borate solutions by formation of complexes. This was for many years the basis for quantitative analysis of boric acid which could be titrated to a phenolphthalein endpoint, which is not possible in the absence of the polyol. The enhancement of the acidity is due to the formation of cyclic borate esters which can be illustrated schematically in Figure 3. The stability of the borate complex formed is strongly dependent on the type of diol used. Borate ester formation with 1,2 diols represented by LH_2 is shown in Equations (5) and (6).

$$B(OH)_4^- + LH_2 \overset{k_1}{\rightleftharpoons} BL(OH)_2^- + 2H_2O \quad (5)$$

$$B(OH)_4^- + 2LH_2 \overset{k_2}{\rightleftharpoons} BL_2^- + 4H_2O \quad (6)$$

Some of the k_1 and k_2 values measured for various polyols are listed in Table 1 (Sanderson, 1980). Clearly, if the diol used involves the OH groups oriented in such a way that they accurately match the structural parameters required by a tetrahedrally coordinated B, a strong complex will be formed. Krantz et al. (1936) showed that erythritan (3,4-*cis*-dihydroxytetrahydrofuran, Figure 4) produced a much lower pK_a with boric acid (more acidity, stronger complex) than did a group of common diols, polyols and sugar alcohols. Later, in 1963, Mazurek and Perlin generalised that the most stable borate esters are formed with *cis*-diols on a furanoid ring. These structures are rare in nature and are limited to apiose and ribose (Figure 4). Although certain sugars can assume a furanose configuration with *cis*-hydroxyls, these sugars are toxic due to Maillard reactions and usually do not occur in high concentrations in plants. This is due to reaction of the aldehydic form with amines, amino acids, and proteins, initially to give Schiff's bases and subsequently leading to other transformations (Hayashi and Namiki, 1986).

There have been numerous studies of the formation of borate complexes by a variety of physical, spectroscopic and chemical techniques. Essentially the esters are formed and dissociate spontaneously on a rapid time scale in a variety of pH dependent equilibria. The amount of acidification produced upon the addition of polyol is proportional to the extent of ester formation. Thus, the monitoring of the electrical conductivity (which is proportional to the acidity) of the solution provides a measure of complex formation. More recently, ^{1}H, ^{11}B and ^{13}C NMR spectroscopy have been increasingly applied to the measurement of solution equilibria in borate complex formation. The use of NMR to monitor such reactions commenced in the early 1960's with ^{1}H NMR spectral studies of the complexation of glucose (Lenz and Heeschen, 1961) and erythritan (Figure 4) by borate (Mazurek and Perlin, 1963). Later work (Gorin and Mazurek, 1973) showed that complex formation between various carbohydrates and borates could be detected by signal broadening and/or chemical shift changes in the ^{13}C NMR spectra of the parent compounds. It was possible in most cases to assign structures to the complexes formed on the basis of the spectral data. However, it should be borne in mind that the ^{13}C nucleus has only 1.1% natural abundance and its relative sensitivity compared to hydrogen is 1.59×10^{-2} (Emsley, 1991). Proton NMR spectroscopy suffers from the disadvantage that the spectrum is often complex owing to the number of hydrogen environments. However, the spectra often become simpler upon complexation to borate and have provided useful data.

More recent work has tended to use ^{11}B NMR spectroscopy to monitor complex formation and the borate structures. One drawback of the ^{11}B NMR studies is that the B must be present in sufficient concentration to allow accurate integration of the signals and calculation of equilibria. For example, in the case of bis(D-

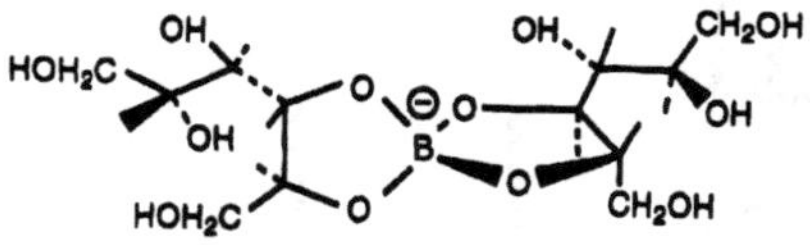

Figure 5. Structural representation of the bis(sorbitol)-2,3,2′,3′-borate complex, one of several isomers identified in phloem of sorbitol-rich plant species (Brown and Hu, 1996).

mannitol)-3,4,3′,4′-borate anion (similar to the sorbitol complex, Figure 5) there is only one B in an ion which has a molecular wt of ca. 367. Below 1 m*M*, accurate integration becomes increasingly difficult particularly when the ^{11}B NMR signal is broad. Thus, the extraction of sufficient quantities of the B complexes from the sample under examination is often the key to obtaining useful spectral data. At near neutral physiological conditions (pH ~7.4), polyborate species can form at concentrations of borate > ca. 25 m*M* which complicates the spectra. In such cases, solution spectroscopy needs to be conducted at relatively low concentrations to avoid this complication. The peaks observed in the ^{11}B spectra of the borates under discussion here fall in the relatively narrow chemical shift range of 0–20 ppm, which is well within the known range (Kennedy, 1987) for a variety of B compounds.

In typical experiments (e.g. Makkee et al., 1985) designed to study complexation, a concentration of at least ~0.1 *M* boric acid has been used and varying concentrations of the complexing polyol have been added while varying the pH of the solution. For instance, for 0.1 *M* boric acid and 0.01 *M* D-mannitol, the monoborate complex is formed and, even at pH values as high as 12, 90% of the boric acid remains uncomplexed. On the other hand, at a mannitol concentration of 0.5 *M*, the boric acid concentration is essentially zero at pH values greater than ca. 8 and the bis(mannitol) ester and the monomannitol ester are present in a ratio of 9:1. More recent studies (Yoshimura et al., 1996) have dealt with borate complexation with linear and cross-linked polysaccharides.

To date, the vast majority of spectroscopic studies have involved the complexation of borates to polyols. This is not surprising in view of the fact that the naturally occurring, biological compounds of B isolated from microorganisms (boromycin, aplasmomycin and tartrolon B) (Irschik et al., 1995) and the sorbitol and rhamnogalacturonan-II complexes from plants (discussed below) involve complexation of this type. Given the ubiquity of –OH groups in biological molecules, more examples will no doubt be found. In addition, it is quite probable that other types of B complexation may be observed, particularly with nitrogen (e.g. as with serine; Woods, 1996).

Speciation of boron in plants

The initial proof of the essentiality of B in plants was published by Warington (1923). Lewis (1980) has theorized that the essentiality of B arose with the evolution of vascular plants, with B playing a primary role in the biosynthesis of lignin (Ascerbo et al., 1973). The biochemical species involved in B's essentiality in plants have not been definitively proven, although many have been hypothesized (Loomis and Durst, 1991).

It is important to keep in mind, since B forms complexes with a variety of polyhydroxy compounds, that a high concentration of B in a particular cellular fraction does "... not necessarily imply an important physiological role at these sites ...-" (Skok and McIlrath, 1958). A critical review of the role of B in higher plants has very recently appeared (Goldbach, 1997).

Boron in cell walls

There is now little doubt that B plays a key role in the development, cell elongation, and structural integrity of dicotyledonous plant cell walls. Proper cell wall function requires a continuous source of B and Warington (1923) stated that "... it is in some way fixed by the plant and not in a state of circulation". Berger (1949) concluded that "The normal development of cell walls is upset when B becomes deficient". Skok (1957) suggested that the function of B is structural rather than metabolic. Lewis (1980) speculated that life forms requiring B have cell walls, matrices and envelopes rich in carbohydrates. A role for B in the synthesis and stability of the cell wall has been suggested (Augsten and Eichhorn, 1977), via ester formation with *cis*-diol groups (Loomis and Durst, 1991). Studies with added mannitol in cultured pine (*Pinus*) cells induced the authors to conclude that B has a primary role in cell wall biosynthesis (Teasdale and Richards, 1990). Brown and Hu (1994) found that the greatest part of B in plants is non-exchangeable and cannot be rinsed out, suggesting very tight complexation. Subsequent work has shown that most of the B resides in the pectic fraction of the cell wall (Hu and Brown, 1994; Hu et al., 1996). Matoh et al. (1992) assigned 95–98% of the total plant B to cell wall under B-limiting conditions, leaving only a small fraction for possible involvement in other plant functions.

Loomis and Durst (1992) discuss the discovery of apiose (Figure 4), a strong furanosidal complexer of borate, as a component of pectic polymers in aquatic monocots, dicots, gymnosperms and graminaceous monocots. In agreement with the acid sensitivity of B complexes, they speculate that the pH drop caused by auxins can result in cross-link breakage and recomplexation to give 'creep', thereby allowing elongation. Hu and Brown (1994) used a tensile strength instrument to show that cell wall extensibility is impaired under B deficiency.

Most of the B in radish (*Raphanus sativum*) root and apple (*Malus pumila*) fruit cell wall fraction has little mobility as shown by very broad ^{11}B NMR signals, indicating that the B is tied up as rigid polysaccharide complexes (Matsunaga and Nagata, 1995). Cell wall polysaccharide/borate complexes have been isolated from tomato *(Lycopersicon esculentum)* leaf (Yamauchi et al., 1986) and from radish roots following enzymatic digestion (Matoh et al., 1993). Boron-11 NMR of the latter indicated the B to be in the form of a tetravalent 1:2 borate:diol complex, and it was reasoned that about 80% of the complex was with one particular polysaccharide. The B was easily removed by acid hydrolysis (1 *M* HCl), but subsequent elevation of the pH with NaOH failed to regenerate the complex. Careful research work on suspension-cultured sycamore (*Acer pseudoplatanus*) cells by Darvill et al. (1978), Stevenson et al. (1988) and Whitcombe et al. (1995) resulted in the isolation and elucidation of structural features of rhamnogalacturonan II (RG-II), a complex polysaccharide of the pectic fraction of primary cell walls. It contains apiose (0.15–0.3% of the total cell wall fraction) and the B in bamboo (*Phyllostachys edulis*) shoot cell walls was shown by ^{11}B NMR to be present as tetrahedral borate–diol diester (Kaneko et al., 1996). The B/RG-II complex also has been identified in pea (*Pisum sativum*), pinto bean (*Phaseolus vulgaris*) and tomato seedlings (Darvill et al., 1978), Douglas fir (*Pseudotsuga menziesii*) (Thomas et al., 1987), rice (*Oryza sativa*) (Thomas et al., 1989), wine (Pellerin et al., 1996), and in cell walls of 24 species (Matoh et al., 1996). The latter authors suggest that the RG-II/borate complex may be present in all higher plants; it appears to contain all of the cell wall B in at least 15 of the species studied. RG-II contains about 30 glycosyl residues and degradation studies revealed it has a highly branched structure based on an α-(1$\rightarrow$4)-linked galactosyluronic acid backbone.

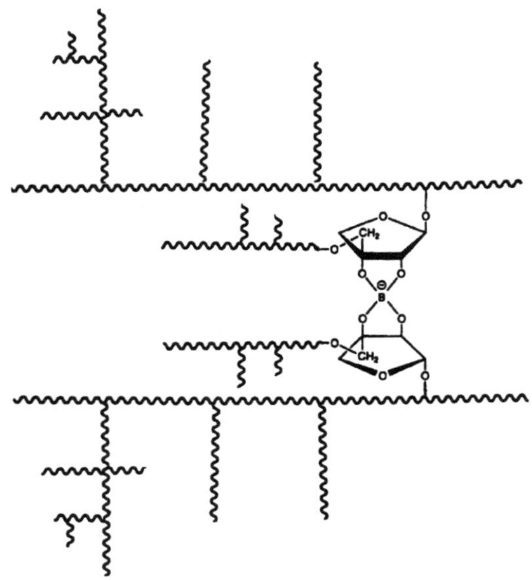

Figure 6. Schematic representation of the bis(rhamnogalacturonan-II)-borate complex, showing the linkage through apiose groups.

Kobayashi et al. (1996) treated radish root tissue with Pectinase-SS and the alkalised digest fractionated on a DEAE-Sepharose column. After purification of the B-containing fractions, acid hydrolysis reduced the molecular weight by half and the B was released, suggesting that two polysaccharide chains were linked by one 1:2 borate:diol complex. Glycosyl and uronic acid analysis indicated the polysaccharide to be RG-II. Ishii and Matsunaga (1996) digested sugar beet (*Beta vulgaris*) cell walls with Driselase enzyme, followed by isolation of an RG-II/borate complex. Boron^{11} NMR clearly showed that the B was tetrahedral in a 1:2 borate–diol complex (δ −9.5; Van Duin et al., 1984). Analysis gave 0.12% B and 4.5 mol% apiose in the complex. The glycosyl and uronic acid composition, and similarity to that of sycamore RG-II (Stevenson et al., 1988), identified the complex as that of an RG-II. O'Neill et al. (1996) used *endo*-α-1,4-polygalacturonase digestion of suspension-cultured sycamore and etiolated pea stem cell walls to give dimeric RG-II crosslinked by borate/apiose complexation. Figure 6 gives a structure for this complex with branching of the RG-II framework, which is not known in detail, indicated in a generic way. In contrast to the earlier Japanese work, these researchers found that the dimer could be formed *in vitro* from RG-II and boric acid, albeit at a slow rate.

The complexation was catalyzed by divalent cations with ionic radii > 0.11 nm (Sr^{2+}, Pb^{2+}, and Ba^{2+}), with an optimum pH in the presence of Sr^{2+} of 3.5–4. In order to account for rapid wall expansion rates, they hypothesised that enzymatic catalysis may be involved *in vivo* in the hydrolysis and reformation of the RG-II/borate complex.

In heterocystous N_2-fixing cyanobacteria, Mateo et al. (1986) discuss the O_2 diffusion barrier, which breaks down when B is deficient, suggesting that B may cause polysaccharide cross-linking or membrane interaction to prevent O_2 diffusion. These heterocyst glycolipids are hexose derivatives of long chain alcohols with *cis*-diol groups ideally suited to B complexing (Lambein and Wolk, 1973). Calcium could be stabilising the heterocyst envelope along with B (Bonilla et al., 1991), perhaps due to a Ca^{2+}/borate polyol complex ion pair.

In pea, Bolaños et al. (1994) found that cell walls varied in thickness and were missing in places, causing cell breakage in B-deficient nodules.

There are arguments against cell walls being the seat of the essentiality of B in plants. Parr and Loughman (1983) argue that the functions of B in the cell wall are not the reason for its essentiality, since it is not essential for fungi. Another point is that the polysaccharides of graminaceous plants are very low in B, pectin, and complexable polysaccharides [e.g. rice cell walls contain only 0.1% RG-II (Thomas et al., 1989)], hardly sufficient for B crosslinks to account for all of the effects described by Loomis and Durst (1992). Goldbach (1997) also discounts the cell wall - B essentiality hypothesis, preferring a mechanism involving $B(OH)_3$ passing through the plasmalemma, complexing in the cytosol, passing out of the cell and being transported in the apoplast, then doing its essential function.

Membrane effects

Extensive studies have supported the idea that B in plants functions at the membrane level (Parr and Loughman, 1983; Blaser-Grill et al., 1989; Ferrol et al., 1993; Shelp et al., 1995). A key paper by Parr and Loughman (1983) makes a convincing case that B deficiency causes impaired membrane function as its *primary* effect. Symptoms of deficiency such as auxin (Robertson and Loughman, 1974) and phenol build-up, as well as increased RNAase activity (Dave and Kannan, 1980), are *secondary* manifestations of the reduced ability of membranes to transport vital nutrients and metabolites. Gauch and Dugger (1953) proposed that B expedites sugar transport across the membrane by temporary formation of a borate complex. Boron also is credited with maintaining membrane integrity (Cakmak et al., 1995). Boron accumulates in the cell wall, which puts the B in an ideal position to affect the plasmalemma.

Membrane uptake of a number of nutrients is inhibited by B deficiency and this can be reversed by as little as $10^{-5}M$ borate solution. For example, leakage of K^+ from sunflower (*Helianthus annuus*) leaves was 35 times higher in B-deficient than in B-sufficient plants (Cakmak et al., 1995), suggesting that B-deficiency changes the permeability of plant membranes to glucose, phosphate, protein, and IAA, in addition to influencing peroxidase activity (Goldbach, 1985; Goldbach and Amberger, 1986). Heyes et al. (1991) indicated that recovery of reduced transport capacity of cultured plant roots for a number of inorganic ions is due to B maintaining a preferred conformation of the active protein in the membrane. It was further concluded that the membrane capacity is under the influence and probable control of the low levels of B in the cell. Interaction of B with the amino, imidazole, serine or other groups on the transmembrane proteins may well be the source of these nutrient effects. It is of interest that B is believed to be involved in the movement of protein into petunia (*Petunia hybrida*) pollen tube membranes (Jackson, 1991). Boron in the media results in a significantly higher accumulation of K^+ (Schon et al., 1990). The overall conclusion was that potassium uptake by plants does not occur in the absence of B (Schon et al., 1991).

Boron is stated to play an important role in phosphate transport across membranes (Loughman, 1977); the absorption of both phosphate and rubidium is lowered in faba bean (*Vicia faba*) roots when B is deficient (Robertson and Loughman, 1974). Boron deficiency also results in effects on the transport capacity of the tonoplast (Parr and Loughman, 1983). Lower ATPase and NAD(P)H-dependent redox activity are seen under B-deficiency; it was concluded that B and Zn are essential for optimal function of plasma membrane ATPase and redox systems (Lawrence et al., 1995). Membrane potential measurements show that B influences the proton gradient (Blaser-Grill et al., 1989). Hyperpolarisation of the plasmalemma points to a key role of B in membrane stability and optimum ion uptake. Glycoprotein complexation by B at the membrane surface creates an additional negative charge which can influence the electrostatics across the membrane. Microsomal studies showed greater ATPase activity and ATP-dependent proton pumping with 100 μM B in the medium vs. no B (Ferrol et al., 1993). Both deficiency and B at toxic levels can inhibit H^+-ATPase (Roldan et al., 1992).

Reduced cellulose content in cell walls under conditions of B deficiency led to speculation that boric acid influences cell wall deposition by changing membrane properties (Goldbach and Amberger, 1986). The filamentous cyanobacterium *Anabaena* has an inner membrane layer in its heterocysts made up of glycolipid. This membrane prevents the passage of oxygen which can interfere with the function of the N_2-fixing nitrogenase in the interior of the cyst. B deficiency prevents the formation of this glycolipid membrane (Garcia-Gonzalez et al., 1991). These heterocyst glycolipids are hexose derivatives of long chain alcohols with *cis*-diol groups which can complex with B. Boric acid stabilisation of heterocyst glycolipid has been reported (Lambein and Wolk, 1973).

It has been hypothesised that membranes have a high permeability to uncharged free boric acid and that B uptake by plants is primarily by passive transport (Raven, 1980; Goldbach, 1997). Bingham et al. (1970) concluded that B absorption is a physical, non-metabolic process with $B(OH)_3$ moving in response to its concentration gradient; similar conclusions were reached based on work at pH's from 6 to 10 by Oertli and Grgurevic (1975). Brown and Hu (1994) conducted kinetic studies which support this view (Hu and Brown, Chapter 4).

It was speculated (Parr and Loughman, 1983) that membrane conformation can be influenced by B complexation with component glycoproteins and/or glycolipids. Boron may be maintaining the active transmembrane protein in a preferred conformation (Heyes et al., 1991), perhaps through borate cross-linking. Both glycolipids in the bilayer, as well as transmembrane and surface glycoproteins (Alberts et al., 1994), have oligosaccharide side chains capable of forming borate complexes and crosslinks. This could change the membrane permeability, surface charge and rigidity leading to many of the observed effects listed above.

The high B demand of diatoms has been interpreted by Goldbach et al. (1991) as due to the ordered deposition of silica on mannose-rich membranes.

Enzyme interactions

A variety of enzymes in microorganisms, plants, animals and humans interact with B compounds resulting in stimulation, stabilisation, or inhibition. The enzyme

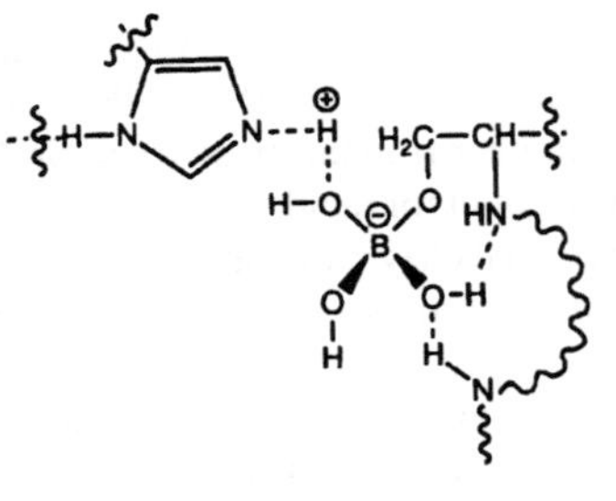

Figure 7. Schematic drawing of the active site of a serine protease, illustrating borate inhibition (Kliegel, 1980).

urease immobilised on a membrane was found to be inhibited by boric acid – the only inhibitor effective on the immobilised enzyme (Zaborska, 1995). Succinic dehydrogenase of *Neurospora* is inhibited by borate buffer and phosphoglucomutase of pea seeds is inhibited by boric acid (Shepherd, 1951). Alcohol dehydrogenase of yeast is competitively inhibited by borate, with a K_i of $3.2 \times 10^{-3} M$ (Weser, 1968). In animals, borate and boronic acid inhibition of a variety of enzyme systems has been demonstrated (Kettner et al., 1990; Kliegel, 1980; Woods, 1996). This inhibition is attributed to borate occupying the active site where it displaces the normal substrate. An example is illustrated for a generic serine protease in Figure 7.

In plants, B deficiency releases the inhibition of 6-phosphogluconate- and of glucose-6-phosphate dehydrogenases, resulting ultimately in increased production of phenols (Gomez-Rodriguez et al., 1987; Heyes et al., 1991; Lee, 1967; Shkolnik, 1974; Shkolnik and Ilinskaya, 1975; Smyth and Dugger, 1981). The mechanism may involve direct inhibition of 6-phosphogluconate dehydrogenase *via* a borate complex (Lee, 1967). Alcohol dehydrogenase is inhibited by borate (Weser, 1968; Smith and Johnson, 1976). These competitive inhibitions likely involve complexation with the ribityl hydroxyls of the coenzyme NAD, slowing its conversion to NADH (Figure 8). Further evidence of this is found in the fact that excess ribose, sorbitol and mannitol can negate the inhibition, presumably by competing with the NAD for the borate (Roush and Gowdy, 1961). A study of the interaction of borate with NAD+ and NMN+ sulfite adducts concluded that borate prefers to complex with the ribose near the positive charge (Johnson and Smith, 1976). In muscle, borate is believed to induce conformational changes in the protein along with NAD complexation (Wolny, 1977).

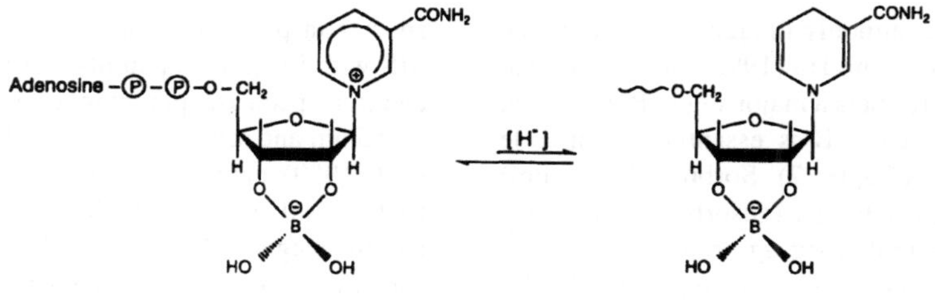

NAD-borate NADH-borate

Figure 8. Schematic of the coenzyme system NAD and NADH, showing shift of the equilibrium towards the electrostatically more stable NAD-borate complex (Johnson and Smith, 1976).

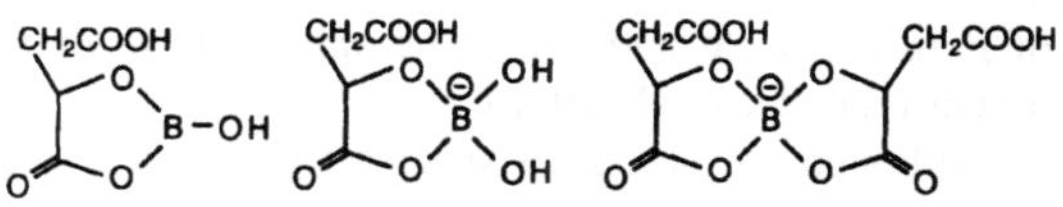

Figure 9. Schematic drawings (l. to r.) of the malic acid neutral borate ester, the monomalic acid borate complex, and the bis(malic acid) borate complex found in apple juice (Matsunaga and Nagata, 1995) and hydroxyacid complexes in wine (Lutz et al., 1991).

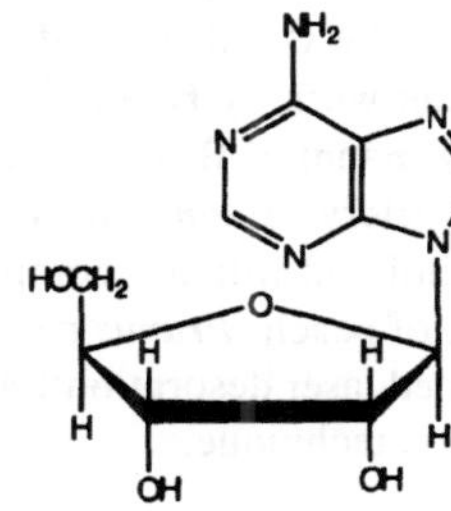

Figure 10. Structural representation of the nucleoside adenosine, capable of forming a stable borate complex.

Blevins and Lukaszewski (1994) have demonstrated that boric acid inhibits allantoate amidohydrolase, a manganese-containing enzyme. It was suggested that many of the functions of B in plants may be due to its regulation of metal-containing enzymes.

Sugar, polyol and hydroxy acid transport

Berger (1949) suggested that B plays an important role in the translocation of sugars. It may be involved in a number of metabolic pathways due to the formation of complexes with a variety of hydroxylated molecules (Augsten and Eichorn, 1976; Dugger, 1983). Gauch and Dugger (1953) showed that B speeds up the uptake and transport of sugars in normal vs. B deficient plants.

Matsunaga and Nagata (1995) recently described *in vivo* detection of B in radish, apple, cabbage (*Brassica oleracea*) and komatsuna (*Brassica campestris*) plant tissue by ^{11}B NMR. The spectra showed free boric acid plus peaks for borate mono- and diesters of diols and of hydroxy acids (Figures 3 and 9). At the pH of radish juice (6.2), a boric acid solution with a 100 fold excess of fructose showed a ^{11}B NMR spectrum similar to that of the radish juice; the same was true of a malic acid solution at pH 4.0 vs. apple juice. It was speculated that fructose was the complexing diol for boric acid in radish root and malic acid in apple juice. Based on the data in the paper, the B concentration in the juices and tissues ranged from 4 to $20\times10^{-5}M$. Subsequent research to better define the radish root juice complexes used size exclusion HPLC/ICP-MS (Matsunaga et al., 1996). The chromatograms indicated the presence of low concentrations of macromolecular B complexes, with most of the B complexed with low molecular wt components. The retention time of the macromolecular complex agreed with that of an RG-II/borate (see section "Boron in cell walls").

A similar study in wine (2–5 mg B L^{-1}) showed three peaks, corresponding to the chemical shifts at the pH of wine (3.5) of a neutral hydroxyacid monoborate ester (or free boric acid), a bis-hydroxyacid–borate complex, and a monohydroxyacid–borate complex (Lutz et al., 1991) (Figure 9). Freshly prepared grape juice gave a similar set of three peaks, as did artificial solutions of 11 ppm B and malic, lactic and tartaric acids. Variations in peak intensities and line widths in the spectra of the ten red and five white wines studied were speculated as being due to different levels of B and proportions of hydroxyacids.

Although B is commonly thought to be immobile in the phloem, Brown and Hu (1996) have found that species in which sorbitol is a major sugar, B is mobile; in sorbitol-poor species, B is essentially immobile (Brown and Shelp, Chapter 7). Sorbitol (D-glucitol) is known to form six stable 1:2 (B:sorbitol) complexes because of the favourable zigzag arrangement of the 2-, 3-, and 4-hydroxyls; the structure of the bis(D-sorbitol) (R)-2,3,2′,3′-borate is shown in Figure 5. The stability of the sorbitol complexes exceeds that of the 3,4,3',4 complex of mannitol (Makkee et al., 1985; Nose and Zenki, 1991) (see section "Borate esters and borate complexes"). At the high sugar:B ratios found in the high sorbitol genera (*Pyrus, Malus, Prunus*), more than 99% of the B will be tied up as 1:2 complexes. Subsequent work by Hu et al. (1997) directly identified the 3,4-mannitol-B-3′,4′-mannitol complex in phloem sap of celery (*Apium graveolens* var. *dulce*) and similar sorbitol and fructose complexes from the extrafloral nectar of peach (*Prunus persica*) by an elegant matrix assisted laser desorption/fourier transform mass spectrometric technique.

Other effects and analytical methods

The variety of biomolecules with which B can interact is illustrated by affinity chromatography on solid or gel media, where fractionation has been demonstrated with nucleotides (Bergold and Scouten, 1983), glucuronides (Yugnait et al., 1994), catecholamines (Higa et al., 1977), membrane proteins (Williams et al., 1982), glycated platelets (Muruganandam et al., 1994), medicinal plant residues in urine (Bongartz and Hesse, 1995), apolipoproteins (Panteghini et al., 1994), and haemoglobins (Middle et al., 1983).

Goldbach et al. (1991) hypothesised that the most likely site for excess borate to interfere with protein synthesis is to esterify the *cis*-diol of the ribose group of RNA (Goldbach et al., 1991); this may also be involved in the proposed essential role of B in protein synthesis (Krueger et al., 1987). Nucleosides such as adenosine (Figure 10) form strong borate complexes (Weith et al., 1970). DNA-B interaction was indicated in work on the fluorescence *in situ* hybridisation (FISH) method of mapping DNA sequences, where alkaline borate treatment gave increased nuclei diameter and interphase distance changes between chromosomes (Yokota et al., 1995).

Analytical methods for B in soils and plants are discussed by Sah and Brown in Chapter 2. Non-destructive techniques suitable for characterising B species in plants include ion microscopy (secondary MS with imaging capability) (Ausserer et al., 1989), electron spectroscopic imaging with resolution of 0.3 to 0.5 nm and sensitivity of 20 to 50 atoms (Bendayan et al., 1989), neutron-capture radiography with a sensitivity of 0.03 mg kg^{-1} B (Martini and Thellier, 1993), and the elegant MALDI-TOF MS technique in which species of interest are analysed by MS after desorption with a nitrogen laser in high vacuum (e.g. O'Neill et al., 1996). $Boron^{11}$ NMR has been mentioned in the previous discussion and, along with ^{13}C and ^{1}H NMR, has been invaluable in characterising B complexes. Magnetic resonance imaging (MRI) using free induction decay of ^{11}B has been used to show spatial distribution of B in a live rat and should be applicable to plant tissues (Kabalka et al., 1992).

Although the precise B specie(s) involved in the essentiality of B in plants has not been definitively established, recent research and new techniques applied to the problem are showing very promising progress. Involvement of complexation with polyhydroxy compounds appears to be involved in all known, isolated, naturally occurring B compounds in microorganisms and plants, with the extent of such complexing being dependent on the concentration of hydroxy species [e.g. sorbitol (Brown and Hu, 1996)].

Acknowledgment

The authors thank Dr Kevin Simpson for assistance with the figures.

References

Alberts B, Bray D, Lewis J, Raff M, Roberts K and Watson J D 1994 Molecular Biology of the Cell. 3rd ed. Garland Publishers, New York, p 503

Ascerbo S, Kastori R, Sochtig H, Harms H and Haider K 1973 Effect of boron on synthesis and transformation of lignin precursors in *Zea mays*. Z. Pflanzenphysiol. 69, 306–317.

Augsten H and Eichhorn B 1976 Biochemistry and physiology of the effect of boron in plants. Biol. Rundsch. 14, 268–285.

Ausserer W A, Ling Y-C, Chandra S and Morrison G H 1989 Quantitative imaging of boron, calcium, magnesium, potassium and sodium distributions in cultured cells with ion microscopy. Anal. Chem. 61, 2690–2695.

Barr R D, Clarke W B, Clarke R M, Venturelli J, Norman G R and Dowing R G 1993 Regulation of lithium and boron levels in normal human blood: Environmental and genetic factors. J. Lab. Clin. Med. 121, 614–619.

Bendayan M, Barth R F, Gingras D, Londono I, Robinson P T Alam F, Adams D M and Mattiazzi L 1989 Electron spectroscopic imaging for high-resolution immunochemistry: use of boronated protein A. J. Histochem. Cytochem. 37, 573–580.

Berger K C 1949 Boron in soils and crops. *In* Advances in Agronomy, vol. 1. Ed. A G Norman. Academic Press, New York.

Bergold A and Scouten W H 1983 Borate chromatography. Chem. Anal. 66, 149–187.

Bingham F T, Elseewi A and Oertli J J 1970 Characteristics of boron absorption by excised barley roots. Soil Sci. Soc. Am. Proc. 34, 613–617.

Blaser-Grill J, Knoppik D, Amberger A and Goldbach H 1989 Influence of boron on the membrane potential in *Elodea densa* and *Helianthus annuus* roots and H^+ extrusion of suspension cultured *Daucus carota* cells. Plant Physiol. 90, 280–284.

Blevins D G and Lukaszewski K M 1994 Proposed physiologic functions of boron in plants pertinent to animal and human metabolism. Environ. Health Persp. 102, Suppl. 7, 31–33.

Bolaños L, Esteban E, de Lorenzo C, Fernandez-Pascual M, de Felipe M R, Garate A and Bonilla I 1994 Essentiality of boron for symbiotic dinitrogen fixation in pea (*Pisum sativum*) rhizobium nodules. Plant Physiol. 104, 85–90.

Bongartz D and Hesse A 1995 Selective extraction of quercetrin in vegetable drugs and urine by off-line coupling of boronic acid affinity chromatography and HPLC. J. Chromatogr. B 673, 223–230.

Bonilla I, Mateo P and Bolaños L 1991 A double interaction B$\leftrightharpoons$Ca in cyanobacteria. *In* Current Topics in Plant Biochem. and Physiol., vol. 10. Eds. D D Randall, D G Blevins and C D Miles. P 295. University Missouri, Columbus.

Brown P H and Hu H 1994 Boron uptake by sunflower, squash and cultured tobacco cells. Physiol. Plant. 91, 435–441.

Brown P H and Hu H 1996 Phloem mobility of boron is species dependent: evidence for phloem mobility in sorbitol-rich species. Ann. Bot. 77, 497–505.

Cakmak I, Kurz H and Marschner H 1995 Short-term effects of boron, germanium, and high light intensity on membrane permeability in boron deficient leaves of sunflower. Physiol. Plant. 95, 11–18.

Cotton F A and Wilkinson G 1989 Advanced Inorganic Chemistry, 5th ed. Wiley, New York.

Crosswell K 1992 Star's Death Boron's Birth? J. Boron and Borate Industry 12, 14–17.

Darvill A G, McNeil M and Albersheim P 1978 Structure of plant cell walls. Plant Physiol. 62, 418–422.

Dave I C and Kannan S 1980 Boron deficiency and its associated enhancement of RNAase activity in bean plants. Z. Pflanzenphysiol. 97, 261–263.

Dugger W M 1983 Boron in plant metabolism. *In* Inorganic Plant Nutrition. Encyclopedia of Plant Physiology, vol 15. Eds. A Läuchli and R L Bieleski. pp 626–650. Springer-Verlag, Heidelberg.

Emsley J 1991 The Elements. 2nd ed. Oxford University Press, UK.

Erd R C 1980 The Minerals of Boron. *In* Supplement to Mellor's Comprehensive Treatise on Inorganic and Theoretical Chemistry, vol. V, Part A: Boron-Oxygen Compounds. Longman, New York.

Ferrol N, Belver A, Roldan M, Rodriguez-Rosales M P and Donaire J P 1993 Effects of boron on proton transport and membrane properties of sunflower (*Helianthus annuus* L.) cell microsomes. Plant Physiol. 103, 763–769.

Garcia-Gonzalez P M, Mateo P and Bonilla I 1991 Boron requirement for envelope structure and function in*Anabaena* PCC 7119 heterocysts. J. Exp. Bot. 42, 925–929.

Gauch H G and Dugger W M 1953 The role of boron in the translocation of sucrose. Physiol. Plant. 28, 457.

Goldbach H 1985 Influence of boron nutrition on net uptake and efflux of ^{32}P and ^{14}C-glucose in *Helianthus annuus* roots and cell cultures of *Dauca carota*. J. Plant Physiol. 118, 431–438.

Goldbach H E 1997 A critical review on current hypotheses concerning the role of boron in higher plants: suggestions for further research and methodological requirements. J. Tr. Microprobe Tech. 15, 51–91.

Goldbach H E and Amberger A 1986 Influence of boron deficiency on ^3H indol-3-yl-acetic acid uptake and efflux in cell cultures of *Daucus carota* L. Plant Growth Regul. 4, 81–86.

Goldbach H E, Blaser-Grill J B, Lindemann N, Portzelt M, Horrmann C, Lupp B and Gessner B 1991 Influence of boron on net proton release and its relation to other metabolic processes. *In* Current Topics in Plant Biochem. and Physiol, vol. 10. Eds. D D Randall, D G Blevins and C D Miles. pp 195–200. University Missouri, Columbus.

Gomez-Rodriguez M V, Luna del Castillo J and Alvarez-Tinaut M C 1987 The evolution of glucose-6P-dehydrogenase and 6P-gluconate-dehydrogenase activities and the ortho-diphenolic content of sunflower leaves cultivated under different boron treatments. J. Plant Nutr. 10, 2211–2229.

Gorin P A J and Mazurek M 1973 ^{13}C magnetic resonance spectroscopic evidence for formation of borate complexes of polyhydroxo compounds. Carbohydr. Res. 27, 325–339.

Greenwood N N and Earnshaw A 1984 Chemistry of the Elements. Pergamon Press, New York.

Hawthorne M F 1993 The role of chemistry in the development of boron neutron capture therapy of cancer. Angew. Chem. Int. Ed. Engl. 32, 950–984.

Hayashi T and Namiki M 1986 Role of sugar fragmentation in the Maillard reaction. In Amino-Carbonyl Reactions in Food and Biological Systems. Eds. M Fujimaki, M Namiki and H Kato. pp 2–38. Elsevier, New York.

Heyes J A, White P J and Loughman B C 1991 The role of boron in some membrane characteristics of plant cells and protoplasts. *In* Current Topics in Plant Biochem. and Physiol. vol 10. Eds. D D Randall, D G Blevins and C D Miles. pp 179–194. University Missouri, Columbus.

Higa S, Suzuki T, Hayashi A, Tsuge I and Yamamura Y 1977 Isolation of catecholamines in biological fluids by boric acid gel. Anal. Biochem. 77, 18–24.

Hu H and Brown P H 1994 Localisation of boron in cell walls of squash and tobacco and its association with pectin; evidence for a structural role of boron in cell wall. Plant Physiol. 105, 681–689.

Hu H, Brown P H and Labavitch J M 1996 Species variability in boron requirement is correlated with cell wall pectin. J. Exp. Bot. 47, 227–232.

Hu H, Penn S G, Lebrilla C B and Brown P H 1997 Isolation and characterisation of soluble B-complexes in higher plants; the mechanism of phloem mobility of boron. Plant Physiol. 113, 649–655.

Irschik H, Schummer D, Gerth K, Hofle G and Reichembach H 1995 The tartrolons, new boron-containing antibiotics from a myxobacterium, *Sorangium cellulosum*. J. Antibiotics 48, 26–30.

Ishii T and Matsunaga T 1996 Isolation and characterisation of a boron-rhamnogalacturonan-II complex from cell walls of sugar beet pulp. Carbohydr. Res. 284, 1–9.

Jackson J F 1991 Borate control of energy-driven protein secretion from pollen and interaction of borate with auxin or herbicide – a possible role for boron in membrane events. *In* Current Topics in Plant Biochem. and Physiol., vol. 10. Eds. D D Randall, D G Blevins and C D Miles. pp 221-229. University Missouri, Columbus.

Johnson S L and Smith K W 1976 The interaction of borate and sulfite with pyridine nucleotides. Biochem. 15, 553–559.

Kabalka G W, Cheng G Q and Bendl P 1992 *In vivo* boron-11 MRI in the rat. *In* Synthesis and Applications of Isotopically Labelled Compounds. Eds. E Buncel and G W Kabalka. pp 363–368. Elsevier, New York.

Kaneko S, Ishii T and Matsunaga T 1996 A boron-rhamnogalacturonan II complex from bamboo shoot cell walls. Phytochem. 44, 243–248.

Kennedy J D 1987 *In* Multinuclear NMR. J Mason Ed, Plenum, New York, Chap 8.

Kettner CA, Mersinger L and Knabb R 1990 Selective inhibition of thrombin by peptideboroarginine. J. Biol. Chem. 265, 18289.

Kliegel W 1980 Bor in biologie, medizin und pharmacie. Springer-Verlag, New York.

Kobayashi M, Matoh T and Azuma J-I 1996 Two chains of rhamnogalacturonan II are cross-linked by borate-diol ester bonds in higher plant cell walls. Plant Physiol. 110, 1017–1020.

Krantz J C, Carr C J and Beck F F 1936 A further study of the effect of sugar alcohols and their anhydrides on the dissociation of boric acid. J. Phys. Chem. 40, 927–936.

Krueger U R, Lovatt J C and Albert L S 1987 Metabolic requirements of *Curcubita pepo* for boron. Plant Physiol. 83, 254–258.

Lambein F and Wolk C P 1973 Structural studies on the glycolipids from the envelope of the heterocyst *Anabaena cylindrica*. Biochem. 12, 791–798.

Lawrence K, Bhalla P and Misra P C 1995 Changes in NAD(P)H-dependent redox activities in plasmalemma-enriched vesicles isolated from boron- and zinc-deficient chick pea roots. J. Plant. Physiol. 146, 652–657.

Lee S 1967 Boron in plants: a biochemical role. Science 158, 798–799.

Lenz R W and Heeschen J P 1961 The application of NMR to structural studies of carbohydrates in aqueous solution. J. Polymer Soc. 51, 247–255.

Lewis D H 1980 Boron, lignification, and origin of vascular plants-a unified hypothesis. New Phytol. 84, 209–229.

Loomis W D and Durst R W 1991 Boron and cell walls. *In* Current Topics in Plant Biochem. and Physiol., vol. 10. Eds. D D Randall, D G Blevins and C D Miles. pp 149–178. University Missouri, Columbus.

Loomis W D and Durst R W 1992 Chemistry and biology of boron. BioFactors 3, 229–239.

Loughman B C 1977 Metabolic factors and the utilisation of phosphate by plants. Ciba Foundation Symposium 57, 155–174.

Lutz O, Humpfer E and Spraul M 1991 Ascertainment of boric acid esters in wine by ^{11}B NMR. Naturwissen. 78, 67–69.

Makkee M, Kieboom A P G and van Bekkum H 1985 Studies on borate esters III. Borate esters of D-mannitol, D-glucitol, D-fructose and D-glucose in water. Recl. Trav. Chim. Pays-Bas 104, 230–235.

Martini F and Thellier m 1993 Boron distribution in parenchyma cells of clover leaves. Plant Physiol. Biochem. 31, 777–786.

Massey AG 1983 Boron subhalides. Adv. Inorg. Chem. Radiochem. 26, 1–54.

Mateo P, Bonilla I, Fernandez-Valiente E and Sanchez-Maeso E 1986 Essentiality of boron for dinitrogen fixation in *Anabaena* sp. PCC 7119. Plant Physiol. 81, 17–21.

Matoh T, Kawaguchi S and Kobayashi M 1996 Ubiquity of a borate-rhamnogalacturonan II complex in the cell walls of higher plants. Plant Cell Physiol. 37, 636–640.

Matoh T, Ishigaki K-I, Ohno K and Azuma J-I 1993 Isolation and characterisation of a boron-polysaccharide complex from radish roots. Plant Cell Physiol. 34, 639–642.

Matoh T, Ishigaki K-I, Mizutami M, Matsunaga W and Takabe K 1992 Boron nutrition of cultured tobacco BY-2 cells I. Requirements for an intracellular localisation of boron and selection of cells that tolerate low levels of boron. Plant Physiol. 33, 1135–1141.

Matsunaga T and Nagata T 1995 *In vivo* ^{11}B NMR observation of plant tissue. Anal. Sci. 11, 889–892.

Matsunaga T, Ishii T and Watanabe H 1996 Speciation of water-soluble boron compounds in radish roots by size exclusion HPLC/ICP-MS. Anal. Sci. 12, 673–675.

Mazurek M and Perlin A S 1963 Borate complexing by five-membered-ring *vic*-diols. Can. J. Chem. 41, 2403.

Middle F A, Bannister A, Bellingham A J and Dean P D G 1983 Separation of glycosylatedhaemoglobins using immobilised phenylboronic acid. Biochem. J. 209, 771–779.

Morgan V 1980 Boron Geochemistry. *In* Supplement to Mellor's Comprehensive Treatise on Inorganic and Theoretical Chemistry, vol. V, Part A: Boron-Oxygen Compounds. Longman, New York.

Muruganandam A, Tannous M and Mutus B 1994 ELISA for *in vivo* assessment of nonenzymatically glycated platelet glutathione peroxidase. Clin. Biochem. 27, 293–298.

Nose K and Zenki M 1991 Flow injection spectrophotometric determination of boron with D-sorbitol using methyl orange as an indicator. Analyst 116, 711–714.

Oertli J J and Grgurevic E 1975 Effect of pH on the absorption of boron by excised barley roots. Agron. J. 67, 278–280.

O'Neill M A, Warrenfeltz D, Kates K, Pellerin P, Doco T, Darvill A G and Albersheim P 1996 Rhamnogalacturonan-II, a pectic polysaccharide in the walls of growing plant cells, forms a dimer that is covalently cross-linked by a borate ester. J. Biol. Chem. 271, 22923–22930.

Paine R T and Narula C K 1990 Synthetic routes to boron nitride. Chem. Rev. 90, 73–91.

Panteghini M, Bonora R and Pagani F 1994 Determination of glycated apolipoprotein B in serum by a combination of affinity chromatography and immunonephelometry. Ann. Clin. Biochem. 31, 544–549.

Parr A J and Loughman B C 1983 Boron and membrane function in plants. *In* Metals and Micronutrients, Uptake and Utilization by Plants. Eds. D A Robb and W S Pierpoint. pp 87–107. Academic Press, New York.

Pellerin P, Doco T, Vidal S, Williams P, Brillouet J-M and O'Neill M A 1996 Structural characterisation of red wine rhamnogalacturonan II. Carbohydr. Res. 290, 183–197.

Power P P 1992 Synthesis and characterisation of compounds with boron-boron double bonds. Inorg. Chim. Acta 198-200, 444–447.

Raven J A 1980 Short- and long-distance transport of boric acid in plants. New Phytol. 84, 231–249.

Reeves H 1974 On the origin of the light elements. Ann. Rev. Astron. Astrophys. 12, 437–469.

Robertson G A and Loughman B C 1974 Modification of phosphate transport in *Vicia faba* by boron deficiency, growth regulators, and metabolic inhibitors. *In* Membrane Transport in Plants. Eds. U Zimmerman and J Dainty. pp 444–449. Springer-Verlag, New York.

Roldan M, Belver A, Rodriguez-Rosales P, Ferrol N and Donaire J P 1992 *In vivo* and *in vitro* effects of boron on the plasma membrane proton pump of sunflower roots. Physiol. Plant. 84, 49–54.

Roush A H and Gowdy B B 1961 The inhibition of yeast alcohol dehydrogenase by borate. Biochem. Biophys. Acta 52, 200–202.

Sanderson B R 1980 Coordinated compounds of boric acid. *In* Supplement to Mellor's Comprehensive Treatise on Inorganic and Theoretical Chemistry, vol V, Part A. Boron-Oxygen Compounds. Longman, New York.

Schon M K, Blevins D G and Novacky A 1991 Boron: from cell membranes to soybean branching. *In* Current Topics in Plant Biochem. and Physiol., vol. 10. Eds. D D Randall, D G Blevins and C D Miles. pp 230–239. University Missouri, Columbus.

Schon M K, Novacky A and Blevins D G 1990 Boron induces hyperpolarization of sunflower root cell membranes and increases membrane permeability to K^+. Plant Physiol. 93, 566–571.

Shelp B J, Marentes E, Kitheka A M and Vivekanandan P 1995 Boron mobility in plants. Physiol. Plant. 94, 356–361.

Shepherd C J 1951 The enzymes of carbohydrate metabolism in *Neurospora*. Biochem. J. 48, 483–486.

Shkolnik M Y 1974 General conception of the physiological role of boron in plants. Fiziolog. Rast. 21, 174–186.

Shkolnik M Y and Ilinskaya N L 1975 Effect of a boron deficiency on activity of glucose-6-phosphate dehydrogenase in plants with different boron requirements. Fiziol. Rast. 22, 801–805.

Skok J 1957 The substitution of complexing substances for boron in plant growth. Plant Physiol. 32, 308–312.

Skok J and McIlrath W J 1958 Distribution of boron in cells of dicotyledonous plants in relation to growth. Plant Physiol. 33, 428–431.

Smith K W and Johnson S L 1976 Borate inhibition of yeast alcohol dehydrogenase. Biochem. 15, 560–565.

Smyth D A and Dugger W M 1981 Cellular changes during boron deficient culture of the diatom*Cylindrotheca fusiformis*. Plant Physiol. 51, 111–117.

Stevenson T T, Darvill A G and Albersheim P 1988 Structural features of the plant cell-wall polysaccharide rhamnogalacturonan-II. Carbohydr. Res. 182, 207–226.

Stock A 1933 Hydrides of Boron and Silicon. Cornell Press, Ithaca, New York.

Teasdale R D and Richards D K 1990 Boron deficiency in cultured pine cells. Plant Physiol. 93, 1071–1077.

Thomas J R, Darvill A G and Albersheim P 1989 Isolation and structural characterisation of the pectic polysaccharide rhamnogalacturonan II from the walls of suspension-cultured rice cells. Carbohydr. Res. 185, 261–277.

Thomas J R, McNeil M, Darvill A G and Albersheim P 1987 Structure of plant cell walls XIX. Isolation and characterisation of wall polysaccharides from suspension-cultured Douglas fir cells. Plant Physiol. 83, 659–671.

Thompson R 1980 Inorganic boron chemistry: an introduction and background. *In* Supplement to Mellor's Comprehensive Treatise on Inorganic and Theoretical Chemistry, vol. V, Part A: Boron-Oxygen Compounds. Longman, New York.

Van Duin M, Peters J A, Kieboom A P G and Van Bekkum H 1984 Tetrahedron 40, 2901–2911.

Warington K 1923 The effect of boric acid and borax on the broad bean and certain other plants. Ann. Bot. 40, 629–671.

Weith H L, Wiebers J L and Gilham P T 1970 Synthesis of cellulose derivatives containing the dihydroxyboryl group and a study of their capacity to form specific complexes with sugars and nucleic acid components. Biochem. 9, 4396.

Wells A F 1984 Structural Inorganic Chemistry, 5th ed. p. 1052. Oxford University Press, UK.

Weser U 1968 Aktivitatshemmung der hefe-alcohol-dehydrogenase in gegenwart von germanat und borat. Hoppe-Seyler's Zeit. Physiol. Chem. 349, 1479–1482.

Whitcombe A J, O'Neill M A, Steffan W, Albersheim P and Darvill A G 1995 Structural characterisation of the pectic polysaccharide, rhamnogalacturonan II. Carbohydr. Res. 271, 15–29.

Williams G T, Johnstone A P and Dean P D G 1982 Fractionation of membrane proteins on phenylboronic acid-agarose. Biochem. J. 205, 167–171.

Wolny M 1977 Effect of borate on the catalytic activities of muscle glyceraldehyde-3-phosphate dehydrogenase. Eur. J. Biochem. 80, 551–556.

Woods W G 1996 A review of possible boron speciation relating to its essentiality. J. Tr. Elem. Exp. Med. 9, 153–163.

Yamauchi T, Hara T and Sonoda Y 1986 Distribution of calcium and boron in the pectin fraction of tomato leaf cell wall. Plant Cell Physiol. 27, 729–732.

Yokota H, Van Den Engh G, Mostert M and Trask B J 1995 Treatment of cells with alkaline borate buffer extends the capability of interphase FISH mapping. Genomics 25, 485–491.

Yoshimura K, Mayazaki Y, Sawada S and Waki H 1996 ^{11}B NMR studies on complexation of borate with linear and crosslinked polysaccharides. J. Chem. Soc. Faraday Trans. 92, 651–656.

Yugnait M, Wilson I D and Nicholson J K 1994 High resolution ^{1}H NMR spectroscopic monitoring of extraction model glucuronides on phenylboronic acid and C_{18} bonded phases. *In* Sample Prep. Biomed. Environ. Anal. (Proc. Chromatogr. Soc. Int. Symp. 1991). Eds. D Stevenson and I D Wilson. pp 127–138. Plenum Press, New York.

Zaborska W 1995 Competitive inhibitors of free and chitosan-immobilised urease. Acta Biochem. Pol. 42, 115–118.

Plant and Soil **193:** 15–33, 1997.

Chapter 2

Techniques for boron determination and their application to the analysis of plant and soil samples

Ram N. Sah and Patrick H. Brown
*Department of Pomology, University of California, Davis, CA 95616, USA**

Abstract

This paper reviews techniques for determining B concentration and isotopic ratio and their application to soil and plant samples. Boron concentration has been determined utilising spectrophotometry, potentiometry, chromatography, flame atomic emission and absorption spectrometry, inductively coupled plasma (ICP) optical emission (OES) and mass spectrometry (MS), and neutron activation analysis using neutron radiography and prompt-γ activation analysis. Isotopic ratios of B have been measured by ICP–MS, thermal ionisation mass spectrometry (TIMS) and secondary ion mass spectrometry (SIMS). For isotopic measurements, TIMS and SIMS are more sensitive and provide higher degrees of accuracy and resolution than ICP–MS, however, extensive sample preparation and purification, and time-consuming measurements limit their usefulness for routine analyses.

While the spectrophotometric technique using a colorimetric reaction of B with azomethine-H has been the most extensively applied B determination method for soil and plant samples, colorimetric methods, in general, suffer from numerous interferences and have poor sensitivity and precision. The prompt-γ method can determine B concentration in intact samples which enables this method to be especially useful for some applications in agriculture. Research involving B behaviour in plant and soil environments would benefit from this technology. In recent years, the use of ICP–OES and ICP–MS for B determination in plant and soil samples has grown tremendously. The application of ICP–OES brought a significant improvement in B analysis because of its simplicity, sensitivity and multielement detection capability. However, besides matrix interferences, the two most sensitive emission lines for B suffer strong spectral interference from Fe. The ICP–OES is not adequately sensitive for some nutritional work involving low B concentrations and B translocation studies using the isotope tracer technique.

Plasma is one of the most effective analyte ionisers and MS is the most sensitive ion detector. Coupling of plasma with MS resulted in the development of plasma source MS technology (ICP–MS) which has outperformed all previous analytical methods for trace element determination. Boron determination by ICP–MS suffers no spectroscopic interferences, and is considered the most practical and convenient technique for B isotope determination. The ability of ICP–MS to measure isotopic ratios as well as B concentration enables: (1) B concentration determination by the isotope dilution method, (2) verification of B concentration by isotope fingerprinting in routine analysis and (3) determination of total B concentration as well as B isotope ratio in the same run for biological tracer studies. Therefore, ICP–MS is the method of choice among the present-day technologies for determining B concentration and a convenient method for B isotope determination. In recent years, new generations of plasma-source MS instruments have been developed using alternative plasma generation methods and high-resolution mass spectrometers. These instruments are expected to bring further improvements in accuracy, sensitivity and precision of B determination.

Introduction

Boron is an essential element for plants and its deficiency affects plant growth and yield in many parts of the world. Boron is a beneficial micronutrient for animal and humans (Nielsen, Chapter 13), however, its essentiality has not yet been established. Excessive amounts of B are toxic to plants and limit crop productivity in many environments (Nable et al., Chapter 12). Determination of the forms and functions in plants involves the use of B isotope tracer studies and requires accurate measurements of B concentration and isotopic

* E-mail:phbrown@ucdavis.edy

ratios (^{11}B:^{10}B). Thus the ability to precisely resolve small quantities of B and determine isotopic compositions are essential for B research in plants and soils.

A number of analytical techniques ranging from a simple potentiometric method to a method requiring a nuclear reactor have been used for B determination. The evolution of the B determination methodology has followed the course of advancement in analytical instrument technology. Most of the earlier methods involved spectrophotometric B determination utilising colour-forming reactions and, to a limited extent, fluorimetric reactions of B. The invention of flame atomic emission spectrometry (AES) and atomic absorption spectrometry (AAS) initiated the age of the atomic spectrometric methods of B determination. However, sensitivity and accuracy of B determination remained less than desirable because of memory effects and interferences in these methods. Parallel development in the application of neutron activation analysis (NAA) has special significance in B determination in intact biological materials. The ^{10}B isotope (approx. 20% abundance) undergoes a neutron capture reaction when placed in a beam of thermal neutrons producing a characteristic α particles and γ rays, both of which have been used for B determination.

The introduction of plasma as an ionisation source and the resulting development of inductively coupled plasma (ICP) optical emission spectrometry (OES) resulted in a significant improvement in B determination technology. The plasma source OES provided higher sensitivity and lower detection capability for B determination than was possible by spectrophotometric, flame AES/AAS and neutron activation methods. The levels of detection and accuracy further improved as a result of coupling of a quadrupole mass spectrometer (MS) as a detector with plasma (ICP) into a new instrument, ICP–MS. The ICP–MS is also a practical and convenient technique for B isotope determination. In recent years, new generations of plasma-source MS instruments have been developed using alternative plasma generation methods and high resolution mass spectrometers. These instruments are expected to bring further improvements in plasma source methods of B determination.

Despite the above mentioned developments in B determination techniques, there are still some problem areas where improvement is needed. Boron tends to adhere to the sample introduction system of the analytical instruments and hence raise the background, affecting subsequent determinations. This phenomenon, called "memory effect" presents a major problem in B analysis, especially if a sample containing low B concentration follows a high concentration sample. Several suggestions have been forwarded in recent years to manage the memory effects. These include, running only relatively dilute concentrations of B and increasing wash time (Ward et al., 1990), running a sodium fluoride (NaF) wash solution between samples (Evans and Krahenbuhl, 1994b; Luguera et al., 1991), frequent measurement of a known reference sample to correct preceding and following B values (Kucharkowski et al., 1996) and the use of direct injector nebuliser (DIN) (Smith et al., 1991; Vaughan and Claassen, 1996). Use of DIN may be the most effective method at present, but its cost may be prohibitive for many laboratories. Instrument memory may have serious implications for the determination of isotope ratios when there is a wide variation in the isotopic composition of samples. However, B isotope ratio determination by ICP–MS may also be affected by the nature and source of sample material (Bassett, 1990; Vanderpool et al., 1994), instrument parameters (Gregoire, 1990), and sample matrices (Gregoire, 1987, 1990). Since B has only two stable isotopes, instrumental isotope fractionation during mass spectrometry is difficult to correct (Aggrawal and Palmer, 1995). Lastly, instruments are only the means of achieving good analytical results. The dexterity of the analyst and sample preparation/handling methods employed for B determination have significant roles in the accuracy and precision of analytical results.

The purpose of this article is to examine the methods for B determination and their application to the analysis of plant, soil and water samples considering sample preparation methods as well as strengths and weaknesses of various B analytical techniques. This article presents information in the following sequence: (1) a survey and brief description of techniques used for B determination in various fields, (2) description of sample extraction/dissolution, matrix reduction and analyte preconcentration techniques applicable to soil, plant and water samples, and (3) application of B analytical techniques for the determination of B concentration and isotope ratios in plant and soil samples.

Boron determination techniques

In this section we will give a brief description of various analytical methods for B determination. Application of these methods to plant and soil samples will follow.

Spectrophotometric techniques

Colorimetric methods

This approach uses specific reactions of B with compounds capable of forming colour-producing B complexes. Organic dyes such as azomethine-H (Ciba and Smolec, 1994; Roignavarro et al., 1996; Wolf, 1974), curcumin (Rand, 1975; Williams, 1979), carmine (Bingham, 1982; Rand, 1975), quinalizarine, arsenazo, crystal violet (Garcia et al., 1985), and methylene blue (Williams, 1979) produce coloured complexes with B, and have been used for colorimetric B determination.

The azomethine-H method, perhaps the most commonly used spectrophotometric method, is based on the formation of a coloured complex with boric acid at pH 5.1. The absorbance at 420 nm (absorption maximum) is linear between the concentration range of 0.5 and 10 mg B L^{-1}. This method suffers fewer interferences (Lopez et al., 1993) and is more reliable, fast, simple, sensitive and convenient than other colorimetric methods (Bingham, 1982). A modification of this method using azomethine-HR (1-(2,4-dihydroxy-benzylideneamino)-8-hydroxynaphthalene-3,6-disulfonic acid, a derivative of azomethine-H) increased the sensitivity 3.5-fold relative to the azomethine-H method, but the azomethine-HR method suffered interferences from Al, Cu, Fe, Ti, and Zr (Zenki et al., 1989).

Two B-curcumin complexes, rubocurcumin and rosocyanin, are of practical significance for B analysis. Commonly used curcumin methods employ the reddish-brown rosocyanin complex which has a 545 nm absorption maximum at about pH 1.0. Parashar et al. (1989) developed the rosocyanin complex by adding a mixture of sulfuric acid (H_2SO_4) and acetic acid (1:1) and curcumin to the sample, allowing the mixture to stand in darkness for two hours followed by the addition of acetate buffer and acetone. A variation of this method described by Bingham (1982) is based on the complexing of boric acid with 2-ethyl-1, 3-hexanediol (in chloroform) and conversion to rosocyanin complex using a solution of curcumin in glacial acetic acid followed by the addition of H_2SO_4. The carmine method uses the complex of carmine (an anthraquinone dye) with boric acid in concentrated H_2SO_4 which has a 585 nm absorption maximum and 0.5 – 10 mg B L^{-1} linear detection range. The direct (without matrix separation) determination of less than 0.5 mg B L^{-1} by the carminic acid method is almost impossible in the presence of dissolved substances (Bebek et al., 1996). The chemical reagents required for B determination by the curcumin and carmine methods are hazardous, limiting their utility.

Fluorimetric method

The fluorimetric method of B determination is based on the formation of fluorescent complexes of B and the measurement of fluorescence induction at a specific wavelength. The difference between the various reported fluorimetric methods is mainly due to the choice of reagents to form fluorescent compounds (Chimpalee et al., 1993). Examples from the published reports include the use of Alizarin Red S (the sodium salt of 1,2-dihydroxyanthraquinone-3-sulfonic acid) (Blanco et al., 1993; Campana et al., 1992; Chimpalee et al., 1993), chromotropic acid (Capitan et al., 1991) and dibenzoylmethane-isobutyl methyl ketone (Aznarez et al., 1983). The reported detection limits are 0.34 μg B mL^{-1} (Chimpalee et al., 1993), 0.3-0.7 μg B mL^{-1} (Blanco et al., 1993) and 7.2 ng B mL^{-1} (Campana et al., 1992). Though more sensitive than colorimetric techniques, this method suffers interferences from a number of chemical species and is sensitive to pH and temperature. Therefore, fluorimetric methods are not as commonly used for B determination as colorimetric methods.

Potentiometric methods

Potentiometric methods involve separation of B from the sample matrix, treatment with HF and development of the resulting tetrafluoroborate ion (BF_4^-) which is measured potentiometrically with a suitable BF_4^- selective electrode (Carlson and Paul, 1968, 1969). A detailed description of the use of resin columns for the matrix separation and BF_4^- generation is given by Bingham (1982). Potentiometric methods that do not require B separation from the sample have also been reported (Imato et al., 1990; Yakimova and Markova, 1992). Potentiometric methods, however, are severely affected by the sample matrix and have not been widely utilised. Two polarographic methods were recently reported for the determination of trace amounts of B in foods (Lu et al., 1994), soil and plant samples (Zheng, 1994), but their usefulness has not been widely tested or adopted. Potentiometric methods have not been very popular for plant and soil samples, possibly due to the complex matrices of these samples, however, interest in this technology remains (Borovskii et al.,

1991; Dolaberidze et al., 1989; Wood and Nicholson, 1994; Zheng, 1994).

Atomic spectrometric methods

Flame atomic spectrometric methods
Flame atomic spectrometry involves two methods: (1) flame atomic emission spectrometry (AES) and (2) flame atomic absorption spectrometry (AAS). Flame AAS is based on the principle that free atoms of an element (e.g. B) at their ground state absorb photons of discrete energy values (a characteristic wavelength) generated by a hollow cathode lamp containing that element (e.g. B). The AES method measures emission from the atomised and excited species when they return to ground state. Samples are generally introduced into a flame (usually the combustion of acetylene–air or acetylene–N_2O–air) where elements of the sample are atomised. Boron determination by flame AAS/AES methods was recently described by Usenko and Prorok (1992) and Zakhariya et al. (1991). These methods generally have poor sensitivity for B (Papaspyrou et al., 1994) and often require separation of B from the sample matrix for acceptable results (Botelho et al., 1994).

The electrothermal vaporisation (ET)–AAS method can determine B in solid or liquid samples without requiring sample digestion and atomisation at high temperatures (Szydlowski, 1979; Botelho et al., 1994; Wiltshire et al., 1994). The ET–AAS method without the use of chemical modifiers has poor detection limits and sensitivity due to inefficient thermal dissociation of B-containing species and severe memory effects (Luguera et al., 1991; Wiltshire et al., 1994). Use of chemical modifiers (e.g. nickel+zirconium salt, Ca–Mg; a list is compiled by Botelho et al., 1994), coating of the pyrolytic graphite tube with chemical agents (e.g. lanthanum carbide, tungsten carbide, zirconium solution) and increasing the pyrolysis temperature are recommended for improving the performance of the ET–AAS method (Barnett et al., 1991; Gong et al., 1995; Liu et al., 1994; Wiltshire et al., 1994). The use of two chemical modifiers for the ET vaporisation, (1) diammonium hydrogen phosphate (to suppress matrix interferences) and (2) NaOH (to retain analyte on the surface of the vaporiser) were reported to be effective in suppressing the loss of B, enhancing detection limits and reducing matrix interferences (Okamoto et al., 1994). The usefulness of this procedure has not yet been confirmed by other researchers. Because of the lack of adequate sensitivity and precision and the need for additional preparation procedures, AES/AAS and ET–AAS methods have not been extensively used for B determination in soil and plant samples.

Plasma-source methods
Plasma is a mass of ionised gas in which numbers of electrons and positive ions are approximately equal (zero total discharge). Argon is most commonly used to generate the plasma, but a number of other gases or their mixtures have also been used (Evans and Giglio, 1993). Most commercial plasma-source instruments use an argon ICP (i.e. ICP generated from argon) for the analyte ionisation, however, other types of plasma, such as direct current plasma (DCP) (Brennan and Svehla, 1989; Urasa, 1984), microwave induced plasma (MIP) (Evans and Caruso, 1993) and glow discharge plasma (GDP) (Sheppard and Caruso, 1994) have also been reported. An ICP provides high temperatures (8000 to 10 000 K), and is an excellent ionisation source for optical emission (ICP–OES) as well as mass spectrometric (ICP–MS) methods.

Generally, samples are transformed (digested) to a liquid state, converted to fine mists using a system composed of a nebuliser and a spray chamber, and introduced into the plasma using a sample injector. Several alternate modes of sample introduction (e.g. slurry, powder, gases, laser ablation, electrothermal vaporisation (ETV) etc.) are used for specific purposes (Evans and Giglio, 1993). Sample introduction into the plasma as a slurry, powder or gas avoids the digestion step. The laser-ablation method uses a laser beam for the ablation of solid samples before it is introduced to the plasma. In the ETV method, heating of the sample can be controlled for selective dissociation and removal or introduction of various components (e.g. solvents, matrices, analytes) of a sample. A detailed discussion of various sample introduction methods is given by Sah (1995). ETV has a significant advantage over other sample introduction methods as it can selectively volatilise and remove solvents from the sample analytes. The addition of mannitol to the B analyte solution significantly improved B detection limits by ETV ICP–MS probably due to improved B transport to the plasma as a B-mannitol complex (Wei et al., 1995). The presence of HF in the sample also increased B transport into the plasma as borotrifluoride and improved the emission signal (Fucsko et al., 1993). The presence of HF in methanolic solutions, however, decreased the ICP emission signals by replacing the

B–OCH_3 bond with B-F bonds and the formation of a less volatile B–fluoro complex (Canals et al., 1985, 1986; Canals and Hernandis, 1987). The presence of HF in the sample also increases contamination from glass and causes damage to the sample introduction system and the ICP–MS interface, thereby limiting its utility.

Plasma-source OES. The development of ICP–OES revolutionised the analysis of several so called "problem elements" such as B, S, Mo and all hard to detect trace elements by virtue of its low detection limits, large linear range, and multi element detection capability. The detection limit and precision for B determination by ICP–OES are better than all previous methods. The basic procedure involves the introduction of solid, liquid or gaseous samples into the plasma and the measurement of atomic emission from the analytes of interest. The high energy atmosphere of the plasma pulverises the sample (if solid), breaks chemical bonds and ionises the sample analytes. The signal is generally linear over a large range. Duffy and Thomas (1996) reported the simultaneous determination of B, P and S using a dual-view ICP–OES. The ICP–OES also allows flow injection (FI) sample introduction to deal with small amounts of samples and samples with difficult matrices. The FI ICP–OES resulted in a near unity dispersion ratio, high linearity of the concentration–peak height relationship and increased the sample throughput (Kempster et al., 1989).

Plasma-source MS. The ICP–MS is the result of coupling an extremely efficient ion source (ICP) with an extremely sensitive ion detection technique (MS). Both ICP–OES and ICP–MS share ICP as the ionisation source and both have the same systems for introduction. In the case of ICP–MS, ions generated in the plasma (at atmospheric pressure) pass through an assembly of a sampler and a skimmer cone to a highly evacuated MS area. An ion optic system focuses ions to the MS which measures mass-to-charge ratios of the ion(s) of interest. The commonly used quadrupole MS acts as a mass filter that allows only a given mass-to-charge ratio of ions to pass to the detector. The ions passing through the MS are deflected to an ion detector that converts the ionic energy to electric energy which forms the basis for the measurement of the analyte concentration. For multielement analyses, the parameters are sequentially changed to allow the passage of other ions of differing mass-to-charge ratios to the detector. Thus, multielement analysis by ICP–MS is, in fact, sequential single element analysis. However, the high speed scanning capability of the quadrupole MS allows the measurement of each element (mass-to-charge ratio) in a few milliseconds thus a suite of several elements and their isotope compositions can be measured in a matter of minutes. The ICP–MS also provides a fast semiquantitative analysis of the entire range of elements of the Periodic Table in two-to-three minutes. The advantages of ICP–MS over other methods are higher sensitivity, lower detection limits and simultaneous measurement of B concentrations and B isotope ratios (^{11}B to ^{10}B). The quadrupole MS can operate at relatively higher pressure but has lower sensitivity and mass resolution than high-resolution MS systems such as magnetic sector MS, time of the flight MS or Fourier transform ion-cyclotron resonance (FTIR) MS. High-resolution ICP–MS using magnetic sector MS is now commercially available. An ICP–MS MS uses a two-step MS system, a quadrupole MS as an ion filter, and a high resolution MS for ion detection. An ICP–OES–MS determines an analyte by simultaneous OES and MS measurements.

The ability of ICP–MS to measure the B-isotope ratio makes it especially suitable for biological B tracer studies (Brown and Hu, 1996). The reported detection limits for B are one μg B L^{-1} (Smith et al., 1991) to three μg B L^{-1} in digests of biological materials (Evans and Krahenbuhl, 1994b), 0.15 μg B L^{-1} in saline waters (Gregoire, 1990) and 0.5 μg B L^{-1} in human serum (Vanhoe et al., 1993). The ICP–MS can carry out B determination by the isotope dilution method which is considered the most precise for quantitative analysis. Some researchers have used FI technique for separation/preconcentration of B and determination by the isotope dilution method to improve sensitivity and precision and detection capabilities (Coedo et al., 1996; Jin et al., 1996).

The external calibration procedure with an internal standard is the most widely used method for plasma-source OES and MS analysis because of its simplicity and labor efficiency. Beryllium (Be, mass = 9) being close in mass to B is an ideally suited internal standard for B determination by ICP–MS (Vanhaecke et al., 1991). Sample introduction by nebulisation for the external calibration method, using Be as the internal standard, works well for most liquid samples with simple matrices. Standard addition and isotope dilution methods are less common and are generally employed to deal with difficult sample matrices or to improve precision. Alternate sample introduction methods and calibration methods are generally employed for non-liquid samples.

Other mass spectrometry. Thermal ionisation mass spectrometry (TIMS) has been used for isotopic determination of BO_2^+ or BO_2^- salts of Na, Cs or Rb (Beary and Xiao, 1990; Ding et al., 1994; Nakamura et al., 1992). Both negative and positive TIMS methods provide a high degree of accuracy and precision (Eisenhut et al., 1996; Hemming and Hanson, 1994; Vengosh et al., 1989; Ziao, 1991; Xiao et al., 1988) but laborious sample preparation steps (Bassett, 1990; Xiao, 1991) and long sample determination times (Johnson, 1989) make this technique less desirable for routine B isotope determination. Isobaric interference may affect the accuracy of isotope ratios if the sample is not well purified and/or the instrument is not sufficiently cleaned to remove B contamination from the previous sample.

All high-resolution MS methods require decomposition of the source material and analyte purification. For geological and other complex materials this is not only difficult and time consuming but may also result in the loss of B and isotope fractionation. Secondary ion mass spectrometry (SIMS) can directly determine B and its isotope ratio in solid samples (Chu et al., 1994). It employs sputtering of the surface atoms from small areas of the sample by an energetic primary ion beam of Ga^+, Cs^+ or O^- and the detection of the secondary ions at mass 10 or 11 by a mass analyser. The disadvantages of this method, besides the high cost, are (1) difficulty in determining absolute B concentration as different matrices may provide different ion yields (2) poorer resolution than TIMS for isotopic ratio measurements, and (3) difficulty in the analysis of volatile phases of B (Aggrawal and Palmer, 1995).

Ion chromatography

Ion chromatography involves separation of ions of interest on an exchange column and analysis after their elution from the column. Determination of B in extracts of gypsiferous soils by a single-column high-performance liquid chromatography (HPLC) with a resin-based anion exchange column using an eluent containing sodium hydroxide and sodium benzoate and a conductivity detector suffered interference from cations and individual simple organic acids (Dudley, 1989). For a satisfactory B analysis by this technique, cation and sulfur interferences had to be removed from the solution being determined. Determination of B in soils, sediments, and water samples by ion-exclusion chromatography using D-sorbitol as the mobile phase was comparable to values obtained using the colorimetric azomethine-H method (Frankenberger et al., 1990). For a successful ion chromatographic method, B must be present in an acceptable ionic form which is far from the reality for many soil and plant extracts and digests. There is no evidence of the adoption of this technique for plant and soil B analysis in any significant extent.

Neutron activation analysis

Neutron activation analysis (NAA) is performed by placing the sample in a beam of thermal neutrons (of about 0.025 eV associated energy with no charge) and measuring the reaction product(s). When B in a sample is placed in an incident beam of thermal neutrons, the ^{10}B isotope (approx. 20% abundance in natural samples) undergoes the following neutron-capture reaction (Moore, 1990).

$$
\begin{aligned}
{}^{10}\mathrm{B} + \mathrm{neutron} \;&=>{}^{7}\mathrm{Li} + 4\mathrm{He} \\
&\qquad (\alpha\,\mathrm{particles}, 2.31\,\mathrm{MeV}) \\
&\qquad +\mathrm{gamma\ ray}\ (478\mathrm{keV})
\end{aligned}
$$

All NAA-based B determination methods are based on the measurement of the reaction products of this reaction. The method based on the measurement of α particles is called neutron capture radiography while that based on the measurement of gamma (γ) ray is called prompt-γ activation analysis. These techniques can determine B in intact samples, but require access to a nuclear reactor for the production of thermal neutrons. Nuclear methods involving non-neutron high-energy beams have also been used for B analysis, but discussion on these methods is beyond the scope of this article. Readers are referred to a recent review of these methods by Sah and Brown (1997).

Neutron capture radiography
For this technique, also called α-track etching, samples containing ^{10}B (e.g. biological materials) in contact with a detector film are placed in a neutron beam. Subsequently, the film is stained, reversed and etched with KOH or NaOH to image the microscopic distribution of the ^{10}B in the sample (Laurent-Pettersson et al., 1992). The quantification of B is possible using an image analyser. To avoid the detection loss caused by self-absorption, the thickness of the B-containing material should be less than the alpha particles range,

typically 1 to 10 micrometers (Lelental, 1972). Other methods under this category include neutron activation mass spectrometry (Clarke et al., 1987; Iyengar et al., 1990) and neutron depth profiling (Lamaze et al., 1993). The advantage of this method is that it allows the measurement of B distribution in intact tissues.

Prompt-γ activation analysis
Prompt-γ activation analysis is a non-destructive method of B determination in intact samples based on the measurement of the γ ray emitted from the disintegrating ^{10}B nuclei upon neutron capture (Matsumoto et al., 1991; Pillay and Peisach, 1992). The limits of accurate quantification for food materials ranged from 1.0 to 2.5 mg B kg^{-1} (Anderson et al., 1990a) while detection limits were 0.3 to 0.8 mg B kg^{-1} sample depending on the irradiation time (Anderson et al., 1990a; Yonezawa and Wood, 1995). As B concentration in the sample decreases, counting time to achieve desired precision increases logarithmically (Rogus et al., 1992). This method has found extensive use for B determination in medical science, especially in the area of B neutron capture therapy. The single most significant drawback of this method is that it requires access to a nuclear reactor for the source of thermal neutrons. The neutron radiography method also requires extensive sample preparation.

Sample preparation considerations

Extraction of boron from soils

Boron present in soils is commonly divided into three categories: (1) total B (2), acid-soluble B and (3) water-soluble B. In a recent study, soil B was fractionated into five types, namely (1) readily soluble B (0.01 M $CaCl_2$) representing <2% of total soil B, (2) specifically adsorbed B (0.05 M KH_2PO_4) representing <2% of total soil B, (3) oxide bound B (0.20 M acidic NH_4-oxalate) roughly 2.3% of total soil B, (4) organically bound B (0.02 M HNO_3 + 30% H_2O_2) with approximately 8.6% of total soil B and (5) the residual or occluded B (HNO_3 + HCl + HF) (Hou et al., 1994). Total B content of a soil is not a reliable indicator of available B or the need for B. The acid-soluble B content of a soil is a somewhat better indicator but is not a reliable index of plant available B, especially in calcareous soils that neutralise the acidity of the extractant. Berger and Truog (1940) proposed a hot water extraction (HWE) for 5 min (with refluxing) as an index of available B in the soil. The HWE method has been widely used and B extracted by this method often correlates well with plant B uptake (Shuman et al., 1992; Self and Gupta, 1993). The hot-water extraction procedure is, however, difficult to standardise (Novozamsky et al., 1990b), time consuming and tedious for routine and reproducible usage (Deabreu et al., 1994). The amount of B extracted by this method is affected by the reflux time (McGeehan et al., 1989), extraction time and temperature (Spouncer et al., 1992) and B may be resorbed during the cooling period (Gupta, 1967). The coloured hot water extracts in some soils may affect B determination. A treatment with activated charcoal to remove the colour of the extract (Gupta, 1979) may lower B concentration in the extract because of B sorption on charcoal (McGeehan et al., 1989).

Several convenient alternatives to the HWE method have been proposed to estimate the amount of plant available B in soils. These include (1) microwave heating of the soil sample and the extractant in sealed plastic pouches (Mahler et al., 1984), (2) extraction with water or 2 mM DTPA at room temperature (Handreck, 1990), (3) extraction with 0.01 M $CaCl_2$ + 0.05 M mannitol at 20 °C (Cartwright et al., 1983), (4) extraction with water or a solution of $BaCl_2$ and microwave heating (Deabreu et al., 1994), (5) extraction with hot 0.01 M $CaCl_2$ (Aitken et al., 1987), (6) extraction with sorbitol (Vaughan and Howe, 1994) and (7) extraction with 1.0 N NH_4OAc (pH 7.0) (Sakal et al., 1993). Cold-dilute $CaCl_2$ extracts less B than hot water or hot-dilute $CaCl_2$ (Aitken et al., 1987; Cartwright et al., 1983), however, B extracted with cold water or 0.01 M $CaCl_2$ correlates well with the hot extraction method (Kaplan et al., 1990; Novozamsky et al., 1990b). In light of clay dispersion and filtration problems commonly encountered in the extraction of soils with either cold or hot water, the alternative extraction method using $BaCl_2$ (Jeffrey and McCallurn, 1988; Parker and Gardener, 1981; Vaughan and Howe, 1994) appears appropriate as this method would also allow the determination of Ca by ICP–OES or MS in the same extract.

The HWE method has limitations for some soils and B extracted by this method did not correlate with crop responses under some management conditions (Gestring and Soltanpour, 1987; Mustafa et al., 1993; Offiah and Axley, 1988). Extraction with 0.05 M HCl worked well for predicting B availability to crop plant in acid soils (Ponnemperuma et al., 1981; Renan and Gupta 1991). However, Fe extracted with the acid extractant often interferes in B determination

Table 1. Levels of boron in soil extracts considered critical for adequate plant growth and yield

Method	Authors	Crop	Concentration (mg B L^{-1})	Remarks
Hot water	Chude and Obigbesan (1984)	Cocoa trees (*Theobroma cocoa*)	0.44–0.88	Nigeria
0.1% $CaCl_2$	Chude and Obigbesan (1984)	Cocoa trees	0.56–1.12	Nigeria
1 N ammonium acetate (pH 7.0)	Sakal et al. (1993)	chickpea (*Cicer arietinum*)	0.31	Calcareous soil
Hot water	Hue et al. (1988)	*Desmodium cannum*[1]	0.3 – 1.6	Hawaii
Hot water	Rashid et al. (1994)	*Brassica* sp.	0.5 – 0.6	Pakistan
HCl ext.	Rashid et al. (1994)	*Brassica* sp.	0.45 – 0.55	Pakistan
Mannitol ext.	Rashid et al. (1994)	*Brassica* sp.	0.4 – 0.48	Pakistan
Hot water	Chang et al. (1983)	Papaya (*Carica papaya*)	0.28	Black soils
Hot water	Chang et al. (1983)	Papaya	0.15	Other soils

[1] *D. canum* is a high B requiring tropical forage legume plant.

by spectrophotometric and ICP–OES methods (Evans and Krahenbuhl, 1994a; Pougnet and Orren, 1986a, 1986b). Levels of B in soils for different extraction methods considered adequate for a variety of crop plants are reported in Table 1.

Decomposition of soil samples

Complete decomposition of soil, geological and silica-rich materials are generally accomplished either by alkali fusion (Smith et al., 1991) or by wet digestion using HF or a mixture of HF with other acids (Alaimo and Censi, 1992; Ricci et al., 1994; Vanhaecke et al., 1991). The alkali fusion involves the fusion of a small amount (usually <1 g) of finely ground samples with an alkali flux in a platinum or nickel crucible at about 1000 – 1500 °C and dissolving the fusion cake with dilute acid. It allows rapid decomposition of a relatively large amount of samples but extraction of B from the sample may be less than quantitative (Aggrawal and Palmer, 1995). Although Na_2CO_3 is the most extensively used flux for alkali fusion, use of other fluxes such as NaOH, KOH and Cs_2CO_3 are also reported (Hofstetter et al., 1991; Musashi et al. 1990; Smith et al., 1991). Beary and Xiao (1990) noted significant advantages of Cs_2CO_3 over Na_2CO_3 for TIMS determination of B. To avoid interferences from free Fe in Na_2CO_3 fusion in colorimetric and ICP–OES methods, Self and Gupta (1993) suggested a two-step fusion method, preliminary P fusion followed by fusion with KOH.

Wet acid sample decomposition is carried out by heating the sample with a mixture of HF, HNO_3 and $HClO_4$. The digest is usually evaporated to dryness or near dryness (Hu, 1991). This process causes the loss of B owing to the high volatility of BF_3 (Ishikawa and Nakamura, 1990; Pougnet and Orren, 1986a, 1986b; Zarcinas and Cartwright, 1987) and there are chances of isotopic fractionation (Nakamura et al., 1992). Suggested preventive methods are: (1) addition of mannitol (to form B–mannitol complex) during the HF digestion and the control of evaporation temperature (Chen et al., 1991; Nakamura et al., 1992) and (2) use of orthophosphoric acid in the $HNO_3/HClO_4/HF$ digestion mixture and controlled temperature evaporation of the digests (Evans and Meisel, 1994; Xu and Rao, 1986).

Decomposition of biological materials

Dry ashing

In this approach, the sample is commonly ashed in a muffle furnace using suitable containers (e.g. quartz or platinum crucibles) by a method similar to those described by Gaines and Mitchell (1979) or Brown et al. (1992) and dry-ashed samples are dissolved in dilute acid (usually HNO_3) for analysis. The temperature of the muffle furnace during the ashing period is important to avoid B loss from the sample. Alwarthan et al. (1993) decomposed seed and pulp of date palm

(*Phoenix dactylifera*) by treating the sample with saturated $Ca(OH)_2$, drying at 105 °C, volatilisation over a burner and then dry ashing in a muffle furnace for colorimetric determination of B.

Open-vessel wet ashing
Strong mineral acids such as HNO_3, H_2SO_4, $HClO_4$ or their mixtures are added to the sample in digestion containers (ideally B-free) and heated (usually over a hot plate with some reflux) to decompose the sample. Novozamsky et al. (1993) suggested the use of HNO_3, $HClO_4$ and hydrogen peroxide for decomposition of plant materials.

High-pressure dissolution
The sample is treated with some acid (HCl, HNO_3 or HF or their mixtures) and hydrogen peroxide in a high-pressure Teflon (PTFE) vessel and the closed vessel is heated in a microwave oven. The heating source may also be a hot plate or an oven with high-pressure digestion bombs. The microwave dissolution method is also used for the decomposition of non-biological materials (Coedo et al., 1993; Rivoldini and Cara, 1992). Boron in the microwave digests may then be determined by spectrophotometric (Ferrando et al., 1993), ICP–OES (Ziaziaris and Kacprzak, 1995), and ICP–MS (Stotesbury et al., 1989; Vanhoe et al., 1993) methods. The digest of biological materials usually contains a large amount of dissolved carbon and organic compounds which tend to interfere with B isotope determination by the ICP–MS method.

Alkali fusion
The method is essentially the same as used for soil and other materials. The sample is fused with alkali (e.g. Na_2CO_3) and redissolved for the analysis. The alkali fusion of biological materials caused negligible volatilisation loss of B and resulted in 80–95% recovery of spiked ^{10}B (Smith et al., 1991) but the high-salt environment of the fused materials may cause matrix interferences for a number of analytical techniques (Gregoire, 1987). There are a few reports of some unique and uncommon digestion methods such as the low temperature RF ashing (Daughtrey and Harrison, 1974) and extraction of biological samples with HCl or H_2SO_4 (Dawson et al., 1990). These methods have not been adequately assessed.

The use of B-containing digestion containers such as borosilicate glass should be avoided to prevent B leaching from the container. New Pyrex glass digestion tubes contribute more background B than older tubes (Gestring and Soltanpour, 1981b). HNO_3 is preferred for wet ashing for plasma source methods because it provides a simpler matrix than other mineral acids. Some reports have expressed concerns over the accuracy of B determination after dry ashing in a muffle furnace (Wikner, 1986; Schnug and Haneklaus, 1996). During the combustion process, the sample B may be subject to volatilisation loss or contamination from fire-clay inserts that will not be detected in the blank unless they contain an alkaline substance (Schnug and Haneklaus, 1996).

Published reports on the comparative evaluation of various sample decomposition methods are often contradictory. Sample-B concentrations determined by the wet-ashing method were reported equal to (Pennington et al., 1991; Spiers et al., 1990), lower than (Ciba and Chrusciel, 1992) as well as higher than (Bañuelos and Akohoue, 1994; Bañuelos et al., 1992) those by dry ashing and the microwave dissolution methods. We have recently found quantitative recovery of B and no difference in B values in the National Institute of Standard and Technology (NIST) biological Standard Reference materials (SRMs) decomposed by either of the three dry or wet ashing methods (Nyomora et al., 1997). These findings are consistent with recent findings of Ferrando et al. (1993) and Bratter et al. (1995). We also found a quantitative recovery of B in several biological materials extracted with hot 1M HNO_3 and determined by ICP–MS (Nyomora et al., 1997). Dry ashing has been proven highly reliable in many laboratories while it was apparently inadequate in others. The temperature at which the sample is ashed may affect the recovery of B. It is also likely that individual furnaces and other undetermined variables affect the reproducibility of this method. The recovery and reproducibility should therefore be verified for each laboratory and sample type.

Separation and preconcentration of boron

The presence of substances such as high salts, organic substances and other analyte species may cause interference in B determination. There are also chances of matrix-related ionisation suppression and mass discrimination causing errors in B determination by plasma source OES and MS methods (Gregoire, 1987, 1990). Some matrix substances can also affect the precision of B isotope ratio determination (Aggrawal and Palmer, 1995). The analytes of interest are separated from the sample matrix if the matrix interference reduction methods such as matrix matching, standard addition calibration and isotope dilution methods are not adequate to overcome matrix effects. If the concentration of the analyte in the original sample is too low to be accurately determined, then separation is combined with analyte preconcentration in the new matrix. Published B separation and preconcentration methods may be classified into: (1) exchange separation (Smith et al., 1991), (2) chelation (Gregoire, 1987), (3) gaseous separation as methyl borate (Castillo et al., 1985a, 1985b) and (4) solvent extraction (Garcia et al., 1985).

Exchange separation

The B-specific anion-exchange resin Amberlite IRA-743 has been extensively used for the separation and preconcentration of B from difficult sample matrices (Evan and Meisel, 1994; Smith et al., 1991). Amberlite IRA-743 resin is manufactured by Rohm and Hass company (Philadelphia, PA, USA) and marketed by Sigma chemical company (Saint Louis, MO, USA). This resin contains tertiary amine as a functional group on a hydrophobic polystyrene base which binds B by forming a borate polyol complex. The theoretical B binding capacity of this resin isabout 5.7 mg B mL^{-1} resin. There is conflicting information on the species of B bound on Amberlite IRA-743. While some reports suggest that B should be in the anionic $[B(OH)_4^-]$ form (Aggrawal and Palmer, 1995; Coedo et al., 1993), the manufacturer's product information sheet states that Amberlite IRA-743 binds borate, boric acid as well as some related species. For the quantitative separation, Aggrawal and Palmer (1995) suggested adjusting solution pH to 10 or higher before loading on a column of the preconditioned Amberlite IRA-743 resin. For additional purity of B for isotopic determination, the sample may be passed through a cation exchange resin following the exchange separation of B on Amberlite IRA-743 (Aggrawal and Palmer, 1995). The use of Amberlite IRA-743 resulted in quantitative recovery (Hemming and Hanson, 1994) and substantial improvements in the detection limits of B in complex matrices (Bebek et al., 1996; Chapman et al., 1996). However, the requirement of relatively high solution pH for effective complex formation of B on Amberlite IRA-743 limits the application of this method for strongly acidic digests (Coedo et al., 1993).

The exchange separation method has also been used to separate interfering matrix ions from the sample matrix. An FI system utilising a one-step matrix separation method using a non-selective cation exchange resin column has been used successfully to remove interfering cations (Wang et al., 1996). A two-step separation plus preconcentration system utilising (1) a strongly acidic cation exchange non-selective (Type 732) resin to eliminate cations from the sample followed by (2) a microcolumn of Type VS-II strongly basic anion exchange resin for adsorption separation and preconcentration of B as $H_2BO_3^-$ under basic medium, pH 11.2 also shows promise (Jin et al., 1996). An alternative method for the ion exchange separation of B from acid (HF) digests was proposed by Nakamura et al. (1992) for isotopic ratio determination by TIMS. The sample B is loaded on a cation exchange resin column as a B–mannitol complex and eluted with 0.02 N HF solution. Cations are retained on the column and hence removed from the sample, but the B–mannitol complex acts as an anion and is not retained on the cation exchange column. Following the cation separation, the sample is treated with a few drops of 3% phosphoric acid, evaporated to near dryness and treated with 3 N HF and warmed in sealed Teflon beakers at 80 °C for 15 min. Following these treatments, the sample solution is loaded on an anion exchange resin column which retains B as BF_4^- and eluted with a mixture of dilute HCl and HF.

Chromatographic separation of B using an ion-exchange column was carried out after converting B into BF_4^- ion using 10% HF (Hill and Lash, 1980) or as ionic complexes of B with chromotropic acid (Jun et al., 1988) and H-resorcinol (Motomizu et al., 1990). When HPLC separation of inorganic ions is used in conjunction with a conductivity detector, it is commonly called an ion chromatograph, as several ions can be determined simultaneously. Ion chromatography has been used for B determination (Hill and Lash, 1980; Ricci et al., 1994), however, the sensitivity of this method is not adequate for most applications (Carpio and Mariscal, 1992).

Gaseous separation as methyl borate

Boron reacts with methanol to form gaseous methyl borate which is separated from the sample matrix (Molinero et al., 1993). The process may be performed as a manual distillation (Sanz et al., 1990a, 1990b) or an on-line continuous-flow technique (Johnson et al., 1992; Novozamsky et al., 1988). Castillo et al. (1985a, 1985b) generated methyl borate from waters and plant samples in concentrated H_2SO_4 medium to utilise the heat generated by the hydration of H_2SO_4 for methyl borate volatilisation. This technique is used to separate B from a number of other matrices, such as wine (Sanz et al., 1990a), rocks (Musashi et al., 1990) and biological tissues (Johnson et al., 1992).

Solvent extraction

Boron is extracted as its organic complexes such as (1) the B-2-ethyl-1,3-hexanediol complex of B and extracted with chloroform or benzene (Agazzi, 1967), (2) the 2,4-dinitro-1,8-naphthalenediol complex of B and extracted with toluene (Kuwada et al., 1978), and (3) the complex of B with 2,4-dimethyl-2,4-octanediol in isopentanol (Panov et al., 1989). Novozamsky et al. (1990a) suggested a solvent extraction system based on the formation of BF_4^- ions that were extracted with liquid ion exchanger (Aliquot 336, tricapryl-methyl-ammonium chloride) in xylene for ICP–OES determination. On-line solvent extraction has been suggested (Panov et al., 1989). A flow injection (FI) isotope dilution ICP–MS determination method used solvent extraction for the separation of Fe from the digests using acetylacetone-chloroform (Coedo et al., 1996). Presence of solvents in the sample, however, may limit their applicability to plasma-source methods of B determination (Shabanova et al., 1985).

Application of boron determination methods to plant and soil analysis

Colorimetric methods

There are numerous reports of applications of colorimetric methods for B determination in soil, plant and water samples. A partial list of published methods is given in Table 2. The azomethine-H method is the most extensively used colorimetric procedure, but colorimetric methods using curcumin, carmine and other reagents have also been reported for B determination (Aznarez et al., 1983; Hofstetter et al., 1991). The colorimetric methods, in general, suffer several interferences, such as sample pH in the range of 6.4 to 7.0 (Carrero et al., 1993), sample colour (McGeehan et al., 1989; Evans and Krahenbuhl, 1994a), nitrate complexes in the wet HNO_3 acid digests of plants (Gestring and Soltanpour, 1981a) and the presence Fe, Al, Cu, Zn and Mo (Arruda and Zagatto, 1987). These interferences and lack of sensitivity limit the application of these methods for the samples with low B concentrations and complex matrices.

Interference reduction methods for the colorimetric B determination in plant and soil extracts include the addition of thioglycolic acid (Zarcinas, 1995), a buffer-masking solution containing polyphosphate ions, thiourea and ascorbic acid (Ferran et al., 1988) and 2-methylpentane-2,4-diol in isobutyl methyl ketone (Aznarez et al., 1983) mainly to suppress the interference from Fe and other ions. Addition of charcoal to soil extracts prior to filtering reduces sample colour interferences (McGeehan et al., 1989). The FI sample introduction systems were applied to the existing chromotropic acid (Lussier et al., 1992) and azomethine-H methods (Carrero et al., 1993; Chen et al., 1989; Sekerka and Lechner, 1990). Nogueira et al. (1993) used a monosegmented FI method while the FI system of Nose and Zenki (1991) measured B following the change in the methyl orange indicator as a result of the change in acidity of a sorbitol solution in presence of boric acid.

Fluorimetric method

Spectrofluorimetric methods have been applied to agricultural samples (Table 1) using several fluorimetric agents. Examples are, chromotropic acid as a fluorimetric reagent for plant tissues, fertilisers and natural waters (Capitan et al., 1991; Motomizu et al., 1991), dibenzoylmethane-isobutyl methyl ketone and concentrated phosphoric acid for soil and plant extracts (Aznarez et al.,1983), and Alizarin Red S for plant tissues (Blanco et al., 1993). The fluorimetric method for microwave digests of biological tissues using carminic acid suffered from several problematic interferences (Evans and Krahenbuhl, 1994a). Fluorimetric methods for B, in general, suffer problematic interferences and have not been widely used for plant and soil samples.

Plasma-source OES

Some examples of the applications of the ICP–OES method for B determination in various sample types are presented in Table 3. Reported detection limits for B are 10 to 15 μg B L^{-1} in soil solutions and plant digests (Spiers et al., 1990) and 25 μg B L^{-1} in the microwave digest of biological tissues (Pollmann et al., 1993). Iron interferes with the two most sensitive B lines 249.773 nm (B1) and 249.678 nm (B2) in the ICP–OES method (Pritchard and Lee, 1984). If the sample has high iron concentrations (as encountered in the digests of soil, plants, and some geologic materials) then B 249.773 nm and B 249.678 nm lines cannot

Table 2. A partial list of published spectrophotometric methods for boron analysis

Author	Material	Methods	Interference
Ogner (1980)	Soil extract	Fluorimetric method	
Parker and Gardner (1981)	Soil extract	Azomethine-H	
Porter et al. (1981)	Soil extract	Azomethine-H	
Lohse (1982)	Plant	Azomethine-H	
Garcia et al. (1985)	Plant and natural waters	Crystal violet	
Lee et al. (1987)	Plant	Azomethine-H	
Chen et al. (1989)	Soil	Azomethine-FI and carminic acid	
McGeehan et al. (1989)	Soil	Azomethine-H	
Zenki et al. (1989)	Water	Azomethine-HR	Al, Cu, Fe, Ti and Zr
Sekerka and Lechner (1990)	Water	Azomethine-FI + Preconc.	
Kaplan et al. (1990)	Soil	Azomethine-H	Organoborates
Hofstetter et al. (1991)	Geological material	Carminic acid	
Bañuelos et al. (1992)	Plant	Azomethine-H	
Campana et al. (1992)	Soil, water, plant	Fluorimetric	
Al-Warthan et al. (1993)	Fruit (dates)	Quinalizarin	Al, Cu, Na, Mg and analine
Lopez et al. (1993)	Water	Azomethine-H	
Nogueira et al. (1993)	Plant	Monosegmented	
Zarcinas (1995)	Soil, plant, fertiliser.	Azomethine-H	Fe

Table 3. Applications of the ICP–OES methods for boron determination in plant and soil

Authors	Material analysed	Method used	Remarks
Pritchard and Lee (1984)	Plant and soil sample	ICP–OES	Used 249.678 nm wave length
Nilsson and Jennische (1986)	Plant	DCP-OES	Agreed with colorimetric B
Lee et al. (1987)	Plant	ICP–OES	HNO_3 Digest gave low B
Jeffrey and McCallum (1988)	$CaCl_2$ extract of soil	ICP–OES	
Hunt and Shuler (1989)	Biological materials	ICP–OES	Low-temp wet ashing
Novozamsky et al. (1990b)	Plant	ICP–OES	N/A
Barth et al. (1991)	Biological materials	DCP-OES	
Hu et al. (1991)	Plant leaves	ICP–OES	Fluorinating ETV method
Goto et al. (1992)	Hot water extract of soil	ICP–OES	Fe, Ca and S interfering ions
Spouncer et al. (1992)	Soil	ICP–OES	
Ferrando et al. (1993)	Plant, animal	ICP–OES	N/A
Evans and Krahenbuhl (1994a)	Plant, animal	ICP–OES/MS	Fe for ICP–OES

be used because of the overlaps of Fe 249.782 nm on B 249.773 nm and Fe 249.653 nm on B 249.678 nm (Kucharkowski et al., 1996; Xu and Rao, 1986). Large amounts of Fe often associated with acid extracts of soil may cause severe interference with B determination by ICP–OES (Evans and Krahenbuhl, 1994a; Pougnet and Orren, 1986a, 1986b). The most sensitive wavelengths for B determination also suffered severe interferences from the components of fertiliser materials (K, Fe, P) (Matilainen and Tummavuori, 1995). Boron determination by ICP–OES is also affected by other interfering species, for example, Si (Owens et al., 1982; Din, 1984), Ni, Cr, Al, V, Mn, Ti, Mo and high concentrations of Na (Kavipurapu et al., 1993; Pougnet and Orren, 1986b). Suggested interference management techniques of B determination by ICP–OES analysis are (1) conversion of sample-B to gaseous methoxyborate (Johnson et al., 1992), (2) successive fusion of geological materials with potassium dihydrogenphosphate and potassium hydroxide to obtain iron-free solution (Din, 1984), (3) the use of Cu as internal standard in sea water (Kato and Takashima, 1990) and (4) slurry introduction of plant leaves in a fluorinating ETV ICP–OES (Hu et al., 1991). Analyte separation and preconcentration techniques detailed earlier would also be applicable.

Plasma source MS

Boron determination is not affected by isobaric interferences and by spectroscopic interferences from the elements originating from water, acid or plasma gas (Evans and Giglio, 1993). Use of Be as an internal standard corrected for matrix-induced interferences (Evans and Krahenbuhl, 1994a, 1994b) and suppression of ^{10}B signals (Vanhoe et al., 1993). For B determination in microwave digests of biological materials, the external calibration method using Be as internal standard was as good as isotope dilution or standard addition methods (Evans and Krahenbuhl, 1994a, 1994b). However, if samples vary in acidity, the use of Be as an internal standard may result in errors in the determination of B/Be ratios by ICP–MS (Evans and Krahenbuhl, 1994a).

Isotope tracer techniques and B determination by the isotope dilution method require accurate measurement of B isotope ratios. Some instrumental parameters and the sample matrix may cause unequal transmission of ions of differing masses resulting in mass discrimination that affects the accuracy of B isotope ratio determination (Beauchemin et al., 1987; Gregoire, 1987). The matrix-related mass discrimination, in general, results in more severe suppression of ^{10}B relative to ^{11}B thus increasing the $^{11}B/^{10}B$ ratio (Gregoire, 1987, 1990). Spectral overlap of the ^{12}C peak on ^{11}B peak may also affect $^{11}B/^{10}B$ ratio determination if the sample contains high levels of carbon as commonly found in the microwave digest of biological materials (Evans and Krahenbuhl, 1994b). We noted a significant difference in the $^{11}B/^{10}B$ ratios due to sample decomposition methods for biological materials (Nyomora et al., 1997). Where isotope ratio measurement is not required, total B should be computed on the basis of ^{10}B intensity to avoid error from the overlap of the ^{12}C-peak (Vanhoe et al., 1993). Readers are referred to two recent reviews (Evans and Giglio, 1993; Sah, 1995) dealing with interferences in the ICP–MS analysis. Sample purification techniques discussed in the earlier section will be useful in overcoming interferences in the samples with difficult matrices such as soil extracts. Contact of the sample or reagents with borosilicate glass should be avoided as much as possible throughout the sample preparation and the sample introduction. The presence of HF is especially problematic in increasing the background due to the dissolution of glass by HF. If HF or $HClO_4$ are used for wet ashing, the digest should be evaporated to dryness and redissolved in HNO_3 for ICP–MS determination (Alaimo and Censi, 1992). A Galan-type nebuliser made of an organic polymer and a PTFE sample introduction system that eliminates the sample-glass contact has been suggested for HF digests (Vanhaecke et al., 1991).

Neutron-capture radiography

Thellier and his associates pioneered the neutron capture radiography in plants (Hartman et al., 1984; Martini and Thellier, 1993). Jimenez and associates have also applied this technique to study the absorption and transport of ^{10}B-boric acid by leaves and fruit skin of mango (*Mangifera indica*) following the deposition of ^{10}B enriched boric acid drops on the specimens (Jimenez et al., 1988; Loria and Jimenez, 1993; Loria et al., 1992). Neutron radiography is an excellent technique to study microdistribution of B in intact biological materials. This technique has been widely applied in medical science (Moore, 1990), but its use in agricultural fields has been very limited.

Prompt-γ activation analysis

Anderson et al. (1990a) determined B concentrations in biological reference by prompt-γ activation analysis (PGAA) and reported 1.0 to 2.5 $\mu g\ g^{-1}$ limit of quantification, 0.3 to 0.8 $\mu g\ g^{-1}$ limit of detection depending on the irradiation time. The analytical uncertainty for materials with $\geq$30 $\mu g\ g^{-1}$ B concentrations was $\leq$2%. Food and mineral supplements analysed by the PGAA method before and after freeze-drying indicated a significant loss of B for the wet composite (Anderson et al., 1990b). Hight et al. (1993) used PGAA for the analysis of dietary supplements. In a recent study, PGAA was used to determine H, B, Na, S, Cl, and K in several food products (Anderson et al., 1994).

Conclusions

This paper reviews methods of B determination and their application to plant and soil analysis. A number of methods based on spectrophotometric, potentiometric, atomic spectrometric and NAA are reported for the routine B determination. Spectrophotometric methods utilising colorimetric reactions (colorimetric determination) have been most commonly used for B determination in soil and plant samples. Plasma source atomic spectrometric methods, ICP–OES and ICP–MS, have been adopted for soil and plant analysis as the methods of choice and their use has increased over recent years. The NAA method, especially prompt-γ spectrometry, can detect B concentration in intact samples including live plant tissues. This method is potentially very useful for B research but has remained unutilised. However, NAA and plasma source methods require sophisticated and expensive instruments which may be a limitation for many analytical and research laboratories.

Spectrophotometric methods require low-cost instrumentation, but are not capable of detecting low concentrations of B required for many nutritional and other applications. These methods suffer from numerous interferences, have poor precision and often fail to accurately determine B concentrations. Performance of potentiometric, flame AAS and AES and ion chromatographic methods are less than satisfactory for the routine B analysis of soil and plant samples. The loss of B from the sample during the sample preparation process (drying and decomposition) has been subject of continued argument among researchers. Neutron activation methods avoid the necessity of sample preparation. Prompt-γ activation analysis has been used for B determination in intact materials in other biological fields. This technique looks attractive for some applications for plant and soil samples. However, this method requires access to a nuclear reactor for the source of thermal neutrons and the method is not sensitive nor time efficient for low concentrations of B. The ICP–OES provides better detection limits and sensitivity than earlier methods (e.g. spectrophotometric, potentiometric, flame AES and AAS etc.), however, the two most sensitive B emission lines have strong spectral interference from Fe. The ICP–OES method also cannot measure B isotope ratios required for many B-tracer studies in plant and soil. Plasma-source MS methods (e.g. ICP–MS) can detect lower concentrations of B than ICP–OES which may be important in some nutritional research. The superiority of ICP–MS over other methods is also attributed to its capability to measure B isotopes, which has the following implications: (1) B concentration determination by isotope dilution, (2) verification of B concentration by isotope fingerprinting in routine analysis, and (3) determination of total B concentration as well as B isotope ratio in the same run for biological tracer studies. Currently, ICP–MS is the technology of choice for B concentration determinations. It is also the most practical and convenient technique for B isotope determination. New generations of plasma-source MS instruments using alternative plasma generation methods and high-resolution mass spectrometers are expected to bring further improvements in plasma-source MS methods of B determination.

References

Agazzi E J 1967 Extraction-flame photometric determination of boron. Anal. Chem. 39, 223–235.

Aggrawal J K and Palmer M R 1995 Boron isotope analysis – A review. Analyst 120, 1301–1307.

Aitken R L, Jeffery A L and Compton B L 1987 Evaluation of selected extractants for boron in some Queensland soils. Aust. J. Soil Res. 25, 263–273.

Alaimo R and Censi P 1992 Quantitative determination of major, minor and trace elements in USGS rock standards by inductively coupled plasma-mass spectrometry. Atom Spectrom. 13, 113–117.

Alwarthan A A, Alshowiman S S, Altamrah S A and Baosman A A 1993 Spectrophotometric determination of boron in dates of some cultivars grown in Saudi Arabia. J. AOAC Int. 76, 601–603.

Anderson D L, Cunningham W C and Lindstrom T R 1994 Concentrations and intakes of H, B, S, K, Na, Cl and NaCl in foods. J. Food Comp. Anal. 7, 59–82.

Anderson D L, Cunningham W C and Mackey E A 1990a Determination of boron in food and biological reference materials by neutron capture prompt-gamma activation. Fres. J. Anal. Chem. 338, 554–558.

Anderson D L, Cunningham W C and Mackey E A 1990b Neutron capture prompt-gamma activation analysis of foods. Biol. Trace Elem. Res. 26/27, 613–622.

Arruda M A Z and Zagatto E A G 1987 A simple stopped-flow method with continuous pumping for the spectrophotometric flow-injection determination of boron in plants. Anal. Chim. Acta 199, 137–145.

Aznarez J, Bonilla A and Vidal J C 1983 Spectrophotometric and fluorimetric determination of boron in soils plants and water by extraction with 2-methylpentane-2 4-diol in isobutyl methyl ketone. Analyst 108, 368–373.

Bañuelos G S and Akohoue S 1994 Comparison of microwave digestion with block digestion for selenium and boron analysis in plant tissues. Comm. Soil Sci. Plant Anal. 25, 1655–1670.

Bañuelos G S, Cardon G, Pflaum T and Akohoue S 1992 Comparison of dry ashing and wet acid digestion on the determination of boron in plant tissue. Comm. Soil Sci. Plant Anal. 23, 2383–2397.

Barnett N, Ebdon L, Evans E H and Ollivier P 1991 The matrix modification of boron and its determination by graphite furnace atomic absorption spectrometry slurry atomization using totally pyrolytic graphite tubes. Microchem. J. 44, 168–178.

Barth R F, Adams D M, Soloway A H, Mechetner E B, Alam F and Anisuzzaman A K M 1991 Determination of boron in tissues and cells using direct-current plasma atomic emission spectroscopy. Anal. Chem. 63, 890–893.

Bassett R L 1990 A critical evaluation of the available measurements for the stable isotopes of boron. Appl. Geochem. 5, 541–554.

Beary E S and Xiao Y K 1990 Rapid and high-precision determination of boron isotope ratios in boron carbide by thermal ionisation mass spectrometric measurement of the dicaesium metaborate cation. Analyst 115, 911–913.

Beauchemin D, McLaren J W and Berman S S 1987 Study of the effects of concomitant elements in inductively coupled plasma mass spectrometry. Spectrochim. Acta 42B, 467–490.

Bebek M, Mitko K and Kwapulinsk J 1996 Determination of aluminium, barium, molybdenum, scandium, beryllium, titanium, vanadium, fluoride and boron in highly salinated waters. Water Sci. Tech. 33, 349–356.

Berger K C and Truog E 1940 Boron deficiencies as revealed by plant and soil tests. J. Am. Soc. Agron. 32, 297–301.

Bingham F T 1982 Boron. *In* Methods of Soil Analysis – Part 2: Chemical and Microbiological Properties. Eds. A L Page, R H Miller and D Keeney. pp 431–446. American Society of Agronomy and Soil Science Society of America Publishers, Madison, Wisconsin, USA.

Blanco C C, Campana A G, Barrero F A and Ceba M R 1993 Simultaneous spectrofluorimetric determination of traces of molybdenum and boron in plant leaves. Anal. Chim. Acta 283, 213–223.

Borovskii E S, Ragoizha G E, Voitovich A I, Rakhmanko E M and Gulevich A L 1991 Potentiometric determination of boron in borophosphosilicate glasses. Indust. Lab. USSR 57, 683–685.

Botelho G M A, Curtius A J and Campos R C 1994 Determination of boron by electrothermal atomic absorption spectrometry, testing different modifiers atomization surfaces and potential interferents. J. Anal. Atom Spectrom. 9, 1263–1267.

Bratter V E N, Bratter P, Reinicke A, Schulze G, Alvarez W O L and Alvarez N 1995 Determination of mineral and trace elements in total diet by inductively coupled plasma atomic emission spectrometry, comparison of microwave-based digestion and pressurized ashing systems using different ashing mixtures. J. Anal. Atom Spectrom. 10, 487–491.

Brennan M C and Svehla G 1989 Flow injection determination of boron copper molybdenum tungsten and zinc in organic matrices with direct current plasma optical emission spectrometry. Fresenius J. Anal. Chem. 335, 893–899.

Brown P H and Hu H 1996 Phloem mobility of boron is species dependent, evidence for phloem mobility in sorbitol-rich species. Ann. Bot. 77, 497–505.

Brown P H, Picchioni G, Jenkins M and Hu H 1992 Use of ICP–MS and ^{10}B to trace the movement of boron in plants and soil. Comm. Soil Sci. Plant Anal. 23, 2781–2807.

Campana A M G, Barrero F A and Ceba M R 1992 Spectrofluorimetric determination of boron in soils plants and natural waters with Alizarin red-S. Analyst 117, 1189–1191.

Canals A and Hernandis V 1987 Fluoride interference on the boron inductively coupled plasma atomic emission in methanol solutions. J. Anal. Atom Spectrom. 2, 379–381.

Canals A, Hernandis V and Sala V 1985 Fluoride interference in the molecular emission of boron in water/methanol media. Anal. Chim. Acta 169, 377–383.

Canals A, Hernandis V and Sala J V 1986 Flame atomic emission spectrometric determination of boron in methanolic solutions: influence of fluoride on the solute transport efficiency. J. Anal. Atom Spectrom. 1, 277–280.

Capitan F, Navalon A, Manzano E, Capitan-vallvey L F and Vilchez J L 1991 Determination of boron with chromotropic acid by 1st-derivative synchronous spectrofluorimetry. Fres. J. Anal. Chem. 340, 6–10.

Carlson R M and Paul J L 1968 Potentiometric determination of boron as tetrafluoroborate. Anal. Chem. 40, 1292–1295.

Carlson R M and Paul J L 1969 Potentiometric determination of boron in agricultural samples. Soil Sci. 108, 266–272.

Carpio R A, Mariscal R and Welch J 1992 Determination of boron and phosphorus in borophosphosilicate thin films on silicon substrates by capillary electrophoresis. Anal. Chem. 64, 2123–2129.

Carrero P, Burguera J L, Burguera M and Rivas C A 1993 Time-based injector applied to the flow injection spectrophotometric determination of boron in plant materials and soils. Talanta 40, 1967–1974.

Cartwright B, Tiller K G, Zarcinas B A and Spouncer L R 1983 The chemical assessment of B status of soils. Aust. J. Soil Res. 21, 321–332.

Castillo J R, Mir J M, Martinez C and Bendicho C 1985a Determination of boron waste in vegetal matter by atomic absorption spectrometry using direct atomization of methyl borate as volatile phase. Atom Spectrosc. 6, 152–155.

Castillo J R, Mir J M, Martinez C and Bendicho C 1985b Determination of boron in waste waters by using methyl borate generation and flame atomic-emission spectrometry. Analyst 110, 1435–1438.

Chang S S, Hu N H, Chen C C and Chiu T F 1983 The diagnostic criteria of boron deficiency in papaya and the soil boron status of Taitung area. J. Agric. Res. China 32, 238–252.

Chapman V J, Kerven G L, Edwards D G, Asher C J and Blammey F P C 1996 Determination of low boron concentrations in nutrient solution. Plant and Soil 180, 97–99.

Chen D, Lazaro F, Decastro L and Valcarcel M 1989 Direct spectrophotometric determination of total boron in soils with ultrasonic leaching in automatic flow systems. Anal. Chim. Acta 226, 221–227.

Chen J S, Lin H M and Yang M H 1991 Spectrophotometric determination of trace and ultratrace levels of boron in silicon and chlorosilane samples. Fres. J. Anal. Chem. 340, 357–362. Chimpalee N, Chimpalee D, Boonyanitchayakul B and Burns D T 1993 Flow-injection spectrofluorimetric determination of boron using alizarin red s in aqueous solution. Anal. Chim. Acta 282, 643–646.

Chu P K, Bleiler R J and Metz J M 1994 Determination of sub-parts per billion boron contamination in n+ Czochralski silicon substrates by SIMS. J. Electrochem. Soc. 141, 3453–3456.

Chude V O and Obigbesan G O 1984 A comparison of soil test methods for estimating available soil boron in cocoa growing areas of southwestern Nigeria. Beitrage-zur-Tropischen-Landwirtschaft-und-Veterinarmedizin 22, 245–254.

Ciba J and Chrusciel A 1992 Spectrophotometric determination of boron in human hair with azomethine-H. Fres. J. Anal. Chem. 342, 147–149.

Ciba J and Smolec B 1994 Determination of boron in steel by an ICP emission spectrometric technique. Fres. J. Anal. Chem. 348, 215–217.

Clarke W B, Koekebakker M, Barr R D, Downing R G and Fleming R F 1987 Analysis of ultratrace lithium and boron by neutron activation and mass spectrometric measurement of ^3He and ^4He. Appl. Radiat. Isot. (Int. J. Radiat. Appl. Instrum. Part A) 38, 735–747.

Coedo A G, Dorado T, Escudero E and Cobo I G 1993 Boron determination in steels by inductively coupled plasma atomic emission spectrometry – comparative study of spark ablation and pneumatic nebulization sampling systems. J. Anal. Atom Spectrom. 8, 827–831.

Coedo A G, Dorado T, Fernandez B J and Alguacil F J 1996 Isotope dilution analysis for flow injection ICPMS determination of microgram per gram levels of boron in iron and steel after matrix removal. Anal. Chem. 68, 991–996.

Daughtrey E H Jr and Harrison W W 1974 Analysis for trace levels of boron by exchange-hollow cathode emission. Anal. Chim. Acta 72, 225–230.

Dawson B S W, Parker G F, Cowan F J, Croucher M C, Hong S O and Cummins N H 1990 Interlaboratory determination of boron compounds in preservative treated timber. Anal. Chim. Acta 236, 423–430.

Deabreu C A, Deabreu M F, Vanraij B and Bataglia O C 1994 Extraction of boron from soil by microwave heating for ICP-AES determination. Comm. Soil Sci. Plant Anal. 25, 3321–3333.

Din V K 1984 The preparation of iron-free solutions from geological materials for the determination of boron (and other elements) by inductively coupled plasma emission spectrometry. Anal. Chim. Acta 159, 387–391.

Ding T P, Zhao D M and Pan M 1994 Analytic method of boron isotopes by thermal ionisation mass spectrometry of $Rb_2BO_2^+$. Chin. Sci. Bull. 39, 1714–1719.

Dolaberidze L D, Zhgenti K A, Chkhetiani N A and Dzhaliashvili A G 1989 Ionometric determination of boron in rocks and ores. Indust. Lab. USSR 55, 31–32.

Dudley L M 1989 Determination of borate in extracts of gypsiferous soils by single column ion chromatography. Comm. Soil Sci. Plant Anal. 20, 483–499.

Duffy M and Thomas R 1996 Benefits of a dual-view ICP–OES for the determination of boron phosphorus and sulfur in low alloy steels. Atom Spectrosc. 17, 128–132.

Eisenhut S, Heumann K G and Vengosh A 1996 Determination of boron isotopic variations in aquatic systems with negative thermal ionisation mass spectrometry as a tracer for anthropogenic influences. Fres. J. Anal. Chem. 354, 903–909.

Evans E H and Caruso J A 1993 Low pressure inductively coupled plasma source for mass spectrometry. J. Anal. Atom Spectrom. 8, 427–431.

Evans E H and Giglio J J 1993 Interferences in inductively coupled plasma mass spectrometry – a review. J. Anal. Atom Spectrom. 8, 1–18.

Evans S and Krahenbuhl U 1994a Boron analysis in biological material – microwave digestion procedure and determination by different methods. Fres. J. Anal. Chem. 349, 454–459.

Evans S and Krahenbuhl U 1994b Improved boron determination in biological material by inductively coupled plasma mass spectrometry. J. Anal. Atom Spectrom. 9, 1249–1253.

Evans S and Meisel T 1994 Low blank determination of boron in geochemical materials. Anal. Chim. Acta 298, 267–270.

Ferran J, Bonvalet A and Casassas E 1988 New masking agents in the azomethine-H method for boron determination in plant tissues. Agrochimica 32, 171–181

Ferrando A A, Green N R, Barnes K W and Woodward B 1993 Microwave digestion preparation and ICP determination of boron in human plasma. Biol. Trace Elem. Res. 37, 17–25.

Frankenberger W T, Huysmans K D and Mehra H C 1990 Determination of borate at trace levels in environmental samples by ion-exclusion chromatography. J. Chromatography 508, 265–270.

Fucsko J, Tan S H, La H and Balazs M K 1993 Fluoride interference in the determination of boron by inductively coupled plasma emission spectroscopy. Appl. Spectrosc. 47, 150–155.

Gaines T P and Mitchell G A 1979 Boron determination in plant tissue by the azomethine-H method. Comm. Soil Sci. Plant Anal. 10, 1099–1108.

Garcia I L, Cordova M H and Sanchez-Pedrono C 1985 Sensitive method for the spectrophotometric determination of boron in plants and waters using crystal violet. Analyst 110, 1259.

Gestring W D and Soltanpour P N 1981a Boron analysis in soil extracts and plant tissue by plasma emission spectroscopy. Comm. Soil Sci. Plant Anal. 12, 733–742.

Gestring W D and Soltanpour P N 1981b Evaluation of wet and dry digestion methods for boron determination in plant samples by ICP-AES. Comm. Soil Sci. Plant Anal. 12, 743–753.

Gestring W D and Soltanpour P N 1987 Comparison of soil tests for assessing boron toxicity to alfalfa. Soil Sci. Soc. Am. J. 51, 1214–1219.

Gong B, Liu Y, Xu Y, Li Z and Lin T 1995 Direct determination of boron in a cobalt-based alloy by graphite furnace-atomic absorption spectrometry. Talanta 42, 1419–1423.

Goto I, Muramoto J and Ninaki M T I 1992 Application of inductively coupled plasma atomic emission spectrometry (ICP-AES) to soil analysis (Part 5) determination of hot water soluble boron in soils by ICP-AES. Jap. J. Soil Sci. Plant Nutr. 63, 53–57.

Gregoire D C 1987 Determination of boron isotope ratios in geological materials by inductively coupled plasma mass spectrometry. Anal. Chem. 59, 2479–2484.

Gregoire D C 1990 Determination of boron in fresh and saline waters by inductively coupled plasma mass spectrometry. J. Anal. Atom Spectrom. 7, 623–626.

Gupta U C 1967 A simplified method for determining of hot-water-soluble boron on podzol soil. Soil Sci. 103, 424–428.

Gupta U C 1979 Some factors affecting the determination of hot-water-soluble boron from podzol soils using azomethine-H. Can. J. Soil Sci. 59, 241–247.

Handreck K A 1990 Methods of assessing boron availability in potting media with special reference to toxicity. Comm. Soil Sci. Plant Anal. 21, 2265–2280.

Hartman A, Wissocg J C, Thellier M, Borner H, Mampe W, Fourcy A, Fantini M and Larsson B 1984 Use of cold neutrons to increase the sensitivity of detection of stable isotopes with (n, alpha) nuclear reaction for tracer experiments. Physiol. Veg. Paris 22, 887–895.

Hemming N G and Hanson Gg N 1994 A procedure for the isotopic analysis of boron by negative thermal ionisation mass spectrometry. Chem. Geol. 114, 147–156.

Hight S C, Anderson D L, Cunningham W C, Capar S G, Lamont W H and Sinex S A 1993 Analysis of dietary supplements for nutritional, toxic and other elements. J. Food Comp. Anal. 6, 121–139.

Hill C J and Lash R P 1980 Ion chromatographic determination of boron as tetrafluoroborate. Anal. Chem. 52, 24–27.

Hofstetter A, Troll G and Matthies D 1991 Determination of trace amounts of fluorine boron and chlorine from a single sodium carbonate fusion of small geological sample masses. Analyst 116, 65–67.

Hou J, Evans L J and Spiers G A 1994 Boron fractionation in soils. Comm. Soil Sci. Plant Anal. 25, 1841–1853.

Hu B, Jiang Z and Zeng Y 1991 Slurry sample introduction with fluorinating electrothermal vaporisation for the direct ICP-AES determination of boron in plant leaves. Fres. J. Anal. Chem. 340, 435–438.

Hu W D 1991 Determination of boron in high-purity silica using direct current plasma emission spectrometry. Anal. Chim. Acta 245, 207–209.

Hue N V, Hirunburana N and Fox R L 1988 Boron status of Hawaiian soils as measured by B sorption and plant uptake. Comm. Soil Sci. Plant Anal. 19, 517–528.

Hunt C D and Shuler T R 1989 Open-vessel wet-ash low-temperature digestion of biological materials for inductively coupled argon plasma spectroscopy (ICAP) analysis of boron and other elements. J. Micronutr. Anal. 6, 161–174.

Imato T, Yoshizuka T and Ishibashi N 1990 Potentiometric flow-injection determination of boron by using a flow-through tetrafluoroborate ion-selective poly(vinyl chloride) membrane electrode. Anal. Chim. Acta 233, 139–141.

Ishikawa T E and Nakamura E 1990 Suppression of boron volatilisation from hydrofluoric acid solution using boron-mannitol complex. Anal. Chem. 62, 2612–2616.

Iyengar G V, Clarke W B and Downing R G 1990 Determination of boron and lithium in diverse biological matrices using neutron activation–mass spectrometry (NA–MS). Fres. J. Anal. Chem. 338, 562–566.

Jeffrey A J and McCallum L E 1988 Investigation of a hot 0.01 *M* $CaCl_2$ soil boron extraction procedure followed by ICP-AES analysis. Comm. Soil Sci. Plant Anal. 19, 663–673.

Jimenez R, Loria L G, Gallardo M and Thellier M 1988 Detection and imaging of boron in plant tissues for agronomical purposes in Costa Rica using small 252CF neutron sources. Comp. Rendus. Academie Sci. III Sciences de la Vie 306, 583–590.

Jin Q, Zhang H Q, Liang F, Yang W J and Jin Q 1996 Determination of trace amounts of boron by microwave plasma porch atomic emission spectrometry using an on-line separation and preconcentration technique. J. Anal. Atom Spectrom. 5, 331–337.

Johnson P E 1989 Methodology for stable isotope analysis in biological materials (a review). J. Micronutr. Anal. 6, 59–83.

Johnson D A, Siemer D D and Bauer W F 1992 Determination of nanogram levels of boron in milligram-sized tissue samples by inductively coupled plasma atomic emission spectroscopy. Anal. Chim. Acta 270, 223–230.

Jun J, Mitsuko O and Shoji M 1988 Determination of boron with chromotropic acid by high-performance liquid chromatography. Analyst 113, 1631–1638.

Kaplan D I, Burkman W, Adriano D C, Mills G L and Sajwan K S 1990 Determination of boron in soils containing inorganic and organic boron sources. Soil Sci. Soc. Am. J. 54, 708–714.

Kato K and Takashima K 1990 Determination of traces of boron in a mixed solution of heavy water and light water by ICP-AES with internal standardization. Bunseki Kagaku 39, 139–143.

Kavipurapu C S, Gupta K K, Dasgupta P, Chatterjee N N and Pandey L P 1993 Determination of boron in steels by inductively coupled plasma optical emission spectrometry. Analusis 21, 21–25.

Kempster P L, Vlict H R and van Staden J F 1989 Determination of boron in water by flow-injection inductively coupled plasma emission spectrometry. Anal. Chim. Acta 218, 69–76.

Kucharkowski R, Muller E and Wustkamp D 1996 Determination of boron in amorphous and nanostructured soft magnetic alloys ribbons and layered materials by ICP atomic emission spectrometry. Fres. J. Anal. Chem. 355, 256–260.

Kuwada K, Motomizu S and Toei K 1978 Solvent extraction-spectrophotometric determination of boron with 2,4-dinitro-1,8-naphthalenediol and brilliant green. Anal. Chem. 50, 1788–1792.

Lamaze G P, Downing R Gg, Pilione L, Badzian A and Badzian T 1993 Analysis of boron in CVD diamond surfaces using neutron depth profiling. Appl. Surface Sci. 65/66, 587–592.

Laurent-Pettersson M, Delpech B and Thellier M 1992 The mapping of natural boron in histological sections of mouse tissues by the use of neutron-capture radiography. Histochemical J. 24, 939–950.

Lee J J, van-der Walinga I, Manyeki P K, Houba V J G and Novozamsky I 1987 Determination of boron in fresh and in dried plant material by plasma emission spectrometry after extraction with HF-HCl. Comm. Soil Sci. Plant Anal. 18, 789–802.

Lelental M 1972 Determination of submicrogram amounts of boron using the B-10 (n alpha) Li-7 reaction. Anal. Chem. 44, 1270–1272.

Liu Y M, Gong B, Xu Y, Li Z and Lin T 1994 Determination of boron in iron- and nickel-based alloys by graphite furnace atomic absorption spectrometry with a zirconium nickel chemical modifier and a zirconium-pretreated graphite tube. Anal. Chim. Acta 292, 325–328.

Lohse G 1982 Microanalytical azomethine-H method for boron determination in plant tissue. Comm. Soil Sci. Plant Anal. 13, 127–134.

Lopez F J, Gimenez E and Hernandez F 1993 Analytical study on the determination of boron in environmental water samples. Fres. J. Anal. Chem. 346, 984–987.

Loria L G and Jimenez R 1993 Autoradiography by neutron capture (RNC) using the nuclear reaction B-10 (n, alpha) Li7. Tecnologia Marcha 12, 61–68.

Loria G, Jimenez R, Gallardo M, Martini F and Thellier M 1992 A study of the distribution and uptake of boron in mango fruits using neutron-capture radiography. Comptes Rendus Academie Sciences Series-III Sciences de la Vie 315, 255–264.

Lu G, Li X and Deng Y 1994 Polarographic determination of trace boron in foods. Food Chem. 50, 91–93.

Luguera M, Madrid Y and Camara C 1991 Combination of chemical modifiers and graphite tube pre-treatment to determine boron by atomic absorption spectrometry. J. Anal. Atom Spectrom. 8, 669–672.

Lussier T, Gilbert R and Hubert J 1992 Determination of boron in light-water and heavy-water samples by flow injection analysis with indirect UV visible spectrophotometric detection. Anal. Chem. 64, 2201–2205.

Mahler R L, Naylor D V and Fredrickson M K 1984 Hot water extraction of boron from soils using sealed plastic pouches. Comm. Soil Sci. Plant Anal. 15, 479–492.

Martini F and Thellier M 1993 Boron distribution in parenchyma cells of clover leaves. Plant Physiol. Biochem. 31, 777–786.

Matilainen R and Tummavuori J 1995 Determination of boron in fertilisers by inductively coupled plasma-atomic emission spectrometry – studies of some spectral interferences at different wavelengths. J. AOAC Int. 78, 598–604.

Matsumoto T, Aoki M and Aizawa O 1991 Phantom experiment and calculation for *in vivo* boron-10 analysis by prompt gamma-ray spectroscopy. Phys. Medicine Biol. 36, 329–338.

McGeehan S L, Topper K and Naylor D V 1989 Sources of variation in hot water extraction and colorimetric determination of soil boron. Comm. Soil Sci. Plant Anal. 20, 1777–1786.

Molinero A L, Ferrer A and Castillo J R 1993 Volatilisation of methyl borate in iron matrix – determination of boron in steel by ICP atomic emission spectrometry. Talanta 40, 1397–1403.

Moore D E 1990 A review of techniques for the analysis of boron in the development of neutron capture therapy agents. J. Pharmaceut. Biomed. Anal. 8, 547–553.

Motomizu S, Oshima M and Jun Z 1990 Anion-exchange chromatographic behaviour of the complex of boron with H-resorcinol – sensitive determination of boric acid. Analyst 115, 389–392.

Motomizu S, Oshima M and Jun Z 1991 Fluorimetric determination of boron with chromotropic acid by flow-injection analysis. Anal. Chim. Acta 251, 269–274.

Musashi M, Oi T, Ossaka T and Kakihana H 1990 Extraction of boron from GSJ rock reference samples and determination of their boron isotopic ratios. Anal. Chim. Acta 231, 147–150.

Mustafa W A, Falatah A M and El Shall A A 1993 Effect of excess boron fertilization on status and availability of boron in calcareous soils. Fert. Res. 36, 71–78.

Nakamura E, Ishikawa T, Birck J L and Allegre C J 1992 Precise boron isotopic analysis of natural rock samples using a boron mannitol complex. Chem. Geol. 94, 193–204.

Nilsson L G and Jennische P 1986 Determination of boron in soils and plants. Swed. J. Agric. Res. 16, 97–103.

Nogueira A R A, Brienza S M B, Zagatto E A G, Lima J L F C and Araujo A N 1993 Multi-site detection in flow analysis: Part 2. Monosegmented systems with relocating detectors for the spectrophotometric determination of boron in plants. Anal. Chim. Acta 276, 121–125.

Nose K and Zenki M 1991 Flow injection spectrophotometric determination of boron with D-sorbitol using methyl orange as an indicator. Analyst 116, 711–714.

Novozamsky I, Vaneck R, Houba V J G and Vanderlee J J 1990a A new solvent extraction for the determination of traces of boron by ICP-AES. Atom Spectrosc. 11, 83–84.

Novozamsky I, Barrera L L, Houba V J, Vanderlee J J and Eck R 1990b Comparison of a hot water and cold 0.01 *M* $CaCl_2$ extraction procedures for the determination of boron in soil. Comm. Soil Sci. Plant Anal. 21, 2189–2195.

Novozamsky I, Houba V J G, Vanderlee J J, Vaneck R and Mignorance M D 1993 A convenient wet digestion procedure for multi-element analysis of plant materials. Comm. Soil Sci. Plant Anal. 24, 2595–2605.

Novozamsky I, Vaneck R, Vanderlee J J, Houba V J G, Vanderlee J J and Ayaga G O 1988 Continuous flow technique for generation nad separation of methyl borate from iron-containing matrices with susequent determination of boron by ICP-AES. Atom Spectrosc. 9, 97–99.

Nyomora A M, Sah R N, Brown P H and Miller R O 1997 Boron determination in biological materials by inductively coupled plasma atomic emission and mass spectrometry: Effects of sample dissoluiton methods. Fres. J. Anal. Chem. (*In press*)

Offiah O and Axley J H 1988 Improvement of boron soil test. Comm. Soil Sci. Plant Anal. 19, 1527–1542.

Ogner G 1980 Automatic determination of boron in water samples and soil extracts. Comm. Soil Sci. Plant Anal. 11, 1209–1219.

Okamoto Y, Sugawa K and Kumamaru T 1994 Chemical modification for inductively coupled plasma atomic emission spectrometric determination of boron with tungsten boat furnace vaporiser. J. Anal. Atom Spectrom. 9, 89–92.

Owens J W, Gladney E S and Knab D 1982 Determination of boron in geological materials by inductively coupled plasma emission spectrometry. Anal. Chim. Acta 135, 169–172.

Panov V A, Semenko K A and Kuzyakov Y Y 1989 Determination of boron by extraction and molecular emission spectrometry with pulse nebulization of extracts into flame. J. Anal. Chem. USSR 44, 1117–1122.

Papaspyrou M, Feinendegen L E, Mohl C and Schwuger M J 1994 Determination of boron in cell suspensions using electrothermal atomic absorption spectrometry. J. Anal. Atom Spectrom. 9, 791–795.

Parashar D C, Sarkar A K and Singh N 1989 Spectrophotometric determination of traces of boron in high purity silicon. Anal. Letters 22, 1961–1967.

Parker D R and Gardner E H 1981 The determination of hot-water-soluble boron in some acid Oregon soils using a modified azomethine-H procedure. Comm. Soil Sci. Plant Anal. 12, 1311–1322.

Pennington H D, Finch C R, Lyons C C and Littau S A 1991 Microwave digestion of plant samples for boron analysis. HortScience 26, 1496–1497.

Pillay A E and Peisach M 1992 Some studies on nuclear methods for boron determination. Nucl. Instr. & Meth. Phys. Res. Sec. B 66, 226–229.

Pollmann D, Broekaert J A C, Leis F, Tschopel P and Tolg G 1993 Determination of boron in biological tissues by inductively coupled plasma optical emission spectrometry (ICP–OES). Fres. J. Anal. Chem. 346, 441–445.

Ponnemperuma F N, Cayton M T and Lantin R S 1981 Dilute hydrochloric acid as an extractant for available zinc, copper and boron in rice soils. Plant and Soil 61, 297–310.

Porter S R, Spindler S C and Widdowson A E 1981 An improved automated colorimetric method for the determination of boron in extracts of soils soil-less peat-based composts plant materials and hydroponic solutions with Azomethine-H. Comm. Soil Sci. Plant Anal. 12, 461–473.

Pougnet M A B and Orren M J 1986a Determination of boron by inductively coupled plasma atomic emission spectroscopy: Part 1. Method development. Int. J. Environ. Anal. Chem. 24, 253–266.

Pougnet M A B and Orren M J 1986b Determination of boron by inductively coupled plasma atomic emission spectroscopy. Part 2. Applications to South African environmental samples. Int. J. Environ. Anal. Chem. 24, 267–282.

Pritchard M W and Lee J 1984 Simultaneous determination of boron phosphorus and sulphur in some biological and soil materials by inductively-coupled plasma emission spectrometry. Anal. Chim. Acta. 157, 313-326.

Rand M C 1975 Standard methods for the examination of water and wastewater. pp 287–291 American Public Health Assoc. Washington, DC.

Rashid A, Rafique E and Bughio N 1994 Diagnosing boron deficiency in rapeseed and mustard by plant analysis and soil testing. Comm. Soil Sci. Plant Anal. 25, 2883–2897.

Renan L and Gupta U C 1991 Extraction of soil boron for predicting its availability to plants. Comm Soil Sci. Plant Anal. 22, 1003–1012.

Ricci L, Lanza P and Lanzoni E 1994 Determination of boron in nickel-boron alloys. Annali Chim. 84, 261–269.

Rivoldini A and Cara S 1992 Boron determination in sulcis coal ash by inductively coupled plasma optic emission spectrometry. Chem. Geol. 98, 317–322.

Rogus R, Harling O K, Olmez I and Widzek S 1992 Boron-10 prompt gamma analysis using a diffracted neutron beam. *In* Progress in Neutron Capture Therapy for Cancer. Eds. B J Allen, D E Moore and B V Harrington. pp 301–305. Plenum Press, New York.

Roignavarro A F, Lopez F J and Hernandez F 1996 Application of the azomethine-H method to the determination of boron in workplace atmospheres from ceramic factories. Fres. J. Anal. Chem. 356, 103–106.

Sah R N 1995 Plasma source mass spectrometric analysis of biological and environmental samples: dealing with potential interferences. Appl. Spectrosc. Rev. 30, 35–80.

Sah R N and Brown PH 1997 Boron determination: a review of analytical methods. Michrochem. J. (*In press*).

Sakal R, Singh S P, Singh A P and Bhogal N S 1993 Evaluation of soil test methods for response of chickpea to boron in calcareous soil. Anal. Agric. Res. 14, 377–387.

Sanz J, Martin R L, Galban J and Castillo J R 1990a Some obsevations on the determination B in wine by flame emission spectrometry after methyl borate generation in heterogeneous phase. Microchem. J. 41, 164–171.

Sanz J, Martin R L, Galban J and Castillo J R 1990b Use of methyl borate generation – flame emission spectrometry combined technique for boron determination in wine. Analusis 4, 279–283.

Schnug E and Haneklaus S 1996 Parameters influencing the calcination of plant materials in muffle furnaces and their importance for micronutrient analysis. Comm. Soil Sci. Plant Anal. 27, 993–1000.

Sekerka I and Lechner J F 1990 Automated method for the determination of boron in water by flow-injection analysis with in-line preconcentration and spectrophotometric detection. Anal. Chim. Acta 234, 199–206.

Self J R and Gupta U C 1993 Soil testing for boron on alkaline soils. *In* Boron and its Role in Crop Production. Ed. U C Gupta. pp 125–135 CRC Press Inc, Boca Raton, FL, USA.

Shabanova G L, Bukhbinder G I and Gilbert E N 1985 Source of solvent for organic samples in direct analysis by the inductively coupled plasma atomic emission method. J. Anal. Chem. USSR 40, 1563–1569.

Sheppard B S and Caruso J A 1994 Plasma mass spectrometry: consider the source. J. Anal. Atom Spectrom. 9, 145–149.

Shuman L M, Bandel V A, Donohue S J, Isaac R A, Lippert R M, Simes J T and Tucker MR 1992 Comparison of Mehlich-1 and Mehlich-3 extractable soil boron with hot-water extractable boron. Comm. Soil Sci. Plant Anal. 23, 1–14.

Smith F G, Wiederin D R, Houk R S, Egan C B and Serfass R E 1991 Measurement of boron concentration and isotope ratios in biological samples by inductively coupled plasma mass spectrometry with direct injection nebulization. Anal. Chim. Acta 248, 229–234.

Spiers G A, Evans L J, Mcgeorge S W, Moak H W and Su C 1990 Boron analysis of soil solutions and plant digests using a photodiode-array equipped ICP spectrometer. Comm. Soil Sci. Plant Anal. 21, 1645–1661.

Spouncer L R, Nable R O and Cartwright B 1992 A procedure for the determination of soluble boron in soils ranging widely in boron concentrations sodicity and pH. Comm. Soil Sci. Plant Anal. 23, 441–453.

Stotesbury S J, Pickering J M and Grifferty M A 1989 Analysis of lithium and boron by inductively coupled plasma mass spectrometry. J. Anal. Atom Spectrom. 5, 457–460.

Szydlowski F J 1979 Boron in natural waters by atomic absorption spectrometry with electrothermal atomization. Anal. Chim. Acta 106, 121–125.

Urasa I T 1984 Determination of arsenic, boron, carbon, phosphorus, selenium and silicon in natural waters by direct current plasma atomic emission spectrometry. Anal. Chem. 56, 904–908.

Usenko S I and Prorok M M 1992 Determining boron in natural waters by atomic-absorption spectrometry with electrothermal atomization. Indus. Lab. 58, 487–488.

Vanderpool R A, Hoff D and Johnson P E 1994 Use of inductively coupled plasma-mass spectrometry in boron-10 stable isotope experiments with plants, rats, and humans. Environ. Health Perspect. 102 (Suppl. 7), 13–20.

Vanhaecke F, Vanhoe H, Vandecasteele C and Dams R 1991 Precise determination of boron in titanium by inductively coupled plasma mass spectrometry. Anal. Chim. Acta. 244, 115–122.

Vanhoe H, Dams R, Vandecasteele C and Versieck J 1993 Determination of boron in human serum by inductively coupled plasma mass spectrometry after a simple dilution of the sample. Anal. Chim. Acta. 281, 401–411.

Vaughan B and Claassen L 1996 Optimising an inductively coupled plasma emission spectrometer analysis time with the use of a direct injection nebuliser. Comm. Soil Sci. Plant Anal. 27, 819–827.

Vaughan B and Howe J 1994 Evaluation of boron chelates in extracting soil boron. Comm. Soil Sci. Plant Anal. 25, 1071–1084.

Vengosh A, Chivas A R and Mcculloch M T 1989 Direct determination of boron and chlorine isotopic compositions in geological materials by negative thermal-ionisation mass spectrometry. Chem. Geol 79, 333–343.

Wang H, Chen Y and Wang J 1996 Determination of boron in steels by on-line flow injection cation exchange eliminating matrix inductively coupled plasma-atomic emission. Spectrom. Microchem. J. 53, 88–92.

Ward N I, Abu-Shakra F R and Durrant S F 1990 Trace element content of biological materials: A comparison of NAA and ICP–MS analysis. Biol. Trace Element Res. 26, 177–187.

Wei W C, Chen C J and Yang M H 1995 Determination of boron using mannitol-assisted electrothermal vaporisation for sample introduction in inductively coupled plasma mass spectrometry. J. Anal. Atom Spectrom. 11, 955–961.

Wikner B 1986 Pretreatment of plant and soil samples. A problem in boron analysis. Part I: Plants Comm. Soil Sci. Plant Anal. 17, 1–25.

Williams W J 1979 Handbook of anion determination. pp 23–39. Butterworth, London.

Wiltshire G A, Bolland D T and Littlejohn D 1994 Study of the atomization of boron in electrothermal atomic absorption spectrometry and hollow cathod furnace atomic non-thermal excitation spectrometry. J. Anal. Atom Spectrom. 9, 1255–1262.

Wolf B 1974 Improvement in the azomethine-H method for the determination of boron. Comm. Soil Sci. Plant Anal. 5, 39–44.

Wood J and Nicholson K 1994 Determination of boron in environmental samples by ion-selective electrode – an evaluation. Environ. Geochem. Health 16, 87–87.

Xiao Y K 1991 Rapid and high precision determination of boron isotopic ratio in boron carbide using mass spectrometry by the thermal ionisation of the dicesium metaborate cation. Chin. Sci. Bull. 36, 173–175.

Xiao Y K, Beary E S and Fassett J D 1988 An improved method for high-precision isotopic measurement of boron by thermal ionisation mass spectrometry. Internat. J. Mass Spectrom Ion Processes 85, 203–213.

Xu L and Rao Z 1986 Determination of B in soils by sequential scanning ICP-AES using side line indexing method. Fres. J. Anal. Chem. 325, 534–538.

Yakimova V P and Markova O L 1992 Ionometric determination of boron in natural and waste water and in biological material. J. Anal. Chem. USSR 47, 1477–1483.

Yonezawa C and Wood A K H 1995 Prompt gamma-ray analysis of boron with cold and thermal neutron guided beams. Anal. Chem. 67, 4466–4470.

Zakhariya A N, Novak I V, Chebotarev A N and Zhila S I 1991 Atomic absorption determination of boron in some nitrogen-containing organic salts of tetrafluoroboric acid. Indust. Lab. USSR 57, 1130–1132.

Zarcinas B A 1995 Suppression of iron interference in the determination of boron using the azomethine-H procedure. Comm. Soil Sci. Plant Anal. 26, 713–729.

Zarcinas B A and Cartwright B 1987 Acid dissolution of soils and rocks for the determination of boron by inductively coupled plasma atomic emission spectrometry. Analyst 112, 1107–1112.

Ziaziaris J and Kacprzak J L 1995 Microwave digestion inductively coupled plasma atomic emission spectrometric determination of boron in raw noodles. JAOAC Int. 78, 874–877.

Zenki M, Nose K and Toei K 1989 Spectrophotometric determination of boron with an azomethine-H derivative. Fres. J. Anal. Chem. 334, 238–241.

Zheng K 1994 Oscillopolarographic determination of trace amount of boron in soil, water and plant samples. Int. J. Environ. Anal. Chem. 39, 75–79.

Plant and Soil **193:** 35–48, 1997.

Chapter 3

Reactions of boron with soils

Sabine Goldberg
*USDA, US Salinity Laboratory, 450 W Big Springs Road, Riverside, CA 92507, USA**

Abstract

Boron is an essential micronutrient for plants, but the range between deficient and toxic B concentration is smaller than for any other nutrient element. Plants respond directly to the activity of B in soil solution and only indirectly to B adsorbed on soil constituents. Soil factors affecting availability of B to plants are: pH, texture, moisture, temperature, organic matter and clay mineralogy. Boron adsorbing surfaces in soils are: aluminium and iron oxides, magnesium hydroxide, clay minerals, calcium carbonate, and organic matter. Boron adsorption reactions can be described empirically using the Langmuir adsorption isotherm equation, the Freundlich adsorption isotherm equation, and the phenomenological Keren model. Chemical models such as the constant capacitance model, the triple layer model, and the Stern VSC-VSP model can describe B adsorption over changing conditions of solution pH and B concentration. Boron desorption reactions often exhibit hysteresis. The rate of B desorption can be described using the first order rate equation, the Elovich reaction rate equation, and the power function equation.

Introduction

Boron is an essential micronutrient element required for the normal growth of plants. Of all micronutrient deficiencies in plants, B deficiency is the most widespread in the USA (Sparr, 1970). Alfalfa (*Medicago sativa*) is most prone to B deficiency and receives more B fertiliser than any other economic crop (Murphy and Walsh, 1972). Boron deficiency is most likely in coarse textured soils in humid regions. Temporary B deficiency can be triggered by liming of acid soils because of increased B adsorption at higher soil pH (Reisenauer et al., 1973).

Plants vary in their B requirement, but the range between deficient and toxic soil solution concentrations of B is smaller than for any other nutrient element. Both deficient and toxic levels of B in the soil solution can occur during a single growing season (Reisenauer et al., 1973). In arid and semi-arid areas, B toxicity results from high levels of B in soils and from additions of B via the irrigation water (Nable et al., Chapter 12). In saline soils, lack of drainage leads to excessive concentrations of B in the soil solution. Land application of coal ash residues is often limited by their high B content (Adriano et al., 1980). Reclamation of high B soils requires about three times as much water as reclamation of saline soils (Keren and Bingham, 1985).

Compared with other nutrient elements, the chemistry of B in soils is very simple. Boron does not undergo oxidation-reduction reactions or volatilisation reactions in soils. Boric acid is a very weak, monobasic acid that acts as a Lewis acid by accepting a hydroxyl ion to form the borate anion (Figure 1). Boron containing minerals are either very insoluble (tourmaline) or very soluble (hydrated B minerals) and generally do not control the solubility of B in soil solution (Goldberg, 1993). The B concentration in the soil solution is generally controlled by B adsorption reactions, as is the amount of water soluble B available for plant uptake. Plants respond only to the B activity in soil solution; B adsorbed by the soil surfaces is not perceived as toxic by plants (Keren et al., 1985a, 1985b; Ryan et al., 1977). Factors affecting B availability and extent of B adsorption in soils are solution pH, soil texture, soil moisture, and temperature.

* E-mail: S Goldberg@ussl.ars.usda.gov

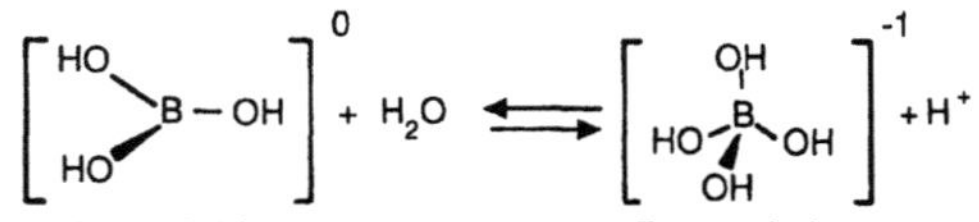

Figure 1. Dissociation of boric acid.

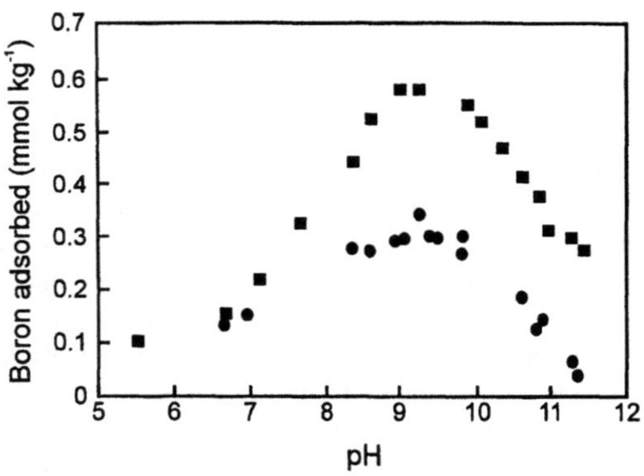

Figure 2. Boron adsorption on a surface (■) and a subsurface (●) sample of an arid zone soil as a function of pH.

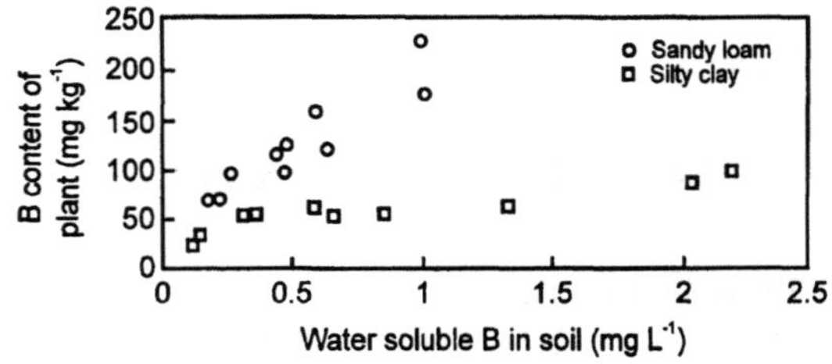

Figure 3. Boron uptake by alfalfa (*Medicago sativa*) as a function of water soluble B for two soil textures. Adapted from Wear and Patterson (1962).

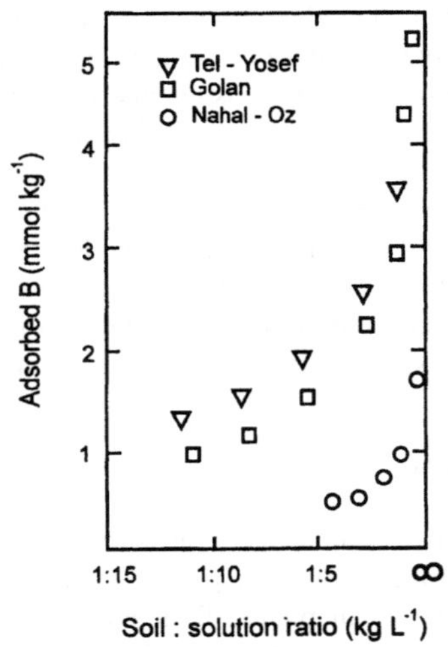

Figure 4. Boron adsorption as a function of soil to solution ratio for three soils. Adapted from Mezuman and Keren (1981).

Factors affecting boron availability in soils

pH

One of the most important factors affecting the availability of B in soils is pH. With increasing soil solution pH, B becomes less available to plants. Therefore, application of lime to acid soils can result in B deficiency symptoms in plants. Soluble B content in soils was highly significantly correlated with solution pH (Berger and Truog, 1945; Elrashidi and O'Connor, 1982). Boron uptake by plants at identical water soluble B content was greater at lower soil solution pH (Wear and Patterson, 1962).

Boron adsorption on soils is very dependent on solution pH (Figure 2). Boron adsorption by soils increased as a function of solution pH in the range of pH 3 to 9 (Barrow, 1989; Bingham et al., 1971; Keren et al., 1985a; Lehto, 1995; Mezuman and Keren, 1981; Schalscha et al., 1973) and decreased in the range of pH 10 to 11.5 (Goldberg and Glaubig, 1986a). Boron adsorption maxima were significantly positively correlated with solution pH (Evans, 1987; Okazaki and Chao, 1968).

Soil texture

Because coarse textured soils often contain less available B than fine textured soils, B deficiency often occurs in plants growing in sandy soils (Fleming, 1980; Gupta, 1968). Native B was significantly positively correlated with soil clay content (Elrashidi and O'Connor, 1982). At identical water soluble B content, plant B uptake was greatest for the most coarse textured soil (Wear and Patterson, 1962; Figure 3). Adsorbed B is dependent on soil texture, and increases with increasing clay content (Bhatnager et al., 1979; Elrashidi and O'Connor, 1982; Mezuman and Keren, 1981; Wild and Mazaheri, 1979). Boron adsorption maxima increased with increasing clay content (Biggar and Fireman, 1960; Goldberg and Glaubig, 1986a; Nicholaichuk et al., 1988; Singh, 1964). A highly sig-

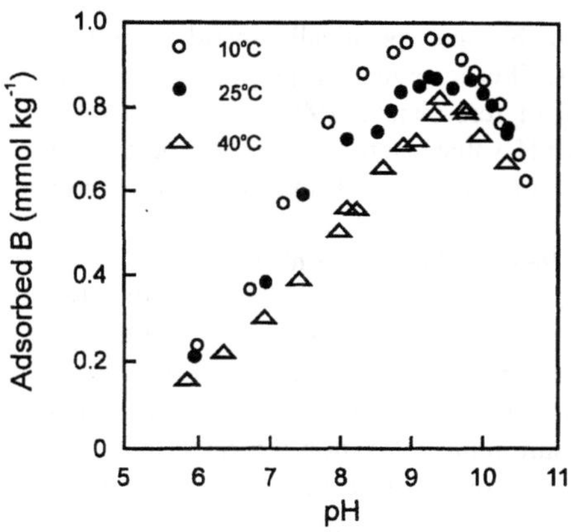

Figure 5. Boron adsorption on an arid zone soil as a function of temperature. From Goldberg et al. (1993a).

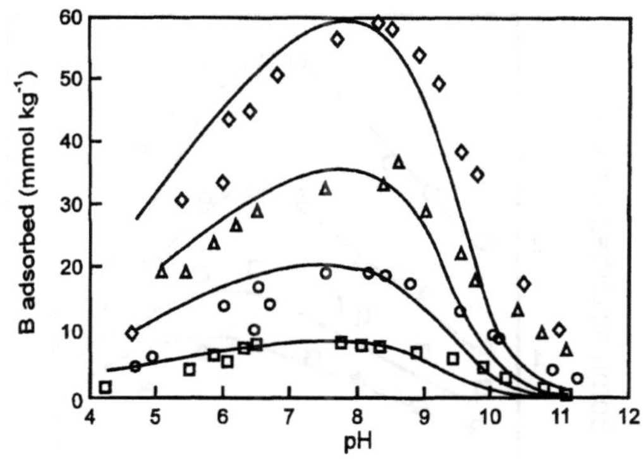

Figure 6. Boron adsorption on an aluminium oxide as a function of pH. Fit of the constant capacitance model to B adsorption from solutions containing 2 mg L^{-1} B (□), 5 mg L^{-1} B (○), 10 mg L^{-1} B (Δ), and 20 mg L^{-1} B (◇). Model results are represented by solid lines. From Goldberg and Glaubig (1988).

nificant correlation was found between soil content of the clay minerals kaolinite, montmorillonite, and chlorite and the B adsorption maximum.

Soil moisture

Boron availability generally decreases as soils dry, making B deficiency in plants more likely (Fleming, 1980). This may be because plants encounter reduced amounts of available B when extracting moisture from lower depths during dry conditions (Fleming, 1980). Diffusivity of B decreased with decreasing water content because drying reduces soil solution mobility and increases the diffusion path length (Scott et al., 1975). However, total diffusible B from repeated extractions was independent of water content (Sulaiman and Kay, 1972). Adsorbed B was independent of variations in soil moisture content from 50 to 100 percent of field capacity in one study (Gupta, 1968) and increased with decreasing soil water content in another (Mezuman and Keren, 1981; Figure 4). Wetting and drying cycles increased the amount of B fixation (Biggar and Fireman, 1960). The effect of drying became more pronounced with increasing additions of B (Biggar and Fireman, 1960).

Temperature

Boron adsorption increases with increasing soil temperature. However, this may be due to an interactive effect of soil temperature with soil moisture since B deficiency is associated with dry summer conditions (Fleming, 1980). Boron adsorption decreased as a function of temperature in the range of 10 – 40 °C on soils dominant in crystalline minerals (Biggar and Fireman, 1960; Goldberg et al., 1993a; Figure 5). In contrast, B adsorption increased slightly as the temperature of an amorphous soil was increased from 10 to 40 °C (Bingham et al., 1971).

Boron adsorbing surfaces in soils

Oxides

Aluminium and iron oxides play an important role in B adsorption behaviour on soils (Bingham, et al., 1971; Elrashidi and O'Connor, 1982; Harada and Tamai, 1968). Boron adsorption is highly significantly correlated with the aluminium oxide content of soils (Bingham et al., 1971; Harada and Tamai, 1968). Percent iron oxide was a significant variable in multiple regression equations explaining the variance in adsorbed B, soluble B, and total B for soils (Elrashidi and O'Connor, 1982).

Numerous studies have investigated B adsorption by various aluminium and iron oxides both crystalline and amorphous (Bloesch et al., 1987; Choi and Chen, 1979; de Bussetti et al., 1995; Goldberg and Glaubig, 1985, 1988; Goldberg et al., 1993a, 1993b, 1996; Hatcher et al., 1967; Keren and Gast, 1983; McPhail et al., 1972; Metwally et al., 1974; Scharrer et al., 1956; Sims and Bingham, 1968; Su and Suarez, 1995;

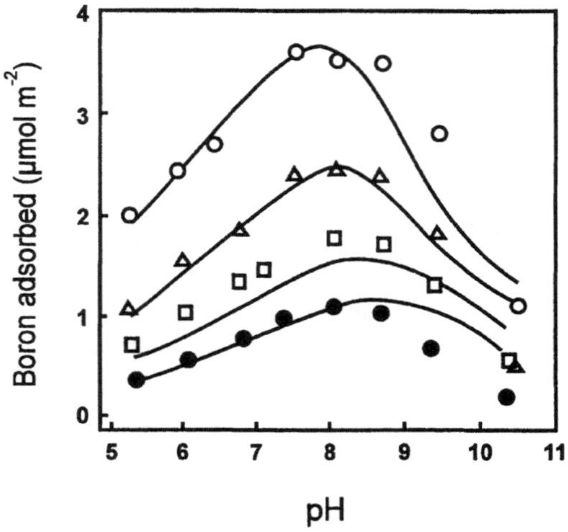

Figure 7. Boron adsorption on an iron oxide as a function of pH. Fit of the Stern VSC-VSP model to B adsorption from solutions containing 139 mg L^{-1} B (●), 270 mg L^{-1} B (□), 525 mg L^{-1} B (Δ), and 921 mg L^{-1} B (○). Model results are represented by solid lines. From Bloesch et al. (1987).

Toner and Sparks, 1995). Boron adsorption on both crystalline and amorphous aluminium and iron oxides increased with increasing pH up to an adsorption maximum at pH 6 to 8 for aluminium oxides and pH 7 to 9 for iron oxide (Bloesch et al., 1987; Choi and Chen, 1979; Goldberg and Glaubig, 1985; McPhail et al., 1972; Metwally et al., 1974; Scharrer et al., 1956; Sims and Bingham, 1968; Su and Suarez, 1995). Above the maximum, B adsorption decreased with increasing pH (Figures 6 and 7). Boron adsorption was greatest on freshly precipitated solids and decreased with aging due to increasing crystallinity (Hatcher et al., 1967; Sims and Bingham, 1968). Boron adsorption per gram was greater for aluminium than iron oxides (Goldberg and Glaubig, 1985; Scharrer et al., 1956; Sims and Bingham, 1968). This is likely due to the higher surface area of aluminium oxides, since adsorption per square metre was similar in magnitude for aluminium and iron oxides (Goldberg and Glaubig, 1985). Boron adsorption on oxide minerals occurred rapidly, being virtually complete after one day of reaction time (Choi and Chen, 1979; Scharrer et al., 1956).

The mechanism of B adsorption on aluminium and iron oxide minerals is considered to be ligand exchange with reactive surface hydroxyl groups (Goldberg et al., 1993b; McPhail et al., 1972; Sims and Bingham, 1968; Su and Suarez, 1995). Ligand exchange with surface hydroxyl groups is a mechanism whereby anions become specifically adsorbed on mineral surfaces. This specific adsorption produces a shift in the point of zero charge (PZC) of the mineral to a more acid pH value. Boron adsorption occurs specifically since it produces a shift in PZC of aluminium and iron oxides (Alwitt, 1972; Beyrouty et al., 1984; Blesa et al., 1984; Fricke and Leonhardt, 1950; Goldberg et al., 1993b; Su and Suarez, 1995). Specifically adsorbed ions are held in inner-sphere surface complexes that contain no water between the adsorbing ion and the surface functional group. Kinetic experiments using pressure jump relaxation indicated that B adsorbs as an inner-sphere surface complex on aluminium oxide via ligand exchange of borate with surface hydroxyl groups (Toner and Sparks, 1995). Fourier transform infrared spectroscopic analyses have shown that B is adsorbed via ligand exchange as both $B(OH)_3^0$ and $B(OH)_4^-$ species on amorphous aluminium and iron oxide (Su and Suarez, 1995).

Boron adsorption on aluminium and iron oxides as a function of pH was exothermic, decreasing with increasing temperature from 5 – 40 °C (de Bussetti et al., 1995; Goldberg et al., 1993a; Su and Suarez, 1995). This effect was only significant near the adsorption maximum. Specific ion adsorption is expected to be exothermic (Helferrich, 1962). Boron adsorption on an iron oxide was independent of solution ionic strength; while B adsorption on an aluminium oxide decreased with increasing ionic strength (Goldberg et al., 1993b).

Competing ions such as silicate, sulfate, phosphate, and oxalate decreased the magnitude of B adsorption on oxides (Bloesch et al., 1987; Choi and Chen, 1979; de Bussetti et al., 1995; Goldberg and Glaubig, 1988; McPhail et al., 1972; Metwally et al., 1974). The effect of competing ions on B adsorption can be slight, in the case of sulfate, or substantial in the case of phosphate (Bloesch et al., 1987; Metwally et al., 1974). The ability of competing anions to leach adsorbed B from oxides increased in the order: chloride < sulfate < arsenate < phosphate (Metwally et al., 1974). Significant silicate adsorption produced only a slight decrease in B adsorption by aluminium oxide suggesting that some B sorption sites show B preference (Goldberg and Glaubig, 1988).

Magnesium hydroxide can remove appreciable amounts of B from solution (Rhoades et al., 1970a). Due to magnesium hydroxide coatings, silicate minerals containing mainly magnesium in their chemical formulas adsorbed more B than a silicate without magnesium in its chemical formula (Rhoades et al., 1970a). The appreciable B sorption capacity of the sand and silt

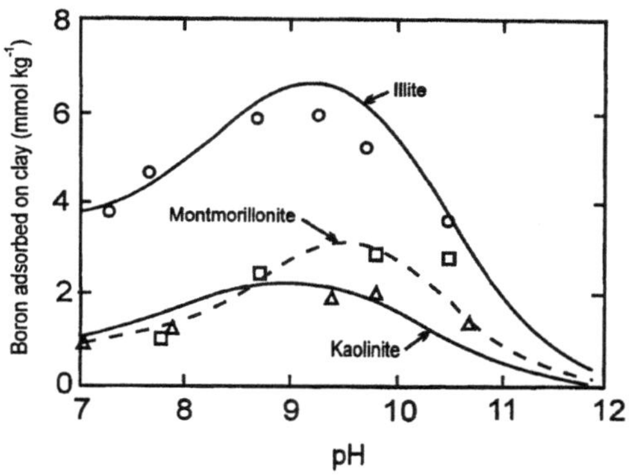

Figure 8. Boron adsorption on clay minerals as a function of pH. Fit of the phenomenological Keren model to B adsorption. Model results are represented by lines. From Keren and Mezuman (1981).

fractions of arid zone soils may result from clusters and coatings of magnesium hydroxide on silicate minerals (Rhoades et al., 1970a).

Clay minerals

Layer silicate clay minerals are important B adsorbing surfaces in soils (Brockamp, 1973; Couch and Grim, 1968; Fleet, 1965; Goldberg and Glaubig, 1986b; Goldberg et al., 1993a, 1993b, 1996; Harder, 1961; Hingston, 1964; Jasmund and Lindner, 1973; Keren and Gast, 1981; Keren and Mezuman, 1981; Keren and O'Connor, 1982; Keren and Sparks, 1994; Keren and Talpaz, 1984; Keren et al., 1981, 1994; Mattigod et al., 1985; Porrenga, 1967; Scharrer et al., 1956; Sims and Bingham, 1967; Singh, 1971). Clay minerals exhibit increasing B adsorption with increasing solution pH with adsorption maxima occurring at pH 8 to 10 (Hingston, 1964; Mattigod et al., 1985; Scharrer et al., 1956; Sims and Bingham, 1967); B adsorption decreases with increasing pH above the maximum (Figure 8). For clay minerals, the order of B adsorption per gram is: kaolinite < montmorillonite < illite (Harder, 1961; Hingston, 1964; Jasmund and Lindner, 1973; Keren and Mezuman, 1981; Scharrer et al., 1956; Sims and Bingham, 1967).

The rate of B adsorption on clay minerals consists of a fast adsorption reaction and a slow fixation reaction. Short term experiments have shown that B adsorption reaches equilibrium in less than one day (Hingston, 1964; Kerren et al., 1981; Scharrer et al., 1956). Long term experiments have shown that fixation of B increased even after six months of reaction time (Harder, 1961; Jasmund and Lindner, 1973). Boron adsorption by clay minerals is considered to be a two step process. Initially, B adsorbs onto the particle edges, subsequently migrates, and incorporates structurally into tetrahedral sites replacing structural silicon and aluminium (Couch and Grim, 1968; Fleet, 1965; Harder, 1961). The mechanism of the B adsorption step is considered to be ligand exchange with surface hydroxyl groups on the clay particle edges (Couch and Grim, 1968; Goldberg et al., 1993b; Keren and Sparks, 1994; Keren and Talpaz, 1984; Keren et al., 1994).

The effect of temperature on B adsorption by clays has been investigated (Couch and Grim, 1968; Goldberg et al., 1993a; Harder, 1961; Jasmund and Lindner, 1973; Singh, 1971). For a very short reaction time of two hours, B adsorption in the pH range of 5.5 – 9.5 decreased with increasing temperature (Goldberg et al., 1993a). Boron adsorption for longer reaction times of twelve hours to sixty days increased with increasing temperature (Couch and Grim, 1968; Jasmund and Lindner, 1973; Singh, 1971). These results suggest that initial adsorption of B is exothermic while the subsequent B fixation reaction is endothermic.

Boron adsorption on clays increased with increasing ionic strength of the solution (Couch and Grim, 1968; Fleet, 1965; Goldberg et al., 1993b; Keren and O'Connor, 1982; Keren and Sparks, 1994). The ionic strength effect was greater for sodium clay than for calcium clay (Keren and O'Connor, 1982) and was greatest around the B adsorption maximum (Goldberg et al., 1993b).

Decreases in water content decreased B adsorption on clays (Keren and Mezuman, 1981). Wetting and drying cycles increased B fixation with the greatest increase occurring during the first wetting and drying cycle (Keren and Mezuman, 1981).

Competing anions affect the magnitude of B adsorption on clays (Brockamp, 1973; Goldberg and Glaubig, 1986b; Goldberg et al., 1996; Jasmund and Lindner, 1973). The presence of chloride, nitrate, and sulfate had little effect on B adsorption on clays; the presence of phosphate, however, appreciably reduced B adsorption (Jasmund and Lindner, 1973). The magnitude of B adsorption on kaolinite was affected slightly while B adsorption on montmorillonite was completely unaffected by substantial silicate adsorption suggesting that the majority of B adsorbing sites are specific to B (Goldberg and Glaubig, 1986b).

The magnitude of B adsorption on clay minerals is affected by the exchangeable cation (Keren and Gast, 1981; Keren and Mezuman, 1981; Keren and

O'Connor, 1982; Mattigod et al., 1985). Calcium clays adsorbed more B than sodium and potassium clays (Keren and Gast, 1981; Keren and O'Connor, 1982; Mattigod et al., 1985). An explanation for this result is that calcium 2:1 clays occur as tactoids consisting of several clay particles while the sodium forms exist in solution as single particles (Keren and Gast, 1981). In tactoids, the diffuse double layer and negative electric field from the planar surfaces is less extensive, making the edge sites of calcium clays more accessible to adsorbing borate anions than those of sodium clays (Keren and Gast, 1981). An alternative explanation for increased B adsorption in the presence of calcium is the formation and adsorption of the calcium borate ion pair (Mattigod et al., 1985).

On a per gram basis clay minerals adsorb significantly less B than do most oxide minerals. It may be more appropriate to compare B adsorption capacity on a surface area basis. However, since B adsorbs on the edges of clay minerals and edge surface areas of clays are difficult to determine, this comparison is not readily available.

Calcium carbonate

Application of lime increases B fixation by soils because it raises the soil solution pH. In addition to its effect on soil pH, calcium carbonate also acts as an important B adsorbing surface in calcareous soils (Elseewi, 1974; Elseewi and Elmalky, 1979; Goldberg and Forster, 1991). Boron adsorption was greater on soils having higher calcium carbonate content (Elrashidi and O'Connor, 1982; Elseewi, 1974). Acid soluble B was highly significantly correlated with calcium carbonate content of soils (Elseewi and Elmalky, 1979). Retention of B on calcium carbonate occurs via an adsorption mechanism (Ichikuni and Kikuchi, 1972). The mechanism could be exchange with carbonate groups.

The magnitudes of the B adsorption maxima for soil samples treated to remove calcium carbonate were statistically significantly lower than those for untreated soil samples indicating that calcium carbonate acts as an important sink for B adsorption in calcareous soils (Goldberg and Forster, 1991). Boron adsorption on reference calcites increased with increasing solution pH from pH 6 to 9, exhibited a maximum at pH 9.5, and decreased with increasing solution pH from pH 10 to 11 (Goldberg and Forster, 1991; Figure 9).

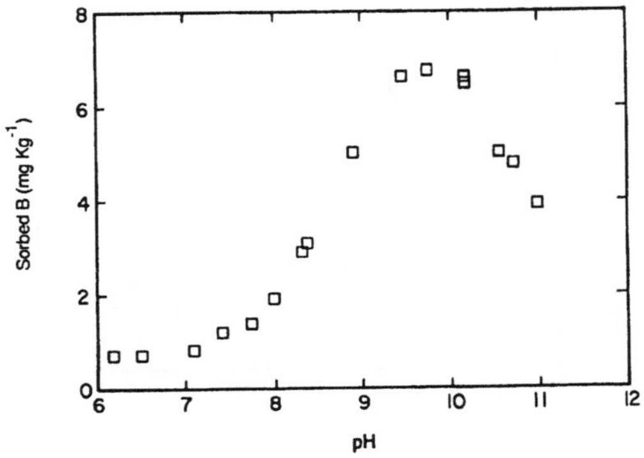

Figure 9. Boron adsorption on a reference calcite as a function of pH. From Goldberg and Forster (1991).

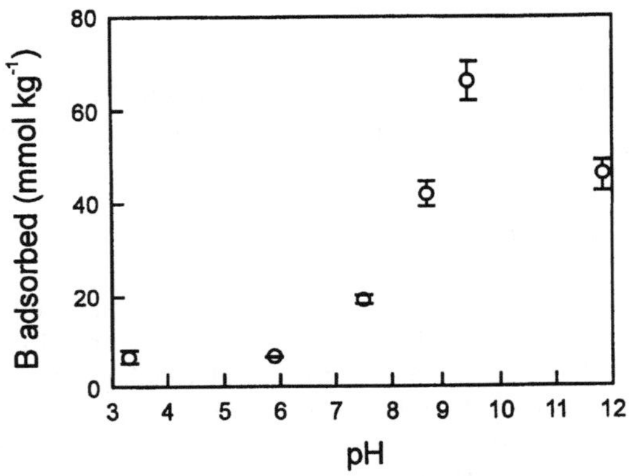

Figure 10. Boron adsorption on a humic acid extracted from soil. From Gu and Lowe (1990).

Organic matter

Organic matter is an important soil constituent affecting the availability of B. Native soil B and hot water soluble B were highly significantly correlated with organic carbon content (Berger and Truog, 1945; Elrashidi and O'Connor, 1982; Gupta, 1968; Miljkovic et al., 1966). Adsorbed B and B adsorption maxima were highly significantly correlated with organic carbon content (Elrashidi and O'Connor, 1982; Evans, 1987; Harada and Tamai, 1968).

Humus extracted from a soil retained significant amounts of B and was considered to play an important role in B adsorption (Parks and White, 1952). Soil organic matter adsorbs more B than mineral soil constituents on a weight basis (Gu and Lowe, 1990; Yermiyaho et al., 1988)). Boron adsorption on a min-

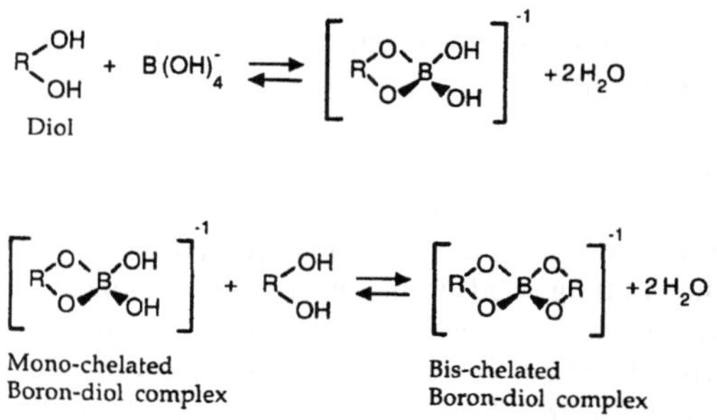

Figure 11. Formation of B-diol complexes.

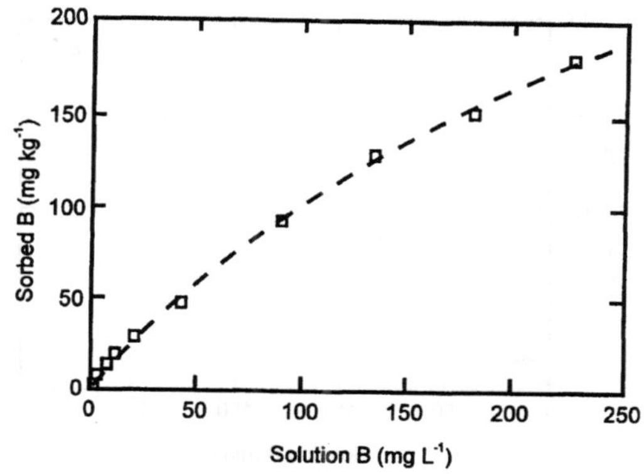

Figure 12. Fit of the Langmuir adsorption isotherm equation to B adsorption on a calcareous soil. Model results are represented by a dashed line. Adapted from Goldberg and Forster (1991).

eral soil increased with increasing additions of composted organic matter (Yermiyahu et al., 1995). The presence of organic material can also occlude B reactive adsorption sites on clays (Gu and Lowe, 1992) and soils (Harada and Tamai, 1968; Marzadori et al., 1991). Boron adsorption on an organic soil (Huettl, 1976; Lehto, 1995) and composted organic matter increased with increasing pH (Yermiyaho et al., 1988). Adsorption on a soil humic acid increased with increasing pH up to a maximum near pH 9, and decreased with increasing pH above 9 (Gu and Lowe, 1990, Figure 10). Boron adsorption on composted organic matter occurred rapidly, reaching equilibrium after three hours and increased with increasing solution ionic strength (Yermiyaho et al., 1988).

Ligand exchange is a possible mechanism for B sorption by organic matter (Yermiyaho et al., 1988). Boron–diol complexes may form with the breakdown products of soil organic matter (Parks and White, 1952); α-hydroxy carboxylic acid groups may act in B sorption on organic matter (Huettl, 1976). Figure 11 shows the formation reactions of mono-chelated and bis-chelated B-diol complexes. The formation of these complexes from reaction of borate with polyhydroxy compounds has been observed using nuclear magnetic resonance spectroscopy (Coddington and Taylor, 1989).

Modeling of boron adsorption

Empirical models

Adsorption isotherm equation

Empirical models provide descriptions of experimental adsorption data without a theoretical basis. Boron adsorption reactions on oxides, clay minerals, and soil materials have historically been described using adsorption isotherm equations (Choi and Chen, 1979; Elrashidi and O'Connor, 1982; Evans, 1987; Goldberg and Forster, 1991; Mondal et al., 1993; Nicholaichuk et al., 1988; Singh, 1971). The most popular are the Langmuir and Freundlich adsorption isotherm equations but Temkin and Brunauer–Emmett–Teller (BET) adsorption isotherm equations have also been used (Mondal et al., 1993).

The Langmuir adsorption isotherm equation was developed to describe gas adsorption onto clean solids, but has often been used to describe B adsorption by soil materials. The equation and its linear forms are provided elsewhere (e.g. Goldberg, 1993). Strictly, the Langmuir adsorption isotherm equation is obeyed only for a uniform adsorbent surface without lateral interactions, implying a constant free energy of adsorption (Brunauer et al., 1967). The Langmuir adsorption isotherm has been used to describe B adsorption on aluminium and iron oxide (McPhail et al., 1972), clay minerals (Gu and Lowe, 1992; Hingston, 1964; Singh, 1971) calcite (Goldberg and Forster, 1991), humic acids (Gu and Lowe, 1990), and soils (Bhatnagar et al., 1979; Biggar and Fireman, 1960; Bingham et al., 1971; Elrashidi and O'Connor, 1982; Evans, 1987; Hatcher and Bower, 1958; Marzadori et al., 1991; Mondal et al., 1993; Nicholaichuk et al., 1988; Okazaki and Chao, 1968; Schalscha et al., 1973; Singh, 1971). For many studies, the Langmuir adsorption isotherm equation was able to describe B adsorption only under conditions of low solution B concentration (Bhatnagar et al., 1979; Biggar and Fireman, 1960; Elrashidi and O'Connor, 1982; Goldberg and Forster, 1991;

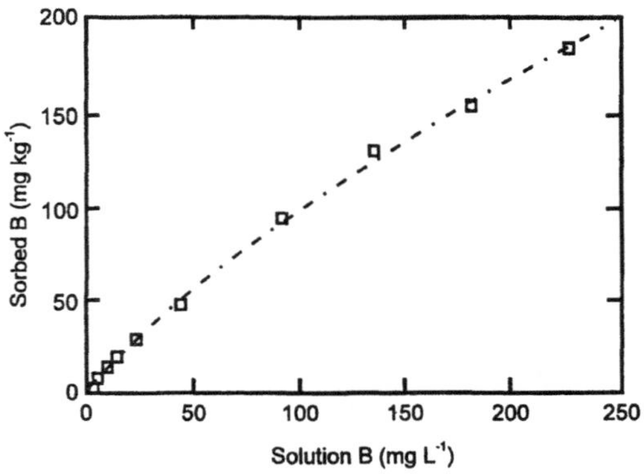

Figure 13. Fit of the Freundlich adsorption isotherm equation to B adsorption on a calcareous soil. Model results are represented by a dashed-dotted line. Adapted from Goldberg and Forster (1991).

Hingston, 1964; McPhail et al., 1972). The ability of the Langmuir adsorption isotherm equation to describe B adsorption on soil is indicated in Figure 12.

The Freundlich adsorption isotherm equation has often been used to describe B adsorption by soil materials. The equation and its linear form are provided elsewhere (e.g Goldberg, 1993). In the Freundlich adsorption isotherm equation, the affinity terms are distributed approximately log normally implying heterogeneity of B adsorbing sites (Sposito, 1984). The Freundlich adsorption isotherm equation is strictly valid only for adsorption data at low concentrations (Sposito, 1984), but has often been used to describe B adsorption by soils over the entire concentration range studied. The Freundlich adsorption isotherm has been used to describe B adsorption on aluminium oxide (Choi and Chen, 1979), clay minerals (Couch and Grim, 1968; Fleet, 1965; Jasmund and Lindner, 1973; Singh, 1971), calcite (Goldberg and Forster, 1991), and soils (Bhatnagar et al., 1979; Elrashidi and O'Connor, 1982; Evans, 1987; Goldberg and Forster, 1991; Lehto, 1995; Nikolaichuk et al., 1988). The ability of the Freundlich adsorption isotherm equation to describe B adsorption on soil is indicated in Figure 13.

Although the adsorption isotherm equations are often excellent at describing B adsorption, they are simply numerical relationships used to describe data (Harter and Smith, 1981). Independent experimental evidence for adsorption must be present before any chemical meaning can be assigned to adsorption isotherm equation parameters (Veith and Sposito, 1977). The adherence of experimental sorption data to the Langmuir or Freundlich equations provides no information about the chemical mechanism of reaction (Sposito, 1982).

In linearising adsorption isotherm equations, all data points are not given equal weighting and the variances of some of the values are disproportionately increased during the transformation (Barrow, 1978). Direct fitting of adsorption data using non-linear least squares methods avoids the difficulties of changes in error distributions and biased parameters associated with linear transformations (Kinniburgh, 1986). Non-linear least squares fits of B adsorption data were much improved over linear transformations as measured by the coefficient of determination, r^2 (Goldberg and Forster, 1991).

Since the use of adsorption isotherm equations is essentially a curve fitting procedure, the parameters obtained are only valid for the conditions under which the experiment was conducted. Prediction of B adsorption behaviour under changing conditions of soil solution B concentration, pH, and ionic strength is impossible.

Keren model

The Keren model is a phenomenological equation developed to describe B adsorption on clay mineral surfaces (Keren and Gast, 1981; Keren and Mezuman, 1981; Keren and O'Connor, 1982; Keren et al., 1981). In this model boric acid, $B(OH)_3^0$, borate ion, $B(OH)_4^-$, and hydroxyl ion, OH^-, all are assumed to compete for the same adsorption sites on the clay surface. Equations for the adsorption of these species and the total amount of B adsorbed are provided elsewhere (Keren et al., 1981).

One of the advantages of the Keren model is that it includes the pH variable and can therefore describe B adsorption behaviour over changing conditions of solution pH (Figure 8). The Keren model was able to fit B adsorption at several pH values and suspension densities using the same set of constants. The Keren model has been used to describe B adsorption by montmorillonite, kaolinite, illite, pyrophyllite, hydroxy aluminium, composted organic matter, and soils (Keren and Gast, 1983; Keren and Mezuman, 1981; Keren and Sparks, 1994; Keren et al., 1981, 1994; Mezuman and Keren, 1981; Yermiyaho et al., 1988, Yermiyahu et al., 1995). No physical significance can be attributed to the absolute values of the constants obtained with the Keren model (Keren et al., 1981; Yermiyaho et al., 1988).

Chemical surface complexation models

Introduction

Chemical models provide a molecular description of adsorption using an equilibrium approach. Chemical models of B adsorption have generally concentrated on the variable charge sites of oxide minerals. Chemical modeling of B adsorption at the oxide solution interface has been successful using various surface complexation models. In these models, B adsorption occurs via complex formation with surface hydroxyl groups on oxide minerals and edges of clay minerals.

Surface complexation models use an equilibrium approach with mass action and mass balance equations (Westall, 1980). Although these models are very successful at describing data over a wide range of experimental conditions, they cannot uniquely describe the physical nature of the interface (Westall and Hohl, 1980). Surface complexation models define surface species, chemical reactions, mass balances, and charge balance and calculate thermodynamic properties such as activity coefficients and equilibrium constants mathematically. The most significant advancement of surface complexation models is the consideration of the charge on both the adsorbate ion and the adsorbent surface. For this reason the models have wide applicability and predictive capability over changing conditions of solution B concentration, pH, and ionic strength.

Constant capacitance model

The constant capacitance model of the oxide aqueous solution interface (Stumm et al., 1980) has been applied to describe B adsorption on aluminium and iron oxides, clay minerals, and soils (Goldberg and Glaubig, 1985, 1986a, 1986b, 1988; Goldberg et al., 1993b; Toner and Sparks, 1995). Surface reactions, equilibrium constants, mass balance, and charge balance for the application of the constant capacitance model to B adsorption are provided elsewhere (Goldberg, 1993). In the constant capacitance model, anion adsorption is assumed to occur via a ligand exchange mechanism with the reactive surface functional group, SOH; no surface complexes are formed with ions in the background electrolyte. Specifically adsorbed ions reside in the surface plane of adsorption along with protons and hydroxyl ions. The constant capacitance model was able to describe B adsorption as a function of pH at four initial solution B concentrations using the same B surface complexation constant (Goldberg and Glaubig, 1988; Figure 6).

The ability of the constant capacitance model to predict B adsorption as a function of solution pH represents a major advantage over the Langmuir and Freundlich adsorption isotherm equations. In the case of oxide minerals, the constant capacitance model has this predictive capability using one less adjustable parameter than the Langmuir and Freundlich isotherm equations. In modeling B adsorption on clay minerals and soils, constants for the protonation and dissociation reactions of the surface functional group were also adjusted (Goldberg and Glaubig, 1986a, 1986b). In some cases, chemically unrealistic values of the protonation-dissociation constants were obtained potentially reducing the chemical significance of the model application to a curve fitting procedure (Goldberg and Glaubig, 1986a, 1986b).

Triple layer model

The triple layer model (Davis and Leckie, 1978, 1980; Davis et al., 1978) has been applied to describe B adsorption on aluminium and iron oxide and kaolinite (Blesa et al., 1984; Singh and Mattigod, 1992; Toner and Sparks, 1995). In contrast to the constant capacitance model, in the triple layer model anion adsorption can occur specifically via ligand exchange or nonspecifically through the formation of outer-sphere surface complexes with the reactive surface functional group, SOH. Outer-sphere surface complexes contain at least one water molecule between the adsorbing ion and the surface functional group. The triple layer model always includes outer-sphere surface complexation reactions for ions of the background electrolyte. Surface reactions, equilibrium constants, mass balance, and charge balances for the application of the triple layer model to B adsorption are provided elsewhere (Goldberg, 1993).

The triple layer model was able to describe B adsorption on aluminium and iron oxide and kaolinite using a ligand exchange mechanism (Blesa et al., 1984; Singh and Mattigod, 1992; Toner and Sparks, 1995; Figure 14). In describing B adsorption on iron oxide (Blesa et al., 1984) and kaolinite (Singh and Mattigod, 1992), additional surface complexes formed between the background electrolyte cations and inner-sphere B surface complexes were included. The model was less successful in describing B adsorption on iron oxide using an outer-sphere adsorption mechanism (Blesa et al., 1984). The triple layer model could describe B adsorption on an aluminium oxide using an outer-sphere adsorption mechanism. Howev-

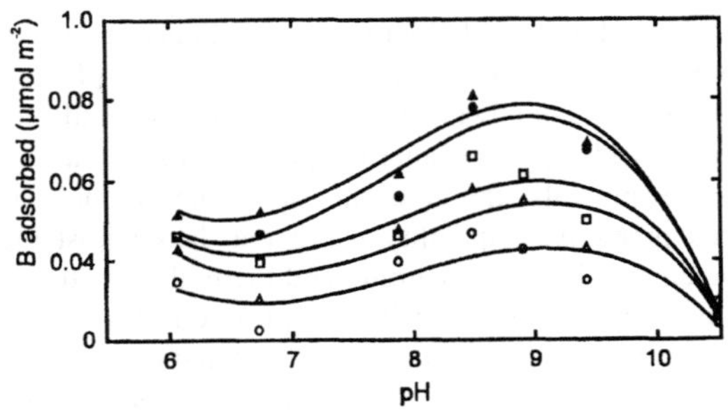

Figure 14. Fit of the triple layer model to B adsorption on a clay mineral from solutions containing 2 mg L^{-1} B (○), 4 mg L^{-1} B (Δ), 6 mg L^{-1} B (□), 8 mg L^{-1} B (●), and 10 mg L^{-1} B (▲). Model results are represented by solid lines. From Singh and Mattigod (1992).

er, only the inner-sphere adsorption mechanism could successfully describe both equilibrium and pressure jump kinetic data (Toner and Sparks, 1995). Experimental surface charge density data on iron oxide as a function of pH could be described using the triple layer model (Blesa et al., 1984). Unlike the constant capacitance model, the triple layer model can describe adsorption as a function of changing solution ionic strength since it considers surface complexation of the background electrolyte.

Stern variable surface charge variable surface potential model

The Stern variable surface charge variable surface potential (VSC-VSP) model (Barrow et al., 1980; Bowden et al., 1977, 1980). has been applied to describe B adsorption on aluminium oxide, iron oxide, and soil (Barrow, 1989; Bloesch et al., 1987). Equations for the application of the Stern VSC-VSP model to B adsorption are provided elsewhere (Goldberg, 1993). In contrast to the previous two models, the surface functional group is defined as $OH{-}S{-}OH_2$, allowing only one protonation or dissociation to occur for every two surface hydroxyl groups (Barrow et al., 1980). Although strongly adsorbed oxyanions form inner-sphere surface complexes, they reside in a separate plane located between the surface plane of protons and hydroxyls and the plane of outer-sphere surface complexation. The Stern VSC-VSP model emphasises parameter optimisation and neglects mass balance and the definition of surface species, surface reactions, and equilibrium constant expressions. The total number of surface sites and the maximum B adsorption are adjustable parameters in the Stern VSC-VSP model.

In describing B adsorption on iron oxide, four adsorbed B species were considered: $B(OH)_4^-$, $B_3O_3(OH)_4^-$, $B_4O_5(OH)_4^{2-}$, and $B_5O_6(OH)_4$-(Bloesch et al., 1987). The fit of the Stern VSC-VSP model to B adsorption as a function of solution pH was good for four initial solution B concentrations optimising binding constants for four adsorbed B species (Bloesch et al., 1987; Figure 7). A comparably good fit was obtained for B adsorption on aluminium oxide using the constant capacitance model by optimising only one binding constant (Figure 6). The Stern VSC-VSP model was applied to B adsorption on a soil by postulating a continuous distribution of binding sites divided into thirty discrete elements (Barrow, 1989). With this very large increase in the number of adjustable parameters, this application of the Stern VSC-VSP model reduces to a curve fitting procedure.

Evaluation of chemical surface complexation models

The major advantage of the chemical models presented is that they consider surface charge resulting from the surface complexation reactions. Chemical surface complexation models provide molecular descriptions of adsorption phenomena. The molecular features of surface complexation models can be given thermodynamic significance (Sposito, 1983). However, as for empirical models, goodness of fit to experimental adsorption data cannot be used as evidence for the actual presence of any of the surface complexes postulated in the models.

The constant capacitance model is the simplest of the surface complexation models. As the complexity of a model increases, the number of adjustable parameters increases; the curve fitting ability of a model improves as the number of adjustable parameters increases. Chemical significance of a model application suffers when parameters whose values are available from independent experiments are adjusted, as is the case in the Stern VSC-VSP model for the maximum surface charge and the maximum B adsorption. In the Stern VSC-VSP model the accuracy of the chemical speciation is also compromised by the lack of mass balance for the surface functional group.

Boron desorption

Boron desorption reactions in soils have been investigated less extensively than B adsorption reactions (Biggar and Fireman, 1960; Elrashidi and O'Connor,

1982; Griffin and Burau, 1974; Hadas and Hagin, 1972; Hatcher and Bower, 1958; Okazaki and Chao, 1968; Rhoades et al., 1970b; Wild and Mazaheri, 1979). Results on the reversibility of B adsorption reactions in soils are contradictory. For some soils, the B desorption isotherm corresponded closely to the B adsorption isotherm (Elrashidi and O'Connor, 1982; Hatcher and Bower, 1958; Wild and Mazaheri, 1979). Other soils exhibited hysteresis, that is the B adsorption isotherm did not correspond to the B adsorption isotherm (Elrashidi and O'Connor, 1982; Okazaki and Chao, 1968; Rhoades et al., 1970b). Hysteresis was not significantly correlated with any of the following soil properties: clay, organic carbon, pH, electrical conductivity, cation exchange capacity, surface area, aluminium oxide content, and iron oxide content (Elrashidi and O'Connor, 1982). Adsorbed soil B was completely desorbed after five sequential extractions (Okazaki and Chao, 1968).

Boron desorption reactions on soil minerals have been found to be reversible in some studies (Bloesch et al., 1987; Harder, 1961; Hingston, 1964; Keren and Gast, 1981) and hysteretic in others (Harder, 1961; Jasmund and Lindner, 1973). The pronounced hysteresis of B desorption from composted organic matter is attributed to the formation of dihydroxy and hydroxy–carboxy bonds with B (Yermiyaho et al., 1988).

Various mechanisms of anion sorption may explain apparent irreversibility of B sorption including: ligand exchange, formation of bidentate surface complexes, and incorporation into clay mineral lattices. Ligand exchange is pH dependent and reversible with respect to pH changes; however, anion desorption at constant pH exhibits varying degrees of irreversibility (Hingston, 1981). "Kinetic irreversibility" of anion adsorption has been attributed to the conversion of reversibly adsorbed, monodentate forms to bidentate forms which are "irreversibly" adsorbed (Kafkafi et al., 1967). The formation of bidentate anion surface complexes thus account for the slowness of the desorption reaction. Adsorbed B may undergo a monodentate to bidentate conversion at the mineral surface (Keren and Gast, 1981). Incorporation of B into the tetrahedral sites of clay lattices may also explain irreversibility of B adsorption (Harder, 1961). Boron diffusion into the particle interior may be responsible for the decrease in B desorption rate from soils after long term reaction (Griffin and Burau, 1974).

Modeling of boron desorption

For non-hysteretic systems, B desorption reactions can be described using the empirical and chemical B adsorption models discussed above. The Langmuir equation was able to describe B adsorption–desorption on soils (Hatcher and Bower, 1958), but due to hysteresis, was unable to describe B leaching from soils (Rhoades et al., 1970b). Boron adsorption and B desorption could be fit with the same Freundlich parameters for non-hysteretic soils, but different sets of Freundlich parameters were necessary to describe B adsorption and B desorption for hysteretic soils (Elrashidi and O'Connor, 1982).

Boron desorption from soils as a function of time can be described by the first order rate equation, the Elovich rate equation, and the power function equation (Griffin and Burau, 1974; Peryea et al., 1985; Sharma et al., 1989). The power function equation provided the best fit to B desorption based on coefficient of determination, r^2 values (Sharma et al., 1989).

Future research activities

In the future, increased use of spectroscopic techniques will provide direct experimental evidence for B adsorption mechanisms. Fourier transform infrared spectroscopy can be used to evaluate B adsorption mechanisms in aqueous systems. The approach of Su and Suarez (1995) may be extended to evaluate B adsorption mechanisms on clay minerals, calcium carbonate, and perhaps organic matter. Nuclear magnetic resonance can be used to distinguish between protonated and unprotonated surface complexes and between inner- and outer-sphere surface complexes. The approach used by Bleam et al. (1991) for phosphate adsorption can be modified to evaluate the configuration of adsorbed B on aluminium oxides. Indirect experimental procedures of establishing B adsorption mechanisms, such as the pressure jump relaxation procedure of Toner and Sparks (1995), may also be applied to B adsorption on clay minerals and calcium carbonate.

It is anticipated that additional advances will be made in surface complexation modeling of B adsorption by incorporating experimental results of B surface configuration obtained with the techniques described above. Surface complexation modeling studies of B adsorption must be expanded to consider the effect of competing ions. Competitive surface complexation modeling approaches such as those of Manning and Goldberg (1996a, 1996b) should be applicable to B adsorption reactions on soil minerals. Incorporation of surface complexation models into transport models is necessary to allow prediction of B adsorption–desorption reactions in natural systems and during reclamation processes.

References

Adriano D C, Page A L, Elseewi A A, Chang A C and Straughan I 1980 Utilization and disposal of fly ash and other coal residues in terrestrial ecosystems: A review. J. Environ. Qual. 9, 333–344.

Alwitt R S 1972 The point of zero charge of pseudoboehmite. J. Colloid Interface Sci. 40, 195–198.

Barrow N J 1978 The description of phosphate adsorption curves. J. Soil Sci. 29, 447–462.

Barrow N J 1989 Testing a mechanistic model. X. The effect of pH and electrolyte concentration on borate sorption by a soil. J. Soil Sci. 40, 427–435.

Barrow N J, Bowden J W, Posner A M and Quirk J P 1980 An objective method for fitting models of ion adsorption on variable charge surfaces. Aust. J. Soil Res. 18, 37–47.

Berger K C and Truog E 1945 Boron availability in relation to soil reaction and organic matter content. Soil Sci. Soc. Am. Proc. 10, 113–116.

Beyrouty C A, van Scoyoc G E and Feldkamp J R 1984 Evidence supporting specific adsorption of boron on synthetic aluminum hydroxides. Soil Sci. Soc. Am. J. 48, 284–287.

Bhatnagar R S, Attri S C, Mathur G S and Chaudhary R S 1979 Boron adsorption equilibrium in soils. Annals Arid Zone 18, 86–95.

Biggar J W and Fireman M 1960 Boron adsorption and release by soils. Soil Sci. Soc. Am. Proc. 24, 115–120.

Bingham F T, Page A L, Coleman N T and Flach K 1971 Boron adsorption characteristics of selected soils from Mexico and Hawaii. Soil Sci. Soc. Am. J. 35, 546–550.

Bleam W F, Pfeffer P E, Goldberg S, Taylor R W and Dudley R 1991 A ^{31}P solid-state nuclear magnetic resonance study of phosphate adsorption at the boehmite/aqueous-solution interface. Langmuir 7, 1702–1712.

Blesa M A, Maroto A J G and Regazzoni A E 1984 Boric acid adsorption on magnetite and zirconium dioxide. J. Colloid Interface Sci. 99, 32–40.

Bloesch P M, Bell L C and Hughes J D 1987 Adsorption and desorption of boron by goethite Aust. J. Soil Res. 25, 377–390.

Bowden J W, Posner A M and Quirk J P 1977 Ionic adsorption on variable charge mineral surfaces. Theoretical-charge development and titration curves. Aust. J. Soil Res. 15, 121–136.

Bowden J W, Nagarajah S, Barrow N J, Posner A M and Quirk J P 1980 Describing the adsorption of phosphate, citrate and selenite on a variable charge mineral surface. Aust. J. Soil Res. 18, 49–60.

Brockamp O 1973 Borfixierung in authigenen und detritischen Tonen. Geochim. Cosmochim. Acta 37, 1339–1351.

Brunauer S, Copeland L E and Kantro D L 1967 The Langmuir and BET theories. *In* The Solid-Gas Interface, Vol 1. Ed. E A Flood. pp 77–103. Marcel Dekker, New York.

Choi W W and Chen K Y 1979 Evaluation of boron removal by adsorption on solids. Environ. Sci. Technol. 13, 189–196.

Coddington J M and Taylor M J 1989 High field ^{11}B and ^{13}C NMR investigations of aqueous borate solutions and borate-diol complexes. J. Coord. Chem. 20, 27–38.

Couch E L and Grim R E 1968 Boron fixation by illites. Clays Clay Miner. 16, 249–256.

Davis J A and Leckie J O 1978 Surface ionization and complexation at the oxide/water interface. II. Surface properties of amorphous iron oxyhydroxide and adsorption of metal ions. J. Colloid Interface Sci. 67, 90–107.

Davis J A and Leckie J O 1980 Surface ionization and complexation at the oxide/water interface. 3. Adsorption of anions. J. Colloid Interface Sci. 74, 32–43.

Davis J A, James R O and Leckie J O 1978 Surface ionization and complexation at the oxide/water interface. I. Computation of electrical double layer properties in simple electrolytes. J. Colloid Interface Sci. 63, 480–499.

de Bussetti S G, Ferreiro E A and Helmy A K 1995 Sorption of boron by hydrous Al-oxide. Clays Clay Miner. 43, 58–62.

Elrashidi M A and O'Connor G A 1982 Boron sorption and desorption in soils. Soil Sci. Soc. Am. J. 46, 27–31.

Elseewi A A 1974 Some observations on boron in water, soils and plants at various locations in Egypt. Alex. J. Agric. Res. 22, 463–473.

Elseewi A A and Elmalky A E 1979 Boron distribution in soils and waters of Egypt. Soil Sci. Soc. Am. J. 43, 297–300.

Evans L J 1987 Retention of boron by agricultural soils from Ontario. Can. J. Soil Sci. 67, 33–42.

Fleet M E L 1965 Preliminary investigations into the sorption of boron by clay minerals. Clay Miner. 6, 3–16.

Fleming G A 1980 Essential micronutrients. I: Boron and molybdenum. *In* Applied Soil Trace Elements. Ed. B E Davies. pp 155–197. John Wiley and Sons, New York.

Fricke F and Leonhardt I 1950 Isoelektrischer Punkt und Pufferlösung. Die Naturwissenschaften. 37, 428.

Goldberg S 1993 Chemistry and mineralogy of boron in soils. *In* Boron and its Role in Crop Production. Ed. U C Gupta. pp 3–44. CRC Press, Boca Raton, USA.

Goldberg S and Forster H S 1991 Boron sorption on calcareous soils and reference calcites. Soil Sci. 152, 304–310.

Goldberg S and Glaubig R A 1985 Boron adsorption on aluminum and iron oxide minerals. Soil Sci. Soc. Am. J. 49, 1374–1379.

Goldberg S and Glaubig R A 1986a Boron adsorption on California soils. Soil Sci. Soc. Am. J. 50, 1173–1176.

Goldberg S and Glaubig R A 1986b Boron adsorption and silicon release by the clay minerals kaolinite, montmorillonite, and illite. Soil Sci. Soc. Am. J. 50, 1442–1448.

Goldberg S and Glaubig R A 1988 Boron and silicon adsorption on an aluminum oxide. Soil Sci. Soc. Am. J. 52, 87–91.

Goldberg S, Forster H S and Heick E L 1993a Temperature effects on boron adsorption by reference minerals and soils. Soil Sci. 156, 316–321.

Goldberg S, Forster H S and Heick E L 1993b Boron adsorption mechanisms on oxides, clay minerals and soils inferred from ionic strength effects. Soil Sci. Soc. Am. J. 57, 704–708.

Goldberg S, Forster H S, Lesch S M and Heick E L 1996 Influence of anion competition on boron adsorption by clays and soils. Soil Sci. 161, 99–103.

Griffin R A and Burau R G 1974 Kinetic and equilibrium studies of boron desorption from soil. Soil Sci. Soc. Am. Proc. 38, 892–897.

Gu B and Lowe L E 1990 Studies on the adsorption of boron on humic acids. Can. J. Soil Sci. 70, 305–311.

Gu B and Lowe L E 1992 Observations on the effect of a soil polysaccharide fraction on boron adsorption by clay minerals. Can. J. Soil Sci. 72, 623–626.

Gupta U C 1968 Relationship of total and hot-water soluble boron, and fixation of added boron, to properties of Podzol soils. Soil Sci. Soc. Am. Proc. 32, 45–48.

Hadas A and Hagin J 1972 Boron adsorption by soils as influenced by potassium. Soil Sci. 113, 189–193.

Harada T and Tamai M 1968 Some factors affecting behaviour of boron in soil. I. Some soil properties affecting boron adsorption of soil. Soil Sci. Plant Nutr. 14, 215–224.

Harder H 1961 Einbau von Bor in detritische Tonminerale. Experimente zur Erklärung des Borgehaltes toniger Sedimente. Geochim. Cosmochim. Acta 21, 284–294.

Harter R D and Smith G 1981 Langmuir equation and alternate methods of studying "adsorption" reactions in soils. *In* Chemistry in the Soil Environment. Am. Soc. Agron. Spec. Pub. no. 40, pp 167–182. Am. Soc. Agron. and Soil Sci. Soc. Am., Madison, Wisconsin, USA.

Hatcher J T and Bower C A 1958 Equilibria and dynamics of boron adsorption by soils. Soil Sci. 85, 319–323.

Hatcher J T, Bower C A and Clark M 1967 Adsorption of boron by soils as influenced by hydroxy aluminum and surface area. Soil Sci. 104, 422–426.

Helferrich F 1962 Ion Exchange. McGraw-Hill Book Company, Inc., New York, USA. 624 p.

Hingston F J 1964 Reactions between boron and clays. Aust. J. Soil Res. 2, 83–95.

Hingston F J 1981 A review of anion adsorption, *In* Adsorption of Inorganics at Solid-Liquid Interfaces. Eds. M A Anderson and A J Rubin. pp 51–90. Ann Arbor Science Publishers Inc., Ann Arbor, Michigan, USA.

Huettl P J V 1976 The pH dependent sorption of boron by soil organic matter. M.Sc. Thesis, University of Wisconsin, Madison, Wisconsin, USA.

Ichikuni M and Kikuchi K 1972 Retention of B by travertines. Chem. Geol. 9, 13–21.

Jasmund K and Lindner B 1973 Experiments on the fixation of boron by clay minerals. Proc. Int. Clay Conf. 1972, 399–412.

Kafkafi U, Posner A M and Quirk J P 1967 Desorption of phosphate from kaolinite. Soil Sci. Soc. Am. Proc. 31, 348–353.

Keren R and Bingham F T 1985 Boron in water, soils, and plants. Adv. Soil Sci. 1, 229–276.

Keren R and Gast R G 1981 Effects of wetting and drying, and of exchangeable cations, on boron adsorption and release by montmorillonite. Soil Sci. Soc. Am. J. 45, 478–482.

Keren R and Gast R G 1983 pH-dependent boron adsorption by montmorillonite hydroxy-aluminum complexes. Soil Sci. Soc. Am. J. 47, 1116–1121.

Keren R and Mezuman U 1981 Boron adsorption by clay minerals using a phenomenological equation. Clays Clay Miner. 29, 198–203.

Keren R and O'Connor G A 1982 Effect of exchangeable ions and ionic strength on boron adsorption by montmorillonite and illite. Clays Clay Miner. 30, 341–346.

Keren R and Sparks D L 1994 Effect of pH and ionic strength on boron adsorption by pyrophyllite. Soil Sci. Soc. Am. J. 58, 1095–1100.

Keren R and Talpaz H 1984 Boron adsorption by montmorillonite as affected by particle size. Soil Sci. Soc. Am. J. 48, 555–559.

Keren R, Bingham F T and Rhoades J D 1985a Plant uptake of boron as affected by boron distribution between liquid and solid phases in soil. Soil Sci. Soc. Am. J. 49, 297–302.

Keren R, Bingham F T and Rhoades J D 1985b Effect of clay content in soil on boron uptake and yield of wheat. Soil Sci. Soc. Am. J. 49, 1466–1470.

Keren R, Gast R G and Bar-Yosef B 1981 pH-dependent boron adsorption by Na-montmorillonite. Soil Sci. Soc. Am. J. 45, 45–48.

Keren R, Grossl P R and Sparks D L 1994 Equilibrium and kinetics of borate adsorption-desorption on pyrophyllite in aqueous suspensions. Soil Sci. Soc. Am. J. 58, 1116–1122.

Kinniburgh D G 1986 General purpose adsorption isotherms. Environ. Sci. Technol. 20, 895–904.

Lehto L 1995 Boron retention in limed forest mor. For. Ecol. Manage. 78, 11–20.

Manning B A and Goldberg S 1996a Modeling competitive adsorption of arsenate with phosphate and molybdate on oxide minerals. Soil Sci. Soc. Am. J. 60, 121–131.

Manning B A and Goldberg S 1996b Modeling arsenate competitive adsorption on kaolinite, montmorillonite, and illite. Clays Clay Miner. 44, 609–623.

Marzadori C, Vittori Antisari L, Ciavatta C and Sequi P 1991 Soil organic matter influence on adsorption and desorption of boron. Soil Sci. Soc. Am. J. 55, 1582–1585.

Mattigod S V, Frampton J A and Lim C H 1985 Effect of ion-pair formation on boron adsorption by kaolinite. Clays Clay Miner. 33, 433–437.

McPhail M, Page A L and Bingham F T 1972 Adsorption interactions of monosilicic and boric acid on hydrous oxides of iron and aluminum. Soil Sci. Soc. Am. Proc. 36, 510–514.

Metwally A I, El-Damaty A H and Yousry M 1974 Anion adsorption as a possible mechanism of boron retention by soils. Egypt J. Soil Sci. 14, 23–31.

Mezuman U and Keren R 1981 Boron adsorption by soils using a phenomenological adsorption equation. Soil Sci. Soc. Am. J. 45, 722–726.

Miljkovic N S, Mathews B C and Miller M H 1966 The available boron content of the genetic horizons of some Ontario soils. I. The relationship between water-soluble boron and other soil properties. Can. J. Soil Sci. 46, 133–145.

Mondal A K, Mandal B and Mandal L N 1993 Boron adsorption characteristics of some acidic alluvial soils in relation to soil properties. Commun. Soil Sci. Plant Anal. 24, 2553–2567.

Murphy L S and Walsh L M 1972 Correction of micronutrient deficiencies with fertilizers. *In* Micronutrients in Agriculture. Eds. J J Mortvedt, P M Giordano and W L Lindsay. pp 347–387. Soil Sci. Soc. Am. Inc., Madison, Wisconsin, USA.

Nicholaichuk W, Leyshon A J, Jame Y W and Campbell C A 1988 Boron and salinity survey of irrigation projects and the boron adsorption characteristics of some Saskatchewan soils. Can. J. Soil Sci. 68, 77–90.

Okazaki E and Chao T T 1968 Boron adsorption and desorption by some Hawaiian soils. Soil Sci. 105, 255–259.

Parks W L and White J L 1952 Boron retention by clay and humus systems saturated with various cations. Soil Sci. Soc. Am. Proc. 16, 298–300.

Peryea F J, Bingham F T and Rhoades J D 1985 Kinetics of post-reclamation boron dissolution. Soil Sci. Soc. Am. J. 49, 836–839.

Porrenga D H 1967 Influence of grinding and heating of layer silicates on boron sorption. Geochim. Cosmochim. Acta 31, 309–312.

Reisenauer H M, Walsh L M and Hoeft R G 1973 Testing soils for sulphur, boron, molybdenum, and chlorine. *In* Soil Testing and Plant Analysis. Eds. L M Walsh and J D Beaton. Rev. ed, pp 173–200. Soil Sci. Soc. Am. Inc., Madison, Wisconsin, USA.

Rhoades J D, Ingvalson R D and Hatcher J T 1970a Adsorption of boron by ferromagnesian minerals and magnesium hydroxide. Soil Sci. Soc. Am. Proc. 34, 938–941.

Rhoades J D, Ingvalson R D and Hatcher J T 1970b Laboratory determination of leachable soil boron. Soil Sci. Soc. Am. Proc. 34, 871–875.

Ryan J, Miyamoto S and Stroehlein J L 1977 Relation of solute and sorbed boron to the boron hazard in irrigation water. Plant and Soil 47, 253–256.

Schalscha E B, Bingham F T, Galindo G G and Galvan H P 1973 Boron adsorption by volcanic ash soils in Southern Chile. Soil Sci. 116, 70–76.

Scharrer K, Kühn H and Lüttmer J 1956 Untersuchungen über die Bindung des Bors durch anorganische Bodenbestandteile. Z. Pflanzenernähr. Düng. 73, 40–48.

Scott H D, Beasley S D and Thompson L F 1975 Effect of lime on boron transport to and uptake by cotton. Soil Sci. Soc. Am. Proc. 39, 1116–1121.

Sharma H C, Pasricha N S and Bajwa M S 1989 Comparison of mathematical models to describe boron desorption from salt-affected soils. Soil Sci. 147, 79–84.

Sims J R and Bingham F T 1967 Retention of boron by layer silicates, sesquioxides, and soil materials: I. Layer silicates. Soil Sci. Soc. Am. Proc. 31, 728–732.

Sims J R and Bingham F T 1968 Retention of boron by layer silicates, sesquioxides, and soil materials: II. Sesquioxides. Soil Sci. Soc. Am. Proc. 32, 364–369.

Singh M 1971 Equilibrium adsorption of boron in soils and clays. Geoderma 5, 209–217.

Singh P N and Mattigod S V 1992 Modeling boron adsorption on kaolinite. Clays Clay Miner. 40, 192–205.

Singh S S 1964 Boron adsorption equilibrium in soils. Soil Sci. 98, 383–387.

Sparr M C 1970 Micronutrient needs – which, where, on what – in the United States. Comm. Soil Sci. Plant Anal. 1, 241–262.

Sposito G 1982 On the use of the Langmuir equation in the interpretation of "adsorption" phenomena. II. The "two-surface" Langmuir equation. Soil Sci. Soc. Am. J. 46, 1147–1152.

Sposito G 1983 Foundations of surface complexation models of the oxide-aqueous solution interface. J. Colloid Interface Sci. 91, 329–340.

Sposito G 1984 The Surface Chemistry of Soils. Oxford University Press, New York, USA. 234 p.

Stumm W, Kummert R and Sigg L 1980 A ligand exchange model for the adsorption of inorganic and organic ligands at hydrous oxide interfaces. Croatica Chem. Acta 53, 291–312.

Su C and Suarez D L 1995 Coordination of adsorbed boron: A FTIR spectroscopic study. Environ. Sci. Technol. 29, 302–311.

Sulaiman W and Kay B D 1972 Measurement of the diffusion coefficient of boron in soil using a single cell technique. Soil Sci. Soc. Am. Proc. 36, 746–752.

Toner C V and Sparks D L 1995 Chemical relaxation and double layer model analysis of boron adsorption on alumina, Soil Sci. Soc. Am. J. 59, 395–404.

Veith J A and Sposito G 1977 On the use of the Langmuir equation in the interpretation of "adsorption" phenomena. Soil Sci. Soc. Am. J. 41, 697–702.

Wear J I and Patterson R M 1962 Effect of soil pH and texture on the availability of water-soluble boron in the soil. Soil Sci. Soc. Am. Proc. 26, 344–346.

Westall J 1980 Chemical equilibrium including adsorption on charged surfaces. *In* Particulates in Water. Eds. M C Kavanaugh and J O Leckie. pp 33–41. Adv. Chem. Ser. no. 189, Am. Chem. Soc., Washington, DC, USA.

Westall J and Hohl H 1980 A comparison of electrostatic models for the oxide/solution interface. Adv. Colloid Interface Sci. 12, 265–294.

Wild A and Mazaheri A 1979 Prediction of the leaching rate of boric acid under field conditions. Geoderma 22, 127–136.

Yermiyaho U, Keren R and Chen Y 1988 Boron sorption on composted organic matter. Soil Sci. Soc. Am. J. 52, 1309–1313.

Yermiyahu U, Keren R and Chen Y 1995 Boron sorption by soil in the presence of composted organic matter. Soil Sci. Soc. Am. J. 59, 405–409.

Plant and Soil **193:** 49–58, 1997.

Chapter 4

Absorption of boron by plant roots

Hening Hu and Patrick H. Brown
*Department of Pomology, University of California, Davis, CA 95616, USA**

Abstract

Experimental evidence suggests that B uptake is the result of the passive assimilation of undissociated boric acid. Boron uptake by a particular species should, therefore, be primarily determined by the B concentration in the soil solution and the rate of water uptake by the plant. This simple explanation of B uptake, however, does not adequately explain field observations where dramatic differences in B concentrations are observed between species, even when these species are grown under similar environmental conditions. The apparent contradiction between experimental results and in field observations, suggests, that B uptake is determined by factors that are as yet unknown. In the following, we discuss experimental and field observations as they relate to B uptake and discuss the mechanisms that may be involved in determining B uptake in diverse species.

Introduction

Boron is an essential micronutrient for higher plants, and B deficiency results in rapid inhibition of plant growth. The rapid and specific inhibition of plant growth that occurs upon removal of B is a consequence of two important features of B physiology: the specific structural role B plays in the cell wall (Hu and Brown, 1994; Loomis and Durst, 1992; Matoh, Chapter 5) and the limited mobility of B in the majority of species (Brown and Shelp, Chapter 7; Oertli and Richardson, 1970). As a result of its critical role in expanding tissues and its limited mobility, B has to be supplied continually throughout the life of the plant, usually through the root. For this reason, knowledge of the physiology of B absorption is essential.

Boron is absorbed from soil solution by roots mainly as the undissociated boric acid. Theoretical considerations predict that undissociated boric acid should be membrane permeable; in agreement with this, recent experimental results suggest that B absorption is a passive, non-metabolic process. Based solely upon these conclusions, it is predicted that B absorption is primarily determined by B concentration in the uptake medium and the transpiration rate of the plant. However, extensive field observations and experimental results demonstrate that species can differ significantly in their rate of B accumulation even when grown under identical environmental conditions. This contradiction cannot be explained adequately by our current understanding of B absorption mechanisms.

In this chapter we review current information on the mechanism of B absorption by plants and describe the biotic and abiotic factors that influence this process. This information is then discussed in light of observed differences in B absorption amongst cultivars and species.

Mechanisms of boron absorption

Properties of boric acid

Boric acid is a very weak acid in aqueous solution: its activity as an acid appears to be related to OH^- acceptance by the $B(OH)_3$ rather than to H^+ donation according to the following reaction:

$$B(OH)_3 + 2H_2O = B(OH)_4^- + H_3O^+ \ (pK_a 9.25)$$

Thus, in neutral or slightly acid soils, commonly occurring in plant growth environments, B exists mainly as undissociated boric acid (Raven, 1980).

* E-mail: hlhu@ucdavis.edu

Boric acid is soluble in water and has an ether–water partition coefficient of 0.035. Raven (1980) computed the permeability of plant cell membranes to $B(OH)_3$ on the basis of the 'lipid solution' mechanism. Using the linear relationship between $PM^{3/2}$ (P = permeability coefficient, M = mol. wt) and the ether–water partition coefficient, the computed value of P for $B(OH)_3$ is 8×10^{-6} cm s^{-1}. In another approach (Raven, 1980), using the linear relationship between log $PM^{1/2}$ and N (the number of potential H-bonding groups, N = 2 for –OH, and N = 6 for $B(OH)_3$), the estimated P for $B(OH)_3$ is 4×10^{-6} cm s^{-1}. This theoretical calculation led Raven (1980) to suggest that the permeability coefficient is high enough to account for the measured magnitude of boric acid fluxes across many plant cell membranes. The use of active transport of boric acid to maintain B distribution across a membrane away from thermodynamic equilibrium is consequently likely to be energetically expensive. On this basis, Raven (1980) predicted that B absorption by plant roots would be a passive process.

The conclusions made by Raven (1980) though sound, may be over simplified. The 'high' cost of maintaining a non-equilibrium concentration of an element, does not preclude the possibility of active absorption. Protons, for example, have an estimated membrane permeability of 10^{-3} to 10^{-4} cm s^{-1} (Dencher et al., 1986), which is 100 to 1000 times more permeable than boric acid. However, active H^+ transport occurs in all living systems (Briskin and Reynolds-Niesman, 1991).

Additionally, calculations of membrane permeability based upon ether:water partition coefficients are undoubtedly over simplified and may be misleading. For example, urea has a similar molecular wt (60) to boric acid (62) and is also uncharged at neutral pH. Finkelstein (1976) reported P_{urea} in an artificial membrane to be 0.6–4.0×10^{-6} cm s^{-1}, while Lee-Standelmann et al. (1991) calculated P_{urea} in barley (*Hordeum vulgare*) to be 2.3–2.6×10^{-5} cm s^{-1}. Hence, membrane permeability to urea in barley was found to be an order of magnitude greater than in an artificial membrane. Similarly, the permeability of water can vary significantly between membranes of different composition and cannot be adequately predicted from theoretical considerations alone. Jansen and Blume (1995) measured the water permeability of lipid vesicles composed of a variety of different phospholipids with different head groups and fatty acyl chains. Depending on the nature of the head group, on the chemical structure of the side chains and on the type of chain linkage, the permeability of water can differ by up to 3-fold from one kind of lipid vesicles to another.

In summary, it is clear that calculations of B fluxes based solely on theoretical lipid permeability are inadequate to predict B absorption by plants. Genotypes within a species could differ several fold in B permeability (see "the paradox of boron uptake"), which may have a physiologically important effect on B uptake by roots.

Role of boron complex formation in boron absorption

Another important property of boric acid that will influence B absorption is its ability to form *cis*-diol complexes with a variety of organic molecules. For example, in celery (*Apium graveolens*), B present in the cell sap and phloem exudates occurs almost entirely as the B–mannitol complex. No free boric acid was detected. The reaction of boric acid with mannitol is described in the following reaction (Hu et al., 1997):

$$H_3BO_3 + \underset{\text{mannitol}}{2C_6O_6H_{14}} =$$
$$\underset{[\text{mannitol}-\text{B}-\text{mannitol}]^-}{[(C_6O_6H_{12})B(C_6O_6H_{12})]^-} + H_3O^+ + 2H_2O$$

The equilibrium constant for the reaction is 10^{-4}. The formation of borate–diol complexes has a much lower pK_a than that of borate. The pK_a of B–mannitol complex (about 5.2) is four units lower than that of free boric acid (9.25) (Raven, 1980), strongly suggesting that at normal B supply free boric acid concentrations inside the cytoplasm of celery (pH about 6.2–6.5) would approach zero. This situation is true for celery, in which mannitol is present at millimolar concentrations, and would also be true for other species where B-chelating diols are present at millimolar or greater concentration.

There are a great number of biological compounds that can form complexes with B, both in the cytoplasm and in the cell wall. Compounds capable of complexing with boric acid include sugars, their derivatives (sugar alcohols), phenols, organic acids, and some polymers (Boeseken, 1949; Raven, 1980). Common examples are sorbitol, mannitol, glycerol, ribose, apiose, nicotinamide-adenine-dinucleotide (NAD) and fructose (Loomis and Durst, 1992; Makkee et al., 1985). Low molecular weight B-complexes with mannitol, sorbitol and fructose have been recently isolated and characterised from plant tissue (Hu et al., 1997;

Penn et al., 1997). In addition, B complexes with rhamnogalacturonan-II in the cell wall have been characterised (Ishii and Matsunaga, 1996; Kobayashi et al., 1996; O'Neill et al., 1996).

The presence of significant amounts of B-complexing compounds in plant cells would have a significant effect on B uptake if uptake is solely determined by the free boric acid concentration gradient across the membrane. Raven (1980) suggested that the total B content of the tissue can greatly exceed the free boric acid concentration in the bathing solution due to the formation of B complexes, so that the demonstration of a net B influx into a tissue, in which the total internal B concentration exceeds that of external boric acid concentration is not adequate to show that active transport is occurring.

The uncharged nature of boric acid, the relatively high (though poorly defined) membrane permeability of boric acid, combined with its propensity to form complexes both in the cell wall and in the cytoplasm greatly complicates the study of B absorption. Classic analysis of elemental uptake based upon determination of transmembrane electrochemical gradients, analysis of absorption kinetics and measurement of electrical changes upon ion uptake do not readily apply to studies of B. Nevertheless, important insights into the mechanism of B absorption can be gained from a review of existing experimental data.

Is boron absorption active or passive?

Over the last 30 years this question has been addressed by many research groups with varied and conflicting results. The reason is partly due to the characteristic permeability of boric acid across cell membranes and the significant effects of *cis*-diol complex formation which adds considerable complexity to the interpretation of uptake experiments. Several of the studies of B absorption were conducted using unrealistically high B concentrations with highly variable absorption or desorption periods, and in only rare cases has the compartmentation of B within the root been adequately considered.

Bingham et al. (1970) conducted one of the first detailed studies of B absorption using excised roots of barley. With a B concentration of 0.93 m*M* and a 4 h absorption period, a strong pH dependency in the alkaline range for B absorption was found. Analysis of the pH effect on the proportion of $B(OH)_3$ and $B(OH)_4^-$ in the absorption solution reveals that B absorption by barley roots was primarily contingent upon the $B(OH)_3$ concentration regardless of pH. In a time study from 1 to 24 h, they did not find B accumulation within the root, nor did they find B accumulation against a concentration gradient in solution B concentration ranges from 0.0185 to 7.39 m*M*. Within 3 to 4 h of transfer to $CaSO_4$ solution, approximately 95% of the B, initially absorbed by the root tissue was lost to the $CaSO_4$ solution. At pH 6.0, changes in substrate temperature (from 10 to 30 °C), salt composition (Ca or K salt) or addition of KCN and 2,4-dinitrophenol (DNP) failed to exert any influence on B absorption. These results led Bingham et al. (1970) to conclude that B absorption is a physical, non-metabolic process acting in response to B concentration gradient and that undissociated boric acid is favoured for absorption.

Wilders and Neales (1971) used sliced disks from carrot (*Daucus carota*) and red beet (*Beta vulgaris*) storage roots suspended in 0.1 m*M* boric acid solutions for absorption or desorption studies. They reported that disks of carrot and red beet absorb B to an equilibrium internal concentration which is greater than that of the external solution and that this accumulation of B in the tissues is inhibited by anoxia, by DNP and also by low temperature, suggesting active uptake. On the other hand, B content of storage tissues, after a period of absorption in 0.1 m*M* boric acid solution, was rapidly and almost completely desorbed into a B free solution, suggesting a passive, diffusion type process. Therefore, Wilders and Neales (1971) concluded that there are two components of B uptake: a passive diffusion of $B(OH)_3$ and an active transport for $B(OH)_4^-$ ion.

Wilders and Neales (1971) based their conclusion of active uptake on 1) the observation that internal B concentrations were greater than the external B concentrations as a result of the transport of the $B(OH)_4^-$ ion; and 2) that uptake was inhibited by anoxia and DNP. Their conclusion of active absorption based upon the occurrence of higher internal B concentrations is not, however, supported by their data. The passive membrane permeability of $B(OH)_3$ (suggested by the authors, and proposed by Raven, 1980) predicts that B can diffuse into both internal and external solution. In other words, the internal B concentration would be equilibrated with the external B concentration. From the desorption data of Wilders and Neales (1971), we can calculate that $B(OH)_3$ passively crosses the membrane at a rate of at least 0.1 μmol h^{-1} g^{-1} fresh wt, while the B accumulation rate was more than 0.1 μmol h^{-1} g^{-1} fresh wt. This means that the cells had to pump B internally at a rate of at least 0.2 μmol h^{-1} g^{-1} fresh wt to obtain a net accumulation of 0.1 μmol h^{-1} g^{-1}

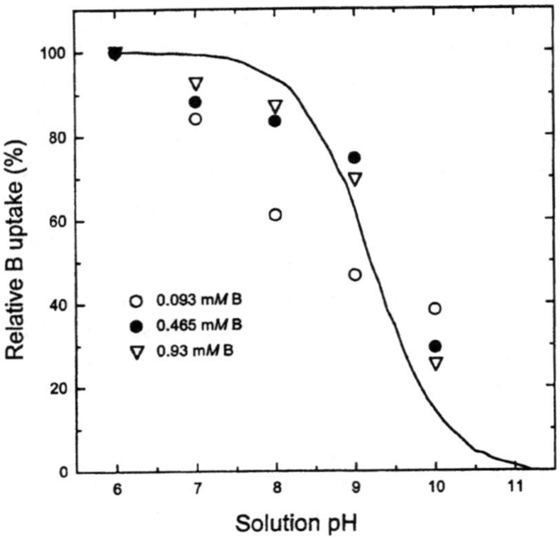

Figure 1. Relative uptake of boron as a function of solution pH. Uptake at pH 6 = 100% at each supply concentration. Solid line: percentage of undissociated H_3BO_3. Adapted from Oertli and Grgurevic (1975).

fresh wt. It is extremely doubtful, that given the limited surface area provided by the storage root disks used, that $B(OH)_4^-$ could be pumped into the cells at a rate of 0.2 μmol h^{-1} g^{-1} fresh wt. This is particularly improbable given the external $B(OH)_4^-$ concentration of 32 n*M* (0.03% of 0.1 m*M* boric acid at pH 5.7) utilised by the authors. It seems clear that Wilders and Neales (1971) conclusion of active uptake was the result of insufficient tissue rinsing and may also have been influenced by the relatively high concentrations of B-binding ligands in carrot (mannitol) and fructose in sugar beet (*Beta vulgaris*).

Oertli and Grgurevic (1975) used excised root sections of 6 day old barley seedlings and studied the pH effect on B absorption. Changes in pH from 6 to 10 and B concentration from 0.093 to 0.93 m *M* suggested that B uptake and the final B contents increased nearly linearly with increasing B supply. The relative B uptake (uptake at pH 6 = 100%) at all supply levels decreased with increasing pH similar to the decrease of the fraction of undissociated $B(OH)_3$ (Figure 1). At pH 6, final B concentrations in roots and treatment solutions were similar. Based on these results, they concluded that B in root tissues and external solution tends to approach a diffusion equilibrium which is governed by the proportion of $B(OH)_3$ vs $B(OH)_4^-$ in the system.

The work of Bingham et al. (1970) and Oertli and Grgurevic (1975) was criticised by Bowen and Nissen (1976), because the barley roots used were only rinsed for 30 s or 1 min after B absorption. As a result, much B remaining in the water free space and free space in the cell wall may have been wrongly considered to be absorbed B. However, it is not a simple matter to determine what is an appropriate time for B desorption. For example, if active uptake is the primary mode of B accumulation, then the desorption period should be long enough to remove all the non-specifically bound B from the free wall space, e.g. 30 min (Bowen and Nissen, 1976). On the other hand, if passive B uptake predominates, then the desorption should be short enough to avoid passive desorption of B across the membrane, while still removing the majority of the apoplastic B. This is clearly an unattainable task for a highly permeable compound such as boric acid. Unfortunately, the significance of the desorption period has not been adequately considered in most studies of B absorption.

Bowen and Nissen (1977) used excised roots from 5 to 7 day old barley seedlings in their short term (hours) B absorption studies. Within the concentration range used, 0.2 – 5 m*M*, they did not find B accumulation against a concentration gradient after 5 h in the absorption solution. However, when they studied B absorption as a function of the external B concentration over the range of 0.05 – 13.3 m*M*, and plotted the data as a double reciprocal plot, they observed that the data could be described by a series of straight lines (phases) separated by sharp transitions or 'jumps', which occurred at B concentrations of 0.16, 0.20, 0.47, 0.55 and 8 m*M*, respectively. Based on these results, they concluded that B absorption in barley is an active transport process since it is compatible with the concept of multiphasic absorption mechanisms outlined by Nissen (1974). To further support their argument, Bowen and Nissen (1977) reported that B uptake from 1 m*M* B solution was strongly dependent upon temperature over 3 – 47 °C range, and also that B uptake from 1 m*M* B solution was inhibited by metabolic inhibitors, including 0.05 m*M* DNP, 0.05 m*M* NaN_3, 1 m*M* NaCN, 5 m*M* arsenate and 0.05 m*M* dicoumarol.

A major concern with Bowen and Nissen's (1977) study is that the B concentrations used were much higher (0.05 – 13.3 m*M*) than is physiologically relevant for barley. It has been reported that barley plants have a low B requirement (Hu et al., 1996; Loomis and Durst, 1992), while sensitive barley cultivars develop B toxicity symptoms after 11 days growth at solution B concentrations of 0.2 m*M*, and moderately resistant cultivars developed toxicity 17 days after exposure to 0.5 m*M* B (Nable, 1988). Bowen and Nissen (1977)

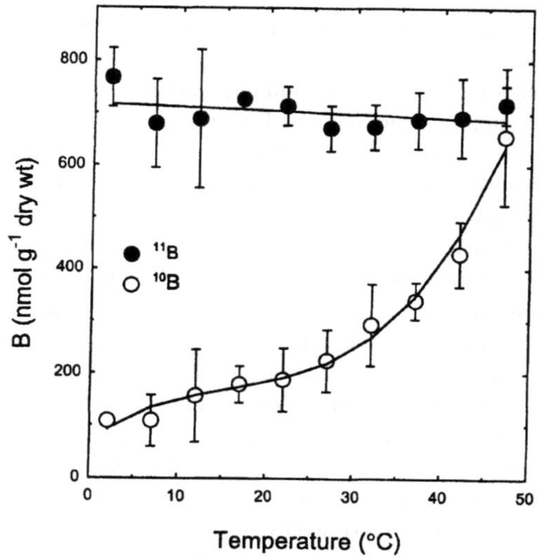

Figure 2. Influence of temperature on the uptake of ^{10}B into ^{11}B precultured intact tobacco cells. Cells were placed in 0.1 m*M* ^{10}B uptake medium for 30 min at various temperatures then transferred to B free desorption solution for 30 min at 5 °C. Values are means ± SE. Adapted from Brown and Hu (1994).

observed transitions in B uptake at concentrations of 0.47, 0.55 and 8 m*M*, which are well in excess of the concentrations that are required for optimal barley growth. Therefore, it is extremely unlikely that the uptake kinetics described by Bowen and Nissen (1977) are of any physiological relevance. The transitions observed by Bowen and Nissen (1977) may be better explained by the saturation of the B-binding capacity of various cellular compartments.

Thellier et al. (1979) conducted a $B(OH)_3$ flux study using isotopes of ^{10}B and ^{11}B in duckweed (*Lemna minor*). After a period in ^{10}B uptake solutions, plants were placed in either ^{11}B boric acid or pure water to determine efflux kinetics. The results suggested that four compartments existed within the cell, i.e. free space, cytoplasm, vacuole and the cell wall. Each of these compartments contained B concentrations much higher than that of the external solution (0.16 m*M*). They did not interpret the occurrence of higher internal than external B concentrations as evidence of active absorption of B, but attributed this to the formation of borate mono- or di-ester within the various compartments and subsequent exchange of B isotopes.

Using stable B isotopes and inductively coupled plasma-mass spectrometry, Brown and Hu (1994) reported that B absorption is linear over a wide range of B concentrations (0–0.2 m*M* or 0.2–10 m*M*) in cultured tobacco (*Nicotiana tabacum*) cells and in the roots of sunflower (*Helianthus annuus*) and squash (*Cucurbita pepo*). There was no indication of saturation kinetics and B absorption was not inhibited by DNP or KCN. Although B absorption did respond to temperature in cultured tobacco cells, absorption was not inhibited by very low (2 °C) temperature, and no inhibition of uptake was reached at temperatures as high as 47 °C (Figure 2). The temperature effect was not typical of temperature dependent kinetics, and the Q_{10} was less than 2. On the basis of these results it was concluded that shifts in B absorption with temperature were associated with changes in membrane conformation (and hence permeability) and the effects of temperature on the relative activity of B molecules in solution.

In these experiments, ^{11}B incorporated during pretreatment growth was not displaced by ^{10}B in the uptake solution even in freeze-thaw killed cells, suggesting that B can be incorporated into some non-exchangeable complexes within the cell (Brown and Hu, 1994). In desorption studies with cultured tobacco cells, after a 30 min uptake period in either 0.1 or 10 m*M*^{10}B solution, it was found that B desorption was almost complete within 1 min of transfer to B free solution. This rapid desorption of B probably represents B within apoplastic spaces. Further loss of recently acquired ^{10}B was very limited. Based on these findings, Brown and Hu (1994) concluded that B absorption is a non-metabolic process, which is controlled, in large, by the formation of non-exchangeable B-complexes in the cytoplasm and cell wall.

In summary, an unequivocal conclusion about B uptake mechanism can not be made at this time. The reason is partly a result of the diverse plant materials and B concentrations used by different authors, the failure to distinguish B compartmentation and the failure to adequately assess B complex formation. It is clear, however, that there is little or no sound evidence supporting active B absorption. Evidence to support a non-metabolic process for B absorption, is somewhat more convincing though far from conclusive. We propose that B absorption can be best explained as a passive diffusion of free boric acid into the cell, followed by a rapid formation of B-complexes within the cytoplasm and the cell wall. The decline in free boric acid within the cell by the formation of B-complexes allows further absorption of B from the external solution and results in tissue B concentration that can greatly exceed the free boric acid concentration in the uptake solution.

Factors affecting boron absorption

Boron absorption by plant roots is affected by various environmental factors, both in the soil and non-soil environment. Important factors influencing B adsorption from solution include the initial B content of the soil, the pH of the soil, the type of exchangeable ions present in the soil solution, the amounts and types of minerals in the soil, the soil organic matter content, the wetting and drying cycles, and the water to soil ratio (Goldberg, Chapter 3; Keren et al., 1985). All these factors affect B absorption by plants. In coarse textured, low organic matter, and high pH soils, low B availability to the plants would be expected.

Among the soil factors, pH is most important in affecting B absorption by plants (Gupta, 1979). In general, plant B absorption is decreased with the increase in soil pH because of two reasons. Firstly, below pH 7, $B(OH)_3$ predominates and since the affinity of soil clay for this species is relatively low, the amount of B adsorption by soil is small. As the pH increases, $B(OH)_4^-$ concentration relative to $B(OH)_3$ increases, and as a consequence of the relatively strong affinity of $B(OH)_4^-$ for clay minerals, the amount of adsorbed B increases (Keren and Bingham, 1985). In other words, with the increase in soil pH, B availability to the root decreases. For more information about reactions of B with soils, the reader is referred to Chapter 3 in this volume. Secondly, Oertli and Grgurevic (1975) found that the absorption of B by solution grown barley roots decreased with the increase of solution pH, this corresponded to the decrease in the fraction of undissociated $B(OH)_3$, demonstrating that $B(OH)_3$ is the form of B available to plants.

In non-soil environments, the transpiration rate seems to be the most important factor governing B absorption. As a result, relative humidity, temperature and light intensity will all alter B absorption. For example, Bowen (1972) made a comprehensive study about the environment factors on B absorption and found that in sugarcane (*Saccharum officinarum*) seedlings, increasing light intensity increased B absorption, while decreasing relative humidity increased B absorption. Similarly, an increase in air temperature increased total B absorption even though the relative humidity was held constant. When other environment conditions were constant, an increase in root temperature increased B absorption.

Similar information was reported by Oertli (1994) in barley seedlings. He found that increasing duration and intensity of illumination significantly increased water consumption as well as B uptake, and that the uptake of water and B was reduced 50% at high humidity in comparison to low humidity. However, the B to water uptake ratio remained constant (about 3.4 μg B mL^{-1} water at 0.93 m*M* external B) irrespective of humidity or temperature treatment. Temperature increased water and B absorption significantly at constant relative humidity (as a consequence of differences in vapour pressure deficits). For example, at an external B concentration of 0.05 m*M*, B absorption was about 0.2 μg per barley seedling at a temperature of 10 °C. When temperature increased to 27 °C, B absorption was about 0.65 μg per barley seedling. However, the ratio of B absorption to water uptake remained constant at 0.39 μg B mL^{-1} water. On the other hand, Oertli (1994) found that the ratio of B absorption to water uptake did not correspond to the external B concentration. For example, at external B concentration of 0.05, 0.23 and 0.93 m*M*, the B to water uptake ratio was 0.39, 1.07 and 3.46 μg B mL^{-1} water, respectively.

Forno et al. (1979) reported that two cassava (*Manihot* sp.) cultivars growing in nutrient solution developed B deficient symptoms when growing at root temperature of 18 °C; their shoot B concentration ranged from 9.7 to 11.9 mg B kg^{-1} dry wt. However, when these plants were grown at 23 °C root temperature, they developed much less severe B deficiency symptoms and symptoms appeared only towards the end of the experiment. These plants accumulated 17 – 19 mg B kg^{-1} dry wt in the shoot. Plants were symptom free at root temperatures of 28 – 33 °C and they accumulated 42 – 55 mg B kg^{-1} dry wt in the shoot.

An exception to the temperature effect on B absorption described above was reported by Nable et al. (1990). These authors reported that nutrient solution temperature had little effect on tissue B concentration within cultivars of barley. For example, at 15 μM B solution, B absorption in 'Sahara 3771' was the same (20 mg B kg^{-1} dry wt) at 5 and 25 °C. At 1 m*M* B in solution, an increase in temperature from 5 to 25 °C resulted in an increase of B absorption from 80 to 90 mg B kg^{-1}. However, different genotypes exhibited differences in B accumulation.

The reported environmental effects on B absorption can largely be explained by changes in transpiration (Raven, 1980). Reduced humidity increases transpiration, resulting in an increase in B absorption. Increased temperature creates water vapour deficits, which will also increase transpiration, resulting in an increase in B uptake. Illumination has the same effect as increased temperature or reduced relative humidity, and as a result, transpiration increases and B uptake increases. However, the observed increases in B uptake with the increase in root temperature can not be explained by transpiration effects. No simple relationship exists between transpiration rates and whole plant B uptake. For example, in wheat (*Triticum aestivum*) cultivars, water use efficiency ranged from 3.1 to 4.0 g dry wt kg^{-1} water. The less than 30% difference in transpiration rate within cultivars of one species can not account for the 400–700% difference in B uptake (Nable, 1988). Further, genotype differences in resistance to B toxicity and B absorption are known to occur in roots assayed without the shoot attached and B uptake in these roots has been frequently observed demonstrating that transpiration is not essential for uptake (Huang and Graham, 1990).

The paradox of boron uptake

The preponderance of available experimental evidence suggests that B uptake is a passive process, the rate of which is influenced by passive diffusion of B through the membrane, the formation of B complexes within the cell and plant water fluxes. If these assumptions are true then one would predict that B uptake would be very closely correlated with water use and that species grown in like environments supplied with non-limiting water would have very similar B uptake rates. Nevertheless, there is clear evidence that B uptake differs dramatically between species and even within genotypes of a single species. For example, Nable (1988) reported that in solution culture experiments, the B concentration or total B content in all organs of five barley and six wheat cultivars differ dramatically. The cultivars resistant to B toxicity always accumulated less B than the sensitive cultivars. When grown in 5 m*M* B solution, barley cultivars 'Sahara 3763' and 'Schooner' accumulated 112 and 710 mg B kg^{-1} dry wt in the youngest expanded leaf blade, respectively, a more than 6-fold difference in B accumulation between genotypes. These differences in B uptake cannot be explained through differences in water use, which has less than 1-fold difference among the cultivars. This apparent contradiction between passive B uptake and significant differences among the genotypes is difficult to reconcile but is of fundamental importance to studies of B nutrition. Several mechanisms have been postulated to explain this apparent paradox (Nable, 1988; Nable et al., 1990).

Firstly, B uptake may be partially under metabolic control. For example, Bowen and Nissen (1977) reported that active uptake accounts for less than 10% of the total B accumulated by barley roots. Alternatively, there may be an active B exclusion mechanism. In this regard, the extent of active uptake may be different from species to species or from genotype to genotype. This mechanism may explain the differences in B absorption among different species (genotypes) and it could also account for the apparently conflicting results obtained when metabolic inhibitors are used. However, in view of the discussion in "Mechanisms of boron absorption", we conclude there is little support for this concept.

Alternatively, one plausible explanation is that the exudation of B complexing agents into the rhizosphere restricts B uptake from the soil. The extent to which B-binding exudates are produced may be species dependent. Organic compounds containing *cis*-diols, as well as some amino acids or some organic acids can complex with B. Because the resultant B-complex is generally an anion, it would have a greatly reduced membrane permeability in comparison to uncharged boric acid, and hence absorption would be reduced. Many compounds present in root exudates, particularly the pectin-rich polysaccharides that are abundant in plant mucilage, would be capable of forming stable B complexes. These complexes could realistically reduce B uptake. However, one might expect that these exudates would be saturated by B at high external B supply and once saturated, uptake should show no difference among genotypes. Even at the very high B application rates (6.4 m*M*) used by Nable et al. (1990), this did not appear to be the case. Also, one would expect that experiments conducted in solution culture (in which rhizosphere exudates would be diluted in the culture medium) would result in diminished genotype differences. In the relatively long term studies of Nable et al. (1990) (see also Nable, 1988; Paull et al., 1988) this did not appear to be true. On the other hand, genotype differences in long term B uptake studies have not been confirmed in short term solution culture experiments (Garnett et al., 1993). Nevertheless, this hypothesis cannot be entirely excluded without more specific information on the amount and localisation of these putative B-binding exudates.

Thirdly, root B-adsorption capacity may differ significantly and hence may affect B uptake. Tanaka (1967) observed a 3- to 10-fold difference in the adsorption capacity of roots between dicots and graminaceous monocots with a much smaller difference within the dicots or within monocots. Yamauchi (1971) demonstrated a highly significant correlation between 'protopectin' (old terminology for water insoluble parent pectic substance) and 0.5 *N* HCl soluble but water insoluble B in 33 species across 13 families. Hu et al. (1996) reported that among the fourteen species studied, there is significant correlation between species B requirement and cell wall pectin. In general, more B is required by dicots than graminaceous monocots presumably because of the higher pectin content in the dicots. Recently, Matoh et al. (1996) demonstrated a high correlation between cell wall B and 2-keto-3-deoxy sugars in the high molecular weight fragments of cell walls. In combination, these results suggest that differences in the cell wall B requirement exist among species and that these differences might affect B uptake. This observation is not adequate, however, to explain the observed differences in B uptake and is not consistent with results from uptake experiments. For example, Hu et al. (1996) report a less than 2-fold difference in cell wall B binding capacity between various monocot species even though differences in B uptake can be as great as 4- to 7-fold between these same species (Nable, 1988). Further, if the cell wall B-binding capacity is the key determinant of the observed differences in B uptake between species, then once the cell wall binding capacity is saturated, uptake should be the same for all species. There is no indication that this is the case.

Fourthly, physical barriers related to root cell wall structure may determine B uptake rates. Some evidence in support of this was provided by Jenkin et al. (1993) who investigated B uptake in protoplasts of two barley varieties with stable B isotope techniques. The two barley varieties used were known to have significant differences in B accumulation on a whole plant and long term basis, with the B toxicity resistant cultivar 'Sahara 3771' accumulating much less B than the sensitive cultivar 'Stirling'. However, in the absence of the cell wall, the two varieties did not show any significant difference in B accumulation (Jenkin et al., 1993). There are three possible explanations for this result: (1), the difference in B accumulation between these two cultivars is determined by the cell wall and, once the cell wall is removed, the cultivar difference disappears; (2), since the membrane integrity was not assessed in the protoplasts used by Jenkin et al. (1993), genotype difference might be overshadowed by membrane leakiness as a consequence of the protoplast isolation technique; or (3), the duration of the uptake experiment was too short to determine species differences. Evidence to support this is provided in the work of Garnett et al. (1993) who reported that in short term B uptake studies (4 h), the wheat cultivar 'Halberd' (resistant to B toxicity) did not differ from 'W1*MMC' (sensitive to B toxicity) in B accumulation. In longer term studies (days) these two varieties did show significant difference in B accumulation. The results of Jenkin et al. (1993) may therefore reflect short term effects and may not be relevant to long term B accumulation. At present, insufficient information is available to determine if there is any relationship between root cell wall structure and B uptake.

If B uptake is a purely passive process, then the differences in B accumulation may be due to differences in membrane permeability to B. Membrane permeability to urea, water and protons may all vary significantly depending upon membrane composition (see "Properties of boric acid"). Boric acid has a lipid solubility of the same order as urea and hence small differences in membrane composition (head groups, fatty acyl chains, etc.) between genotypes may affect B uptake. If passive membrane permeation is the primary means of B uptake, then B uptake would be linear with increasing concentration, and would not be influenced by metabolic inhibitors. This basically agrees with existing information (see "Is boron adsorption active or passive?". If species and genotypes had different membrane permeability coefficients for boric acid, they would also exhibit a linear response to increasing B supply. However, the slope of this response line may differ and the resulting B absorption rate would differ. The genotype comparisons of Nable et al. (1990) can be suitably explained in this manner.

Similarly, small differences in the relative permeability of membranes to B relative to water would alter the ratio of B accumulated to water transpired, though the total amount of water transpired would still directly affect B uptake. This would adequately explain the observed relationship between water uptake and B accumulation while accommodating the observed genotype differences. It is also possible that differences in membrane permeability to B may not be sufficient to influence whole plant B uptake during short term uptake studies, these differences may only become apparent after longer B uptake periods.

Though differences in membrane permeability could theoretically result in the observed differences in B uptake between species (or genotypes), it should be noted that this has not been experimentally verified. Further, we are unaware of any evidence of genotypic differences in membrane composition resulting in the differential membrane permeability of any solute. Finally, experimental evidence suggests that genotypic differences in B uptake are associated with relatively few genes (Nable et al., 1990 and references therein), it would be intriguing to determine if these genes code for subtle differences in membrane composition.

Further research to determine the mechanism of genotype and species differences in B absorption is clearly required.

Conclusions and future directions

Boron uptake in higher plants is probably a passive process acting in response to external boric acid concentration, membrane permeability, internal complex formation and transpiration rates. Species differences in B uptake rate cannot be adequately explained on the basis of current knowledge but may result from differences in membrane permeability, B-complex formation inside or outside the root or some as yet unidentified mechanism.

Future studies on B absorption should include consideration of both short and long term uptake periods, must be conducted at realistic B concentrations and should always consider the impact of B-complex formation in experimental design. Experimental verification of the theoretical membrane permeability coefficient proposed by Raven (1980) remains a priority and may ultimately allow permeability to be compared between species. The recent identification of closely related genotypes differing significantly in B absorption rates has provided invaluable experimental material. These genotypes hold great potential for providing the physiological and genetic information necessary to determine the mechanism of B uptake in plants.

References

Bingham F T, Elseewi A and Oertli J J 1970 Characteristics of boron absorption by excised barley roots. Soil Sci. Soc. Am. Proc. 34, 613–617.

Boeseken J 1949 The use of boric acid for the determination of the configuration of carbohydrates. *In* Advances in carbohydrate chemistry V4. Eds. W W Pigman and M L Wolfrom, pp 189–210. Academic Press, New York.

Bowen J E 1972 Effect of environmental factors on water utilisation and boron accumulation and translocation in sugarcane. Plant Cell Physiol. 13, 703–714.

Bowen J E and Nissen P 1976 Boron uptake by excised barley roots I. Uptake into the free space. Plant Physiol. 57, 353–357.

Bowen J E and Nissen P 1977 Boron uptake by excised barley roots II. Characteristics and kinetics of active uptake. Physiol. Plant. 41, 109–115.

Briskin D P and Reynolds-Niesman I 1991 Determination of H^+/ATP stoichiometry for the plasma membrane H^+-ATPase from red beet (*Beta vulgaris* L.) storage tissue. Plant Physiol. 95, 242–250.

Brown P H, and Hu H 1994 Boron uptake by sunflower, squash and cultured tobacco cells. Physiol. Plant. 91, 435–441.

Dencher N A, Burghaus P A and Grzesiek S 1986 Determination of the net proton-hydroxide ion permeability across vesicular lipid bilayers and membrane proteins by optical probes. Methods in Enzymology 127, 746–760.

Finkelstein A 1976 Water and nonelectrolyte permeability of lipid bilayer membranes. J. Gen. Physiol. 68, 127–135.

Forno D A, Asher C J and Edwards D G 1979 Boron nutrition of cassava, and boron × temperature interaction. Field Crops Res. 2, 265–279.

Garnett T P, Tester M A and Nable R O 1993 The control of boron accumulation by two genotypes of wheat. *In* Plant nutrition – from genetic engineering to field practice. Ed. N J Barrow. pp 397–400. Kluwer Academic Publishers, Dordrecht, The Netherlands.

Gupta U C 1979 Boron nutrition of crops. Adv. Agron. 31, 273–307.

Hu H and Brown P H 1994 Localisation of boron in cell walls of squash and tobacco and its association with pectin. Evidence for a structural role of boron in the cell wall. Plant Physiol. 105, 681–689.

Hu H, Brown P H and Labavitch J M 1996 Species variability in boron requirement is correlated with cell wall pectin. J. Exp. Bot. 47, 227–232.

Hu H, Penn S G, Lebrilla C B and Brown P H 1997 Isolation and characterisation of soluble B-complexes in higher plants. The mechanism of phloem mobility of boron. Plant Physiol. 113, 649–655.

Huang C and Graham R D 1990 Resistance of wheat genotypes to boron toxicity is expressed at cellular level. Plant and Soil 126, 295–300.

Ishii T and Matsunaga T 1996 Isolation and characterisation of a boron-rhamnogalacturonan-II complex from cell walls of sugar beet pulp. Carbohydr. Res. 284, 1–9.

Jansen M and Blume A 1995 A comparative study of diffusive and osmotic water permeation across bilayers composed of phospholipids with different head groups and fatty acyl chains. Biophysical J. 68, 997–1008.

Jenkin M, Hu H, Brown P, Graham R, Lance R and Sparrow D 1993 Investigations of boron uptake at the cellular level. *In* Plant nutrition – from genetic engineering to field practice. Ed. N J Barrow. pp 157–160. Kluwer Academic Publishers, Dordrecht, The Netherlands.

Keren R and Bingham F T 1985 Boron in water, soils, and plants. *In* Advances in soil science VI. Ed. B A Stewart. pp 230–276. Springer-Verlag, New York.

Keren R, Bingham F T and Rhoades J D 1985 Effect of clay content in soil on boron uptake and yield of wheat. Soil Sci. Soc. Am. J. 49, 1466–1470.

Kobayashi M, Matoh T and Azuma J-I 1996 Two chains of rhamnogalacturonan II are cross-linked by borate-diol ester bonds in higher plant cell walls. Plant Physiol. 110, 1017–1020.

Lee-Standelmann O Y, Bushnell W R, Curran C M and Standelmann E J 1991 Permeability to non-electrolytes as an indicator of change in membrane structure in *Hordeum vulgare* cells infected by *Erysiphe graminis* f. sp. *Hordei*. Physiol. Mol. Plant Pathol. 38, 163–178.

Loomis W D and Durst R W 1992 Chemistry and biology of boron. Biofactors 3, 229–239.

Makkee M, Kieboom A P G and van Bekkum H 1985 Studies on borate esters III. Borate esters of D-mannitol, D-glucitol, D-fructose and D-glucose in water. Recl. Trav. Chim. Pays-Bas 104, 230–235.

Matoh T, Kawaguchi S and Kobayashi M 1996 Ubiquity of a borate-rhamnogalacturonan II complex in the cell walls of higher plants. Plant Cell Physiol. 37, 636–640.

Nable R O 1988 Effects of B toxicity amongst several barley wheat cultivars: a preliminary examination of the resistance mechanism. Plant and Soil 112, 45–52.

Nable R O, Lance R C M and Cartwright B 1990 Uptake of boron and silicon by barley genotypes with differing susceptibilities to boron toxicity. Ann. Bot. 66, 83–90.

Nissen P 1974 Uptake mechanisms: inorganic and organic. Ann. Rev. Plant Physiol. 25, 53–79.

Oertli J J 1994 Non-homogeneity of boron distribution in plants and consequences for foliar diagnosis. Comm. Soil Sci. Plant Anal. 25, 1133–1147.

Oertli J J and Grgurevic 1975 Effect of pH on the absorption of boron by excised barley roots. Agron J. 67, 278–280.

Oertli J J and Richardson W F 1970 The mechanism of boron immobility in plants. Physiol. Plant. 23, 108–116.

O'Neill M A, Warrenfeltz D, Kates K, Pellerin P, Doco T, Darvill A G and Albersheim P 1996 Rhamnogalacturonan-II, a pectic polysaccharide in the walls of growing plant cells, forms a dimer that is covalently cross-linked by a borate ester. J. Biol. Chem. 271, 22923–22930.

Paull J G, Cartwright B and Rathjen A J 1988 Responses of wheat and barley genotypes to toxic concentrations of soil boron. Euphytica 39, 137–144.

Penn S G, Hu H, Brown P H and Lebrilla C B 1997 Direct analysis of sugar alcohol borate complexes in plant extracts, by matrix assisted laser desorption fourier transform mass spectrometry (MALDI FTMS). Anal. Chem. (*In press*).

Raven J A 1980 Short- and long-distance transport of boric acid in plants. New Phytol. 84, 231–249.

Tanaka H 1967 Boron adsorption by plant roots. Plant and Soil 27, 300–302.

Thellier M, Duval Y and Demarty M 1979 Borate exchanges of *Lemna minor* L. as studied with the help of the enriched stable isotopes and of a (n,α) nuclear reaction. Plant Physiol. 63, 283–288.

Wilders R A and Neales T F 1971 The absorption of boron by disks of plant storage tissues. Aust. J. Biol. Sci. 24, 873–884.

Yamauchi M 1971 The role of B in higher plants. Relationship between B and Ca on the pectin substances in plants. J. Sci. Soil Manure 42, 207–213.

Plant and Soil **193:** 59–70, 1997.

Chapter 5

Boron in plant cell walls

Toru Matoh
*Laboratory of Plant Nutrition, Kyoto University, Kyoto 606-01, Japan**

Abstract

Boron is an essential element for higher plants, yet the primary functions remain unclear. In intact tissues of higher plants, this element occurs as both water soluble and water insoluble forms. In this review, the intracellular localisation of B and possible function of B in cell walls of higher plants are discussed. The majority of the water soluble B seems to be localised in the apoplastic region as boric acid. The water insoluble B is associated with rhamnogalacturonan II (RG-II) and the complex is ubiquitous in higher plants. In the Brassicaceae, Apiaceae, Chenopodiaceae, Asteraceae, Amaryllidaceae, and Liliaceae, nearly all the cell wall B is associated with RG-II, while in the Cucurbitaceae, only half of the cell wall B is in this complex. In duckweed, a different type of B-polysaccharide complex has been identified.

Analysis of the structure of the B–RG-II complex reveals that the complex is composed of boric acid and two chains of monomeric RG-II. Boric acid does not merely bind to sugars but crosslinks two chains of pectic polysaccharide at the RG-II region through borate-diester bonding, thus forming a network of pectic polysaccharides in cell walls. The B–RG-II complex is reconstituted *in vitro* only by mixing monomeric RG-II and boric acid at pH 4.0. In the *in vitro* reconstitution, germanic acid can substitute for boric acid to some extent. The RG-II epitope, which cross reacts with the antibody toward the B-RG-II complex, is detected in walls of every cell in radish roots. The epitope is also detected in growing pollen tube cell walls, which are claimed to require B.

Whilst it is now clear that boric acid links some cell wall components, it is not yet clear whether there is a structural requirement for B in cell wall function.

Introduction

Warington (1923) conducted convincing experiments that when boric acid was deprived from culture solutions, roots of broad bean (*Vicia faba*) were short and thick with a stunted appearance. The shoots of the plants were also small, stunted, and dark green in colour. The apex of the shoot withered and the stem at this point became blackened. The prevention of, or recovery from, these unhealthy conditions was achieved only by the addition of boric acid to culture solutions. Moreover, the need for the supply was maintained throughout life. Following her work, many growth experiments, such those by Sommer and Lipman (1926) were carried out in order to confirm the essentiality of B for higher plants.

Loomis and Durst (1992) comprehensively reviewed the postulated role of B in higher plants. Almost all the hypotheses were based on observed abnormalities in plants when B was withheld from culture media. Whether the abnormalities are the result of a primary function of the element or merely a secondary consequence of the primary effect has been assessed only by the timing of the noticeable changes in metabolic activities and/or morphology. Therefore, it is most important to detect the earliest manifestation of B deficiency. The timetable of early symptoms of B deficiency are listed by Hirsch and Torrey (1980). However, it is not possible to judge with confidence which event is the earliest, because the physiological changes are not always visible and are quite often subtle. Also, it is likely that the time lag for the manifestation of the deficiency symptom depends on the degree of stringency of B removal, thus adding further ambiguity to the issue of a primary function of B versus a secondary consequence.

Since the specific function of B might be associated with certain cellular constituents, and abnormalities

* E-mail: matoh@kais.kyoto-u.ac.jp

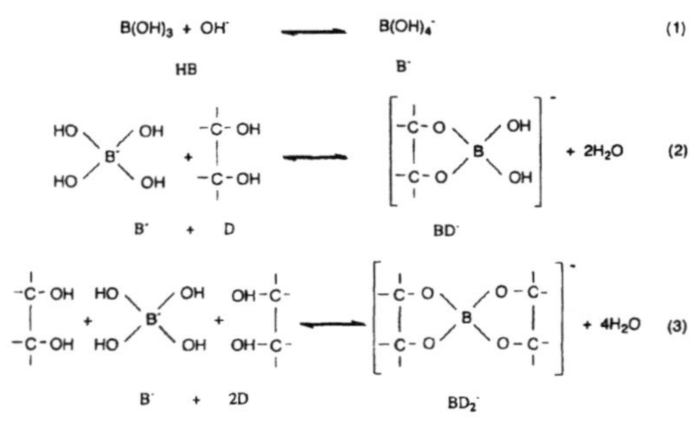

Figure 1. Interconversion of boric acid (HB) and borate anions (B^-), and the interaction of borate with diols (D).

due to B deficiency may develop first where B use is localised, information on the distribution of B within the cell would be of value. Of course, some B may accumulate in regions away from the essential sites. Therefore, both the intracellular localisation of B, and the earliest disorder due to B deficiency should be considered together.

Boric acid interacts spontaneously with diol groups in the *cis* position. The interconversion of boric acid (HB) and borate anions (B^-), and the interaction of borate with diols (D) is represented by the equilibria (1) to (3) in Figure 1. These equilibria in aqueous solution are rapidly established and reversible, and the positions of the equilibria depend upon the pH of the solution and the stereochemistry of the diol (Henderson et al., 1973). Therefore, B in boric acid has been postulated to react and form complexes with sugars and phenolics spontaneously. Many functions of B have been proposed including the activation of enzymes, the maintenance of membrane function, the stimulation of nucleic acid metabolism and so on (summarised by Loomis and Durst, 1992). For each of these proposed functions, the localisation of the element has been neglected and therefore each hypothesis is flawed since they all lack evidence for the presence of the corresponding B-substrate complex.

Relatively little work has been done on the distribution of B within plant cells, therefore, a major part of this review is devoted to a discussion on the intracellular localisation of B and the postulated function of B in cell walls of higher plants. In addition, readers should consult the excellent review articles on this subject, such as those by Loomis and Durst (1992) and Skok (1958), and reports by Kobayashi et al. (1996), Skok and McIlrath (1958), Teasdale and Richards (1990) and Yamanouchi (1971). Boron may have other roles in cellular function apart from those in the cell wall, and these are discussed elsewhere in this book (Cakmak and Römheld, Chapter 6).

Water soluble boron

Marsh (1942) and Marsh and Shive (1941) demonstrated that higher plants contain B both in a water soluble and insoluble forms, and that in broad bean and maize (*Zea mays*), the water soluble B increases according to an increase in B supplied to plants, while the insoluble B does not. The authors suggested a close quantitative relationship between B and pectic materials, and between B and Ca. Eaton (1944), Husa and McIlrath (1965), McIlrath (1965), Parr and Loughman (1983) and Skok and McIlrath (1958) also reported that both forms of B occur in higher plants.

Where does the water soluble B occur in plant tissues? In the symplast, significant amounts of water soluble B were not detected in cultured cells (Durst and Loomis, 1984; Hu and Brown, 1994, Loomis and Durst, 1992; Matoh et al., 1992). Protoplasts prepared from cultured tobacco (*Nicotiana tabacum*) cells did not contain significant amounts of B (Loomis and Durst, 1992; Matoh et al., 1992). Protoplast experiments could not rule out a possibility that the symplasmic B had leaked out during protoplast preparation. Therefore, Matoh et al. (1992) homogenised the tobacco cells in the culture medium in which the cells had been grown and the carry over of the culture medium to the supernatant fluid was assessed by tritiated water and ^{14}C-sorbitol. The concentration of B in the supernatant could be entirely ascribed to the carry over of B from the culture medium, indicating that the symplasmic B did not contribute to any increase in the B concentration of the supernatant. These data suggest that the B concentration in the symplast is negligible in cultured cells.

Boron-11 NMR spectroscopy is certainly one of the most powerful, non-destructive means to study boric acid and borate esters in living tissues. Figure 2 shows ^{11}B-NMR spectra of expressed leaf sap from the middle leaves of Swiss chard (*Beta vulgaris cicla*) grown with 10 m*M* B. Even though the leaves contained 680 mg B kg^{-1} dry wt the plants did not show any symptom of B toxicity, as reported for red beet (*Beta vulgaris*) (Oertli and Roth, 1969). Three major signals were detected at 19.2, 10.6, and 5.8 (Figure 2a), and these peaks were assigned to the B in HB, BD_2^- and BD^- (Figure 1), respectively (Oi et al., 1992). The

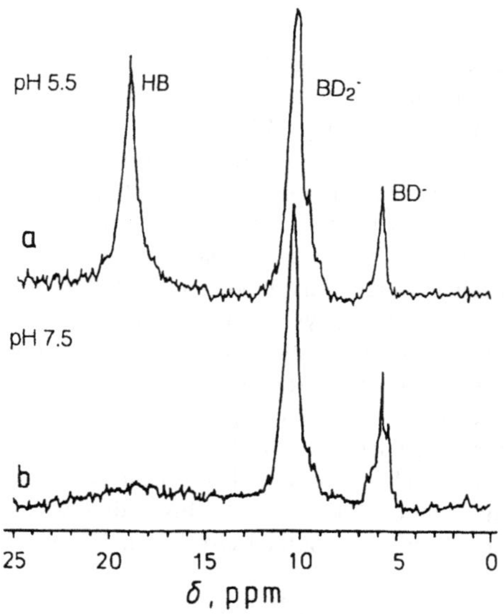

Figure 2. ^{11}B-NMR spectra of the expressed leaf sap of Swiss chard (*Beta vulgaris cicla*) plants. The pH of the sap was 5.5. (a) The spectrum was obtained in 30 min. (b) The pH of the sap was raised to 7.5 with 0.1 M NaOH. The spectrum was obtained in 30 min.

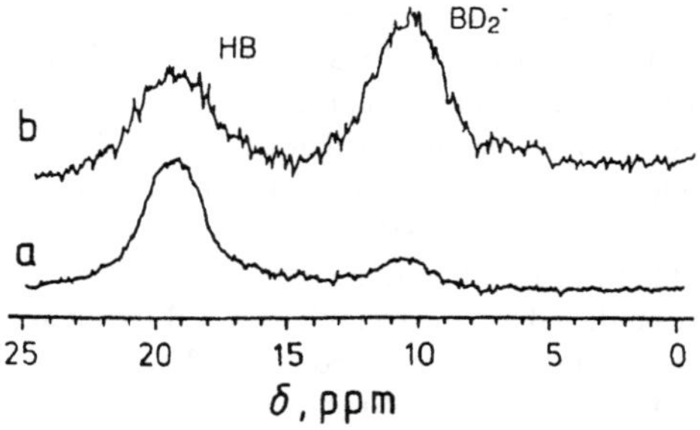

Figure 3. ^{11}B-NMR spectra of intact leaf disks of Swiss chard (*Beta vulgaris cicla*) plants. (a) The spectrum was obtained in 45 min. (b) The leaf disks were frozen by dipping the disks in the NMR tube in liquid N_2 then thawed at room temperature. The spectrum was obtained in 30 min.

pH of the expressed sap was 5.5. When the pH of the sap was raised to 7.5, the signal for HB disappeared and the intensities of the signals due to BD_2^- and BD^- increased (Figure 2b). This suggests that at pH 7.5 almost all the B was engaged in complex formation with soluble ligand. These data suggest that B-ligand interaction is regulated by the pH of the environment, therefore, the compartmentation of B in cells is crucial for the complex formation. When leaf disks (4 mm dia., stacked in an NMR tube having external diameter of 5 mm) were subjected to analysis (Figure 3a), the majority of the B was in a boric acid form (around 19.2) with a minor peak at around 10.6, which is attributable to BD_2^- in a concentration of 3.2% of the HB signal. The signal, which is due to the water insoluble B that is bound inside the cell walls, is likely to be too broad to be detected because of decreasing rates of molecular motion within the native cell walls (Roberts, 1987). When the leaf discs in a sample tube were frozen in liquid N_2 and thawed at room temperature, the HB signal decreased while that of BD_2^- increased (Figure 3b). This indicates that the freeze thaw treatment disrupts the membrane integrity and allows boric acid to bind to water soluble ligand leading to production of BD_2^- complexes. Therefore, the BD_2^- signal detected in such samples does not always indicate that the B occurs in the symplast. Results in Figure 3 suggest that the majority of cellular B occurs as boric acid and that almost all the element is localised where the ligand is absent. Martini and Thellier (1993) employed neutron capture radiography to trace the intracellular localisation of B in leaf parenchyma cells of three week old clover (*Trifolium* sp.). The clover leaves contained 43 mg B kg^{-1} dry wt, the cell wall 82.5 mg B kg^{-1} fresh wt, the cytoplasm 3.78 mg B kg^{-1} fresh wt, and there was practically zero B in the vacuole. Taking into consideration the ^{11}B-NMR data of intact leaves and those of protoplast experiments, it is likely that the majority of cellular B is excluded from the symplast.

Many authors have assumed that B is localised in the cytoplasm, and because the molecular mass of boric acid is small and not ionised at physiological pH, B was considered to readily penetrate the plasma membrane (Oertli and Richardson, 1970; Raven, 1980). However, a survey of the literature reveals that unequivocal evidence of the localisation of this element in the symplast has never been demonstrated. From the results presented here, the author speculates that the majority of the water soluble B may occur as boric acid in the apoplast, that is in the cell walls and intercellular spaces, and not in the symplast in intact plant tissues. Of course, this result does not rule out a possible involvement of B in metabolic reactions in the symplast, and a possibility of a phloem transport of B in some species is postulated (Shelp et al., 1995). Localisation of the water soluble B in intact plant tissues, as well as the membrane transport mechanism for this element, should be examined further.

Cell wall bound boron

Schmucker (1933, 1934) suggested a close relationship between B and pectic substances in growing pollen tubes. Later, Smith (1944) demonstrated in squash (*Cucurbita* sp.) leaves that water insoluble B is cell wall bound and is released with diluted HCl. The author claimed that it is important to identify to which substances B binds. Analogous to Ca, he proposed that pectic polysaccharides are the candidate. Torssell (1956) speculated that B may be involved in gel formation of polysaccharides which were secreted into cell walls. Whittington (1959) suggested that B is involved in the production of pectic polysaccharide. Skok and McIlrath (1958) found that B deficient sunflower (*Helianthus annuus*) plants contain less water soluble and dialysable B than B adequate plants. They also recognised the significance of identifying the structures and compounds which fix B. A decrease in water soluble B during the development of B deficiency was also noted by Shive and Barnett (1973). Teasdale and Richards (1990) found that in the culture medium for pine (*Pinus* sp.) cells the B concentration which gives a maximum growth rate varies considerably according to the concentrations of coexisting Ca and Mg ions in the medium. From these interactions, the authors drew the conclusion that a critical acceptor molecule is activated only by binding both Ca and B at separate sites, and therefore B has a primary role in forming a carbohydrate gel surrounding the cellulose fibrils.

Yamanouchi (1971) was the first to demonstrate that the amount of water insoluble but 0.5 *M* HCl soluble B is highly correlated with that of the water- and oxalate-insoluble but 0.5 *M* HCl soluble pectic polysaccharides in the cell walls of 52 species of higher plants. Hu et al. (1996) obtained similar results. Yamauchi et al. (1986) presented the possibility that cell wall polysaccharides which contain B could be solubilised by hydrolysis of cell walls with pectinase enzymes. Matoh et al. (1993) isolated and purified a particular B-polysaccharide complex (BPC) from radish (*Raphanus sativus*) root cell walls, and subsequently the sugar component of the BPC was analysed (Kobayashi et al., 1995). In the complex, B was present as a tetravalent 1:2 borate–diol complex. Because of the ability of B to form complexes spontaneously with substrates having *cis* diol, it seems possible that B might bind to polysaccharides in cell walls at random. However, nearly 80% of the cell wall B was bound solely to this particular polysaccharide. Acid hydrolysis (0.1 *M* HCl for 15 min at 25 °C) of the BPC released B and decreased its molecular weight by one half with only a slight increase (14%) in the proportion of reducing end groups (Kobayashi et al., 1996). Fragments generated by this mild hydrolysis were identical in terms of size and charge, indicating that B crosslinks two identical polysaccharide chains through B–diol ester bonds. Since the B content was 0.23% and its molecular weight was 9900, 1 mol of BPC contains 2.1 mol of B. Analysis of the glycosyl composition of the complex revealed that the complex contains 2-*O*-methylfucose, 2-*O*-methylxylose, apiose, aceric acid and 2-keto-3-deoxysugars, which are diagnostic of rhamnogalacturonan II (RG-II, Thomas et al., 1989). Glycosyl linkage analysis also indicated that the sugar component of BPC is RG-II. This finding was subsequently confirmed for the RG-II isolated from cultured sycamore (*Acer* sp.) cells and pea (*Pisum sativum*) shoots (O'Neill et al., 1996), red wine (Pellerin et al., 1996) and beet (*Beta vulgaris*) taproots (Ishii and Matsunaga, 1996). In BPC, B crosslinks two chains of the monomeric RG-II. As RG-II is a component of a long chain of pectic polymers (Thomas et al., 1989), the two pectic chains are crosslinked in the RG-II region at two sites with two mol boric acid. That is, B works to form a pectic network in cell walls. In BPC, O'Neill et al. (1996) revealed that the apiosyl residue is responsible for fixing B to the polysaccharide chains. Two apiosyl residues occur in the monomeric RG-II (Thomas et al., 1989), therefore, the BPC could hold two mol of boric acid to form a 1:2 borate-apiosyl complex. Apiose was first predicted to be a candidate as a B fixing agent in higher plants by Loomis and Durst (1992).

O'Neill et al. (1996) showed that dimeric RG-II crosslinked with B is reconstituted spontaneously by only mixing the monomeric RG-II with B, and found that the addition of Sr, Ba and Pb ions to the reconstitution medium stimulates the dimer formation. They discussed that elements having an ionic radius larger than 1.1 Å are stimulative to the dimer formation. However, these elements are not essential, and are seldom found, in higher plants, therefore, these ions may not be functional *in vitro*. Rather, these results suggest that steric factors may in part regulate the dimer formation *in vitro*. Since the RG-II region is only a part of a long chained pectic polysaccharide (Thomas et al., 1989), the dimer formation may be governed not only by the concentration but also by the alignment of the two vicinal chains. Pellerin et al. (1996) demonstrated that red wine RG-II contains Ca ion as a constituent. Recently, our laboratory also confirmed this and found that the addition of Ca ion enhances the stability of the reconstituted B-RG-II complex. Furthermore, the native B–RG-II complex contained two mol Ca (Asaka and Kobayashi, unpublished data). Therefore Ca ion may be a native constituent of the complex. Kobayashi et al. (1997) demonstrated that germanic acid could, to some extent, substitute for B in the dimer formation. Skok (1957) found that germanic acid can substitute for B to some extent in the growth of B deprived sunflower plants. In tomato (*Lycopersicon esculentum*), germanic acid fulfils a similar growth promoting effect by stimulating the upward transport of B (Brown and Jones, 1972). Loomis and Durst (1992) confirmed the effect of germanic acid on B deficiency by establishing cell lines which could grow using germanic acid without boric acid. The ability of germanic acid to form a dimeric RG-II complex *in vitro* may explain the substitution effect of Ge when B is deprived from culture media.

Occurrence of the BPC was surveyed in the cell walls of 24 species of higher plants (Matoh et al., 1996). Boron concentrations in the cell wall, cell wall hydrolysate and high molecular weight (HMW) fragments of the hydrolysate are presented in Table 1. In all the samples apart from soybean (*Glycine max*) leaves, tomato fruits, duckweed (*Lemna pausicostata*) and rice (*Oryza sativa*) leaves, an average of 94% of the cell wall B was solubilised by the Driselase enzyme. In 18 out of 24 samples, an average of 84% of the solubilised B was associated with fragments of HMW. Three cucurbitaceous plants, duckweed, rice, and maize were exceptions. The diagnostic monosaccharide for RG-II, 2-keto-3-deoxysugars, were determined in the HMW fragments and a quantitative relationship between these sugars and B was plotted. As presented in Figure 4, amounts of B were positively correlated (r^2= 0.778) with those of 2-keto-3-deoxysugars. The monomeric RG-II with 30 glycosyl residues contains two residues of 2-keto-3-deoxysugars (Puvanesarajah et al., 1991) and a unit BPC contained 2.1 mol of B. Therefore, the molar ratio of B to 2-keto-3-deoxysugars in a unit BPC should be 0.525. This value is close to the slope of the regression line, 0.535, which represents the molar ratio. The regression line crossed the Y axis close to the origin, suggesting that almost all the B exists as a 2:2 complex with RG-II in the HMW fragments, and moreover that no RG-II exists without B and *vice versa*. In 14 out of 24 species, such as members of the Brassicaceae, Apiaceae, Chenopodiaceae, Asteraceae, Amaryllidaceae, and Liliaceae, an average of 80% of the cell wall B was found in HMW fragments and this B had a close quantitative relationship with RG-II. Therefore, it seems likely that all the cell wall B occurs as a B-RG-II complex, with RG-II being the exclusive carrier of B. Skok and McIlrath (1958) and McIlrath (1965) reported water soluble but non-dialysable B occurs in the leaves of several dicotyledonous plants. We could not confirm the presence of B in such a form in cell walls of radish roots and Swiss chard leaves, but there is a possibility that some of the B–RG-II complex is mechanically fragmented when a fresh tissue is homogenised in a buffer solution.

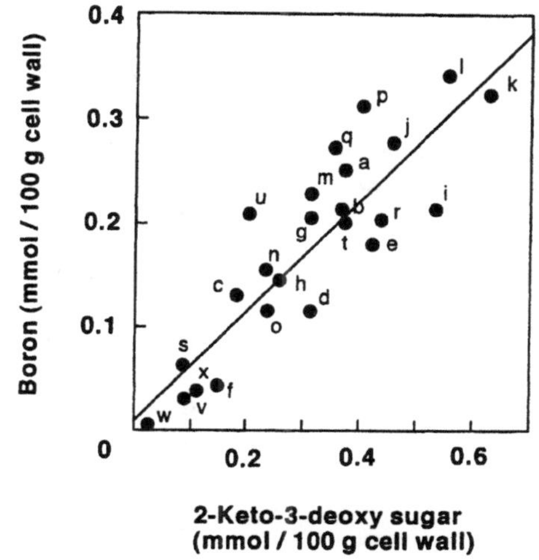

Figure 4. Quantitative relationship of boron and 2-keto-3-deoxy sugars in the high molecular weight fragments of cell walls solubilised by Driselase from 24 species of higher plants. The regression line was $Y = 0.0100 + 0.535\ X$ and the regression coefficient r^2 was 0.778. a: cabbage (*Brassica oleracea capitata*), b: Brussel sprout (*B. oleracea* var.*gemmifera*), c: cucumber (*Cucumis sativus*), d: gourd (*Lagenaria leucantha*), e: towel gourd (*Luffa cylindrica*), f: soybean (*Glycine max*), g: pea (*Pisum sativum*), h: *Vicia sepium*, i: cowpea (*Vigna sinesis*), j: carrot (*Daucus carota*), k: parsley (*Petroselinum sativum*), l: table beet (*Beta vulgaris*), m: spinach (*Spinacia oleracea*), n: sweet pepper (*Capsicum annuum angulosum*), o: tomato (*Lycopersicon esculentum*), p: Garland chrysanthemum (*Chrysanthemum coronarium*), q: lettuce (*Lactuca sativa*), r: lily (*Lilium longiflorum*), s: duckweed (*Lemna pausicostata*), t: leek (*Allium ampelopraum*), u: Welsh onion (*Allium fistulosum*), v: Italian ryegrass (*Lolium multiflorum*), w: rice (*Oryza sativa*) and x: maize (*Zea mays*).

In cucurbitaceous plants, Driselase released B completely from the cell walls, however, only half of the B was recovered in the HMW fragments (Table 1), suggesting that cell wall polysaccharides other than RG-II may be engaged in B fixation in these species. Duckweed cell walls contained a lot of B, as much as

Table 1. Levels of boron in the cell walls (CW) of 24 species of higher plants, and the recovery of B in the hydrolysates of their cell wall with Driselase and in the high molecular wt fragments of each hydrolysate. From Matoh et al. (1996)

Family		B content (mg B kg^{-1} CW)	B in Hydrolysate (% of that in CW)	B in high molecular wt fragments (% of that in CW)
Dicotyledonous plants				
Brassicaceae	Cabbage (*Brassica oleracea*) leaves	27.6	100	89
	Brussel sprout (*B. oleracea* var. *gemmifer*) leaves	26.1	97	86
Cucurbitaceae	Cucumber (*Cucumis sativus*) leaves	36.5	100	35
	Gourd (*Lagenaria leucantha*) leaves	26.5	100	43
	Towel gourd (*Luffa cylindrica*) leaves	46.2	97	42
Legminosae	Soybean (*Glycine max*) leaves	37.2	18	12
	Pea (*Pisum sativum*) leaves	25.9	89	82
	Vicia sepium leaves	20.8	95	75
	Cowpea (*Vigna sinesis*) leaves	30.6	92	75
Apiaceae	Carrot (*Daucus carota*) roots	38.8	87	78
	Parsley (*Petroselinum sativum*) leaves	44.4	92	78
Chenopodiaceae	Table beet (*Beta vulgaris*) leaves	42.1	100	85
	Spinach (*Spinacia oleracea*) leaves	32.4	100	82
Solanaceae	Sweet pepper (*Capsicum annuum angulosum*) fruits	30.9	69	54
	Tomato (*Lycopersicon esculentum*) fruits	31.3	50	41
Asteraceae	Garland chrysanthemum (*Chrysanthemum coronarium*) leaves	40.5	92	83
	Lettuce (*Lactuca sativa*) leaves	42.0	81	70
Monocotyledonous plants				
Liliaceae	Lily (*Lilium longiflorum*) leaves	27.1	100	80
Araeae	Duckweed (*Lemna pausicostata*)	778	30	0.87
Amaryllidaceae	Leek (*Allium ampelopraum*) shoots	26.7	98	86
	Welsh onion (*Allium fistulosum*) shoots	21.4	100	80
Gramineae	Italian ryegrass (*Lolium multiflorum*) shoots	4.0	100	78
	Rice (*Oryza sativa*) shoots	5.0	56	6.0
	Maize (*Zea mays*) shoots	6.8	100	58

780 mg B kg^{-1} cell walls, while only 0.87% of the B was associated with the HMW fragments, and 70% of the B remained in the cell walls. Duckweed cell walls were reported to be a good source of apiose (Duff, 1965). In the cell wall, Hart and Kindel (1970a, 1970b) demonstrated that apiose is a component of apiogalacturonan, which is comprised of an apiosyl disaccharide attached to a main chain of polygalacturonic acid. A borate binding polysaccharide has been purified from the cell wall and the polysaccharide contains apiose, xylose and uronic acid residues (Miyagawa, 1997). The higher capacity of duckweed cell walls for fixing B may be related to the higher content of apiose.

Boron deficiency brings about structural abnormalities, especially in cell walls. Spurr (1957) demonstrated that B apparently affects the rate and process of carbohydrate condensation into wall material. Some collenchyma cells fail to develop typical "corner" thickening, instead forming a uniformly thick, massive layer of wall material. These anatomical changes in cell walls convinced him that B is a morphogenetic agent affecting the development of specific form of the cell walls of higher plants. Starck (1963) observed in B deficient sunflower shoots that the arrangement of microfibrils is impaired and suggested that the stiffening of the pectic matrix caused by B deficiency would probably result in reduced microfibril deposition. Kouchi and Kumazawa (1975, 1976) noted that in B deficient tomato root tips, cell walls suffer an irregular thickening and that the cell wall alterations appeared most distinctly in the longitudinal walls. They also reported the abnormal development of Golgi apparatus. That is, the number and size of the secretory vesicles increased in B deficient root tip cells. Hirsh and Torrey (1980) also demonstrated that cell wall thickening was observed as early as 6 h after the removal of B. They confirmed that cell wall thickening with B deficiency was associated with increased vesicle formation at the cell wall plasma membrane interface, even though the origin of the vesicles were not identified. They speculated that these vesicles are most similar to paramural bodies which are thought to be associated with cell wall synthesis. Specifically, these vesicular aggregations were apparent at the corner edges of three way cell junctions. Selected cells of cultured tobacco BY-2 which had been acclimatised to lower doses of B (1 μg B L^{-1} medium) were characterised as having swollen cell walls and dark stained Golgi vesicles (Matoh et al., 1992; Figure 5). The oven dried cells of the selected and wild cells contained B at 84.3 mg and 2260 mg kg^{-1} dried cells, respectively. All these results suggest that B is functionally impor-

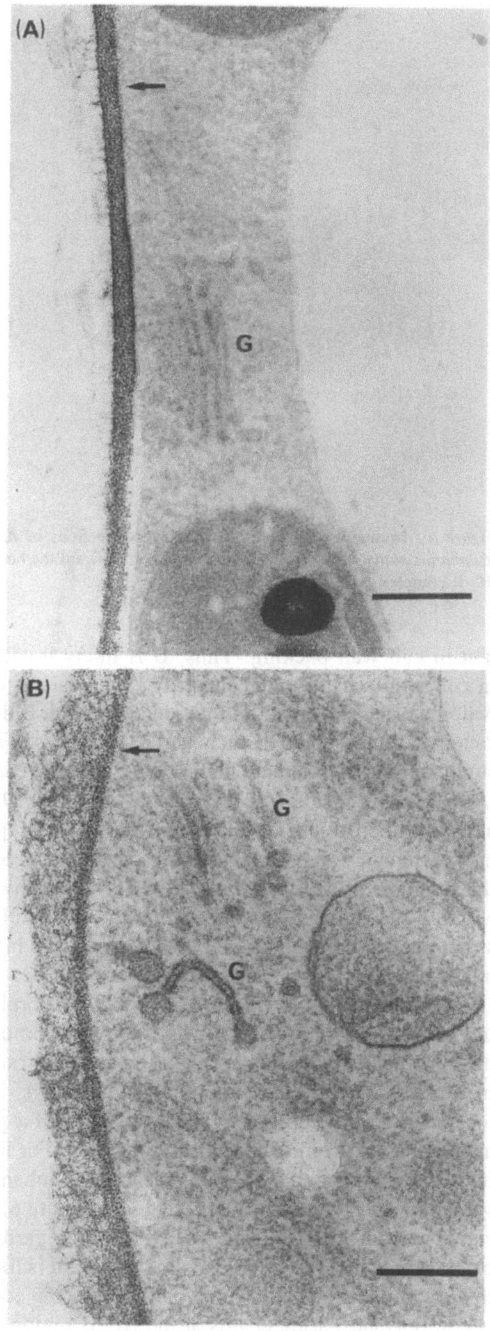

Figure 5. Ultrastructure of the control cells (A) and of the cells that tolerated the reduced concentration of B in the medium (B). The cells were fixed by a rapid freeze and substitution method using liquefied propane gas. To visualise polysaccharides, ultra thin sections were stained with periodic acid-thiocarbohydrazide-silver proteinate. The cell wall (designated by an arrow) is thicker and seemed to be not tightly packed in the selected cells. The Golgi apparatus (G) was accompanied by more strongly staining secretory vesicles. Scale bar is 0.5 μm.

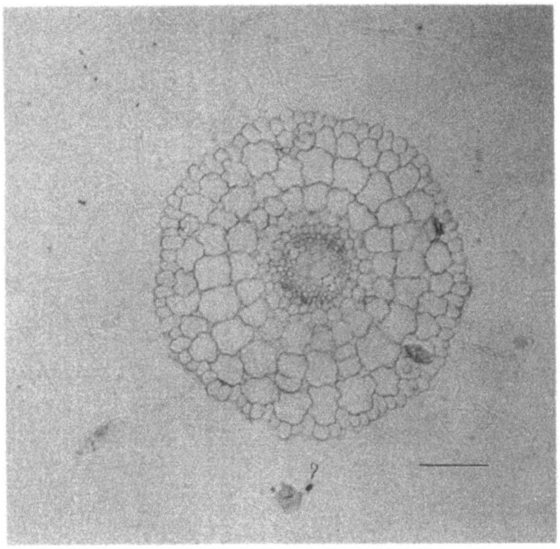

Figure 6. Immunogold labelling of transverse sections of radish (*Raphanus sativus*) root apices with an antibody toward the boron–RG-II complex. Measurement bar is 20 μm.

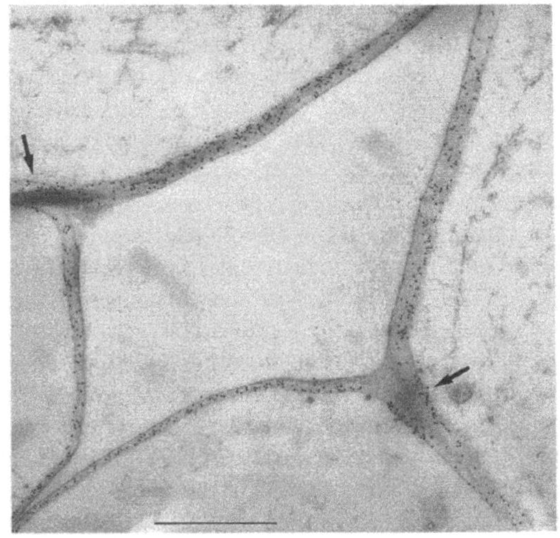

Figure 7. Immunogold labelling with an antibody toward the boron–RG-II complex in radish root apices. Measurement bar is 0.5 μm. Arrowhead indicates a corner close to three way junctions where denser labelling was observed.

tant in cell wall packing. Thus, B is likely involved in cell wall structure and integrity. Synthesis of cell wall materials, however, did not decrease with B deficiency in tomato roots (Kouchi and Kumazawa, 1976), sunflower leaves (Yamanouchi, 1973), cultured carrot (*Daucus carota*) cells (Goldbach and Amberger, 1986) and several crop species (Hu et al., 1996). This apparent discrepancy suggests that B may not play a role in the synthesis of wall materials, which is a symplasmic activity, but rather B may be important for the ultrastructural arrangement of cell wall components in the apoplast.

To understand the effect of B on cell wall morphology, the localisation of BPC was examined immunocytochemically using an antibody toward BPC raised in rabbits (Matoh et al., unpublished results). The localisation of the epitope was visualised with a goat anti-rabbit IgG conjugated to colloidal gold. For light microscopy, the signal on thin sections was enhanced with silver precipitation. As presented in Figure 6, the signal was detected in all tissues of the roots, regardless of the position of the sample along the root. It is likely that B fulfils its essential function through connecting pectic polysaccharide chains at the RG-II region, the distribution of which is ubiquitous. The omnipresence of the epitope contrasts to the RG-I epitope, which was not detected in the stele tissues of the younger part of *Arabidopsis thaliana* roots (Freshour et al., 1996). This result suggests that the RG-I and RG-II region may not share the same pectic polysaccharide chain, at least when cells are immature. In mature cells, the same RG-I antibody labelled the inner quarter of the cell wall, closest to the plasma membrane (Freshour et al., 1996). Knox et al. (1990) noted that a monoclonal antibody which recognises the de-esterified epitope of pectin was located on the inner surface of the primary cell walls adjacent to the plasma membrane in the root apex of carrot. Immunogold labelled antibodies to RG-II were localised primarily in the one third of the cell walls closest to the plasma membrane (Figure 7). Employing a monoclonal antibody to RG-II, Williams et al. (1996) demonstrated that the epitope is localised very close to the plasma membrane. A similar distribution of RG-II, RG-I, and de-esterified epitope of pectic polysaccharides in the primary cell walls suggest that crosslinking with Ca ions and boric acid is important for the network formation of newly produced pectic polysaccharides. The gradient in the distribution of RG-II across the cell wall suggests that following the maturation of a cell wall, the epitope of RG-II may disappear. Kobayashi et al. (1997) detected the monomeric RG-II in the extracellular polysaccharides of cultured tobacco cells BY-2 at the middle log phase when the cells grew vigorously. Together, these results suggest that some of the pectic polysaccharides may be decomposed during the cell wall maturation and the RG-II fragments are released into the surrounding medium, perhaps after removal of B from the complex. A little denser label (arrows in Figure 7) was found in a region of cell corners, where cell wall abnormality was detected in the collenchyma cells of B deficient celery (*Apium graveolens* var. *dulce*) (Spurr, 1957), and where increased vesicle formation at the cell wall plasma membrane interface was observed (Hirsh and Torrey, 1980). Many authors have detected the epitope of RG-I (Moore et al., 1986; Moore and Staehelin, 1988) and of de-esterified pectin (Knox et al., 1990; Liners and Van Cutsem, 1992) in the middle lamella. On the other hand, the RG-II epitope was not detected in the middle lamella at all, suggesting that B may not function to cement cells.

The most dramatic manifestation of B requirement is seen in pollen tube growth. The essentiality of B for pollen tube elongation was first demonstrated by Schmucker (1933, 1934) and thereafter the essentiality has been confirmed by several authors for pollen of a wide range of higher plants. Occurrence of the BPC in growing pollen tubes was confirmed with immunocytochemical analysis (data not presented). Direct evidence is now being prepared by purifying the complex from growing pollen tubes. The other instance which is often noticed for B essentiality is the nodule development and nitrogen fixing activities in the root nodules of leguminous plants (Bolaños et al., 1994). The nodule was also found to be rich in the BPC analysed with an immunocytochemical method (data not presented). This suggests that root nodules may not work fully because the cell walls are not well organised when the B supply was limited.

Function of boron in cell walls

In this review, the author has summarised the topic of B and cell walls with an intention of demonstrating that B functions primarily in cell walls. Skok (1958) discussed the possibility that if B is required for some other more basic physiological function than that of cell wall synthesis, it is reasonable to expect B requirement to be more widespread among living organisms. Loomis and Durst (1992) concluded that the rapid response of pollen tube elongation to withdrawal of B from the culture medium indicates that neither nucleic acid metabolism nor protein synthesis is involved in the function of B. They add that it is most likely that B's role is in the assembly of cell walls. The old concept of the cell wall as a dead box without structure has given way to a new picture of a highly complex structure performing a great diversity of functions in the life of plants. The most recognised function of the cell wall is that related to its rigidity, especially the provision of strength and shape to the cell. Shedletzky et al. (1990) isolated a cell line of tomato which adapted to the presence of 2,6-dichlorobenzonitrile, a potent inhibitor of cellulose synthesis. The adapted cells still form coherent walls, but these contain virtually no cellulose and hemicellulose, instead, the cell walls appear to consist solely of a cross linked meshwork of pectin. Based on this result, Roberts (1990) introduced a notion that a pectin network in the cell wall is separate from the cellulose-hemicellulose network. Hayashi et al. (1980) suggested a better understanding of pectic polymers, namely that they are oriented parallel to the direction of cell elongation. Shea et al. (1989) demonstrated that pectic polysaccharides are secreted into cell walls in advance of cellulose microfibrils during the regeneration of cell walls of carrot protoplasts. A pectic network is no longer compatible with the former notion that pectins are simply in a gel form. Fry (1986) summarised various ways by which pectic polysaccharides such as Ca bridges, ester bonds, and coupled phenols can be cross linked. We now know that borate-diester bonding is also involved. The RG-II regions provide the connecting sites for pectic polysaccharides through B–diester bonding. As the site is specific to boric acid, cells could control the number of knots on the polysaccharide chains by regulating the frequency of RG-II insertion into pectic polysaccharide chains. This may be a merit of borate bridges over Ca bridges. Fry (1986) suggested that loops formed between the knots on polysaccharide chains could envelope microfibrils, hence the loops may anchor the pectic network to cellulose microfibrils. If B is absent from the binding sites, newly secreted pectic chains may not be held by the cell walls. Absence of the monomeric RG-II in cell walls (Kobayashi et al., 1997) may support this notion. What is clarified at present is that B crosslinks two pectic polysaccharide chains at the RG-II region, but this does not directly elucidate the mechanism underlying the abrupt cease in elongation of roots (Kouchi and Kumazawa, 1975) and pollen tubes (Schmucker, 1933) and the immediate death of cultured cells (Matoh et al., 1992) due to withdrawal of B. There must be a link between cell wall integrity and cell viability, which is invisible at present. The study of the function of B has to be considered in terms of the significance of the cell wall on the viability of cells.

It is supposed that the first event following the depletion of B from the surrounding medium is the leakage of newly secreted pectic polysaccharide having RG-II regions from cell walls. Perception of this leakage may trigger off various disorders with impairment of membrane integrity being the most conspicuous one. A great abundance of experiments have demonstrated that deprivation of B results in disturbance of membrane function. Torchia and Hirsh (1982), Tanada (1983) and Parr and Loughman (1983) demonstrated that some of the cellular B is found in a membrane fraction. However, we could not confirm the presence of a significant amount of B in microsomal membranes of cultured tobacco cells (Kobayashi et al., 1997). That is, at most 0.5% of the cellular B is recovered in the 80,000 *g* precipitate of the cell homogenate of tobacco BY-2 cells. Direct evidence for the presence of B in membranes is mandatory, and more experimental evidence for the significant stimulation of boric acid on certain metabolic activities in isolated and purified membrane systems, not in tissues, is necessary.

Boric acid is transported from roots to leaves via xylem vessels (Raven, 1980) and some of the B binds to the RG-II region of newly synthesised pectic polysaccharide chains in cell walls. The majority of the remaining B may accumulate in cell walls as boric acid without complexing with any ligand. Transport of B from roots to shoots increased linearly according to the increase in the concentration of B in the rooting medium (Marschner, 1995). Therefore, if the supply of B to the roots is high, then B is accumulated in leaf cell walls and finally may intrude into cytoplasm and, thus, disturb cytoplasmic metabolism, resulting in B toxicity. To substantiate this working hypothesis, it is necessary to elucidate how B is loaded from root cell walls to xylem vessels.

Acknowledgment

Experiments on B presented here have been carried out in the Laboratory of Plant Nutrition, Faculty of Agriculture, Kyoto University since 1986. I wish to express my profound gratitude to the laboratory members, Professor Emeritus Eiichi Takahashi and Professor Jiro Sekiya, and my colleagues Masaru Kobayashi, Dr. Jun-ichi Azuma, and Dr. Keiji Takabe, and my students. Gratitude is also extended to Ms. Margaret H. Flash for brushing up the English text, and to my family for their steadfast support.

References

Bolaños L, Esteban E, de Lorenzo C, Fernández-Pascual M, de Felipe M R, Gárate A and Bonilla I 1994 Essentiality of boron for symbiotic dinitrogen fixation in pea (*Pisum sativum*) Rhizobium nodules. Plant Physiol. 104, 85–90.

Brown J C and Jones W E 1972 Effect of germanium on utilisation of boron in tomato (*Lycopersicon esculentum* Mill.). Plant Physiol. 49, 651–653.

Duff R B 1965 The occurrence of apiose in *Lemna* (Duckweed) and other Angiosperms. Biochem. J. 94, 768–772.

Durst R W and Loomis W D 1984 Substitution of germanium for boron in suspension-cultured carrot cells. Plant Physiol. 75S, 14.

Eaton F M 1944 Deficiency, toxicity, and accumulation of boron in plants. J. Agric. Res. 69, 237–277.

Freshour G, Clay R P, Fuller M S, Albersheim P, Darvill A G and Hahn M G 1996 Developmental and tissue-specific structural alterations of the cell-wall polysaccharides of *Arabidopsis thaliana* roots. Plant Physiol. 110, 1413–1429.

Fry S C 1986 Cross-linking of matrix polymers in the growing cell walls of angiosperms. Ann. Rev. Plant Physiol. 37, 165–186.

Goldbach H and Amberger A 1986 Influence of boron nutrition on cell wall polysaccharides in cell cultures of *Daucus carota* L. J. Plant Physiol. 123, 263–269.

Hart D A and Kindel P K 1970a A novel reaction involved in the degradation of apiogalacturonans from *Lemna minor* and the isolation of apibiose as a product. Biochemistry 9, 2190–2196.

Hart D A and Kindel P K 1970b Isolation and partial characterization of apiogalacturonans from the cell wall of *Lemna minor*. Biochem. J. 116, 569–579.

Hayashi R, Morikawa H, Nakajima N, Ichikawa Y and Senda M 1980 Oriented structure of pectic polysaccharides in pea epidermal cell walls. Plant Cell Physiol. 21, 999–1005.

Henderson W G, How M J, Kennedy G R and Mooney E F 1973 The interconversion of aqueous boron species and the interaction of borate with diols: A ^{11}B NMR study. Carbohydr. Res. 28, 1–12.

Hirsch A M and Torrey J G 1980 Ultrastructural changes in sunflower root cells in relation to boron deficiency and added auxin. Can. J. Bot. 58, 856–866.

Hu H and Brown P H 1994 Localisation of boron in cell walls of squash and tobacco and its association with pectin. Evidence for a structural role of boron in the cell wall. Plant Physiol. 105, 681–689.

Hu H, Brown P H and Labavitch J M 1996 Species variability in boron requirement is correlated with cell wall pectin. J. Exp. Bot. 47, 227–232.

Husa J G and McIlrath W J 1965 Absorption and translocation of boron by sunflower plants. Botan. Gaz. 126, 186–194.

Ishii T and Matsunaga T 1996 Isolation and characterization of a boron-rhamno galacturonan-II complex from cell walls of sugar beet pulp. Carbohydr. Res. 284, 1–9.

Knox J P, Linstead P J, King J, Cooper C and Roberts K 1990 Pectin esterification is spatially regulated both within cell walls and between developing tissues of root apices. Planta 181, 512–521.

Kobayashi M, Matoh T and Azuma J 1995 Structure and glycosyl composition of the boron-polysaccharide complex of radish roots. Plant Cell Physiol. 36S, 139.

Kobayashi M, Matoh T and Azuma J 1996 Two chains of rhamnogalacturonan II are cross-linked by borate-diol ester bonds in higher plant cell walls. Plant Physiol. 110, 1017–1020.

Kobayashi M, Ohno K and Matoh T 1997 Boron nutrition of cultured tobacco BY-2 cells. II. Characterisation of the boron-polysaccharide complex. Plant Cell Physiol. 38, 676–683.

Kouchi H and Kumazawa K 1975 Anatomical responses of root tips to boron deficiency II. Effect of boron deficiency on cellular growth and development in root tips. Soil Sci. Plant Nutr. 21, 137–150.

Kouchi H and Kumazawa K 1976 Anatomical responses of root tips to boron deficiency III. Effect of boron deficiency on sub-cellular structure of root tips, particularly on morphology of cell wall and its related organelles. Soil Sci. Plant Nutr. 22, 53–71.

Liners F and Van Cutsem P 1992 Distribution of pectic polysaccharides throughout walls of suspension-cultured carrot cells. An immunocytochemical study. Protoplasma 170, 10–21.

Loomis W D and Durst R W 1992 Chemistry and biology of boron. BioFactors 3, 229–239.

Marsh R P 1942 Comparative study of the calcium-boron metabolism of representative dicots and monocots. Soil Sci. 53, 75–78.

Marsh R P and Shive J W 1941 Boron as a factor in the calcium metabolism of the corn plant. Soil Sci. 51, 141–151.

Marschner H 1995 Characteristics of ion uptake by roots. *In* Mineral nutrition of higher plants, 2nd. Ed. pp 30–63. Academic Press, London, UK.

Martini F and Thellier M 1993 Boron distribution in parenchyma cells of clover leaves. Plant Physiol. Biochem. 31, 777–786.

Matoh T, Kawaguchi S and Kobayashi M 1996 Ubiquity of a borate-rhamno-galacturonan II complex in the cell walls of higher plants. Plant Cell Physiol. 37, 636–640.

Matoh T, Ishigaki K, Ohno K and Azuma J 1993 Isolation and characterization of a boron-polysaccharide complex from radish roots. Plant Cell Physiol. 34, 639–642.

Matoh T, Ishigaki K, Mizutani M, Matsunaga W and Takabe K 1992 Boron nutrition of cultured tobacco BY-2 cells. I. Requirement for and intracellular localisation of boron and selection of cells that tolerate low levels of boron. Plant Cell Physiol 33, 1135–1141.

McIlrath W J 1965 Mobility of boron in several dicotyledonous species. Botan. Gaz. 126, 27–30.

Miyagawa I 1997 Purification and characterization of a borate-binding polysaccharide from cell walls of *Lemna minor* L. BSc. thesis, Kyoto University, Kyoto, Japan.

Moore P J and Staehelin L A 1988 Immunogold localisation of the cell-wall matrix polysaccharides rhamnogalacturonan I and xyloglucan during cell expansion and cytokinesis in *Trifolium pratense* L.; implication for secretory pathways. Planta 174, 433–445.

Moore P J, Darvill A G, Albersheim P and Staehelin L A 1986 Immunogold localisation of xyloglucan and rhamnogalacturonan I in the cell walls of suspension-cultured sycamore cells. Plant Physiol. 82, 787–794.

Oertli J J and Richardson W F 1970 The mechanism of boron immobility in plants. Physiol. Plant. 23, 108–116.

Oertli J J and Roth J A 1969 Boron nutrition of sugar beet, cotton, and soybean. Agron. J. 61, 191–195.

Oi T, Takeda T and Kakihana H 1992 Esterification of boric acid with 1,2-propanediol, 3-amino-1,2-propanediol, and (±)-3-dimethylamino-1,2-propanediol as studied by ^{11}B NMR spectroscopy. Bull. Chem. Soc. Jpn. 65, 1903–1909.

O'Neill M A, Warrenfeltz D, Kates K, Pellerin P, Doco T, Darvill A G, and Albersheim P 1996 Rhamnogalacturonan-II, a pectic polysaccharide in the walls of growing plant cells, forms a dimer that is covalently cross-linked by a borate ester. J. Biol. Chem. 271, 22923–22930.

Parr A J and Loughman B C 1983 Boron and membrane function in plants. *In* Metals and Micronutrients. Uptake and utilization by plants. Eds. D A Robb and W S Pierpoint. pp 87–107. Academic Press, New York.

Pellerin P, Doco T, Vidal S, Williams P, Brillouet J-M and O'Neill M A 1996 Structural characterization of red wine rhamnogalacturonan II. Carbohydr. Res. 290, 183–197.

Puvanesarajah V, Darvill A G and Albersheim P 1991 Structural characterization of two oligosaccharide fragments formed by the selective cleavage of rhamnogalacturonan II: evidence for the anomeric configuration and attachment sites of apiose and 3-deoxy-2-heptulosaric acid. Carbohydr. Res. 218, 211–222.

Raven J A 1980 Short- and long-distance transport of boric acid in plants. New Phytol. 84, 231–249.

Roberts J K M 1987 NMR in plant biochemistry. *In* The biochemistry of plants, vol. 13. Ed. D D Davis. pp 181–227. Academic Press, London.

Roberts K 1990 Structures at the plant cell surface. Curr. Opinion Cell Biol. 2, 920–928.

Schmucker T 1933 Zur Blütenbiologie Tropischer Nymphaea-Arten. II. (Bor als entscheidender faktor.) Planta 18, 641–650.

Schmucker T 1934 Über den Einfluss von Borsäure auf Pflanzen, Insbesondere Keimende Pollenkrner. Planta 23, 264–283.

Shea E M, Gilbeaut D M and Carpita N C 1989 Structural analysis of the cell walls regenerated by carrot protoplasts. Planta 179, 293–308.

Shedletzky E, Shmuel M, Delmer D P and Lamport D T A 1990 Adaptation and growth of tomato cells on the herbicide 2,6-dichlorobenzonitrile leads to production of unique cell walls virtually lacking a cellulose-xyloglucan network. Plant Physiol. 94, 980–987.

Shelp B J, Marentes E, Kitheka A M and Vivekanandan P 1995 Boron mobility in plants. Physiol. Plant. 94, 356–361.

Shive J B Jr and Barnett N M 1973 Boron deficiency effects on peroxidase, hydroxyproline, and boron in cell walls and cytoplasm of *Helianthus annuus* L. hypocotyls. Plant Cell Physiol. 14, 573–583.

Skok J 1957 The substitution of complexing substances for boron in plant growth. Plant Physiol. 32, 308–312.

Skok J 1958 The role of boron in the plant cell. *In* Trace Elements. Eds. C A Lamb, O G Bentley and J M Beattie. pp 227–243. Academic Press, London.

Skok J and McIlrath W J 1958 Distribution of boron in cells of dicotyledonous plants in relation to growth. Plant Physiol. 33, 428–431.

Smith M E 1944 The role of boron in plant metabolism. 1. Boron in relation to the absorption and solubility of calcium. Aust. J. Exp. Biol. Med. Sci. 22, 257–263.

Sommer A L and Lipman C B 1926 Evidence on the indispensable nature of zinc and boron for higher green plants. Plant Physiol. 1, 231–249.

Spurr A R 1957 The effect of boron on cell-wall structure in celery. Am. J. Bot. 44, 637–650.

Starck J R 1963 Effect of boron on the cell walls structure of sunflower. Acta Soc. Bot. Pol. 32, 619–623.

Tanada T 1983 Localisation of boron in membranes. J. Plant Nutr. 6, 743–749.

Teasdale R D and Richards D K 1990 Boron deficiency in cultured pine cells. Quantitative studies of the interaction with Ca and Mg. Plant Physiol. 93, 1071–1077.

Thomas J R, Darvill A G and Albersheim R 1989 Isolation and structural characterization of the pectic polysaccharide rhamnogalacturonan II from walls of suspension-cultured rice cells. Carbohydr. Res. 185, 261–277.

Torchia R A and Hirsh A M 1982 Analysis of membrane fractions from boron-deficient and control sunflower root tips by (n, a) nuclear reaction. Plant Physiol. 69S, 44.

Torssell K 1956 Chemistry of arylboric acids. VI. Effects of arylboric acids on wheat roots and the role of boron in plants. Physiol. Plant. 9, 652–664

Warington K 1923 The effect of boric acid and borax on the broad bean and certain other plants. Ann. Bot. 37, 629–672.

Whittington W J 1959 The role of boron in plant growth. II. The effect on growth of the radicle. J. Exp. Bot. 10, 93–103.

Williams M N V, Freshour Gg, Darvill A G, Albersheim P and Hahn M G 1996 An antibody Fab selected from a recombinant phage display library detects deesterified pectic polysaccharide rhamnogalacturonan II in plant cells. Plant Cell 8, 673–685.

Yamanouchi M 1971 The role of boron in higher plants I. The relations between boron and calcium or the pectic substance in plants. J. Soil Sci. Manure Jpn. 42, 207–213.

Yamanouchi M 1973 The role of boron in higher plants (Part 2). The influence of boron on the formation of pectic substances. Bull. Fac. Agri. Tottori Univ. XXV, 21–27.

Yamauchi T, Hara T and Sonoda Y 1986 Distribution of calcium and boron in the pectin fraction of tomato leaf cell wall. Plant Cell Physiol. 27, 729–732.

Plant and Soil **193:** 71–83, 1997.

Chapter 6

Boron deficiency-induced impairments of cellular functions in plants

Ismail Cakmak[1] and Volker Römheld[2]
[1]*Department of Soil Science and Plant Nutrition, Cukurova University, 01330 Adana, Turkey* and* [2]*Institute of Plant Nutrition, University of Hohenheim, 70593, Stuttgart, Germany*

Abstract

The essentiality of B for growth and development of plants is well-known, but the primary functions of B still remain unknown. Evidence in the literature supports the idea that the major functions of B in growth and development of plants are based on its ability to form complexes with the compounds having *cis*-diol configurations. In this regard, the formation of B complexes with the constituents of cell walls and plasma membranes as well as with the phenolic compounds seems to be a decisive step affecting the physiological functions of B. Boron seems to be of crucial importance for the maintenance of structural integrity of plasma membranes. This function of B is mainly related to stabilisation of cell membranes by B association with membrane constituents. Possibly, B may also protect plasma membranes against peroxidative damage by toxic O_2 species. In B-deficient plants, plasma membranes are highly leaky and lose their functional integrity. Under B-deficient conditions, substantial changes in ion fluxes and proton pumping activity of the plasma membranes were noted. Impairments in phenol metabolism and increases in levels of phenolics and polyphenoloxidase activity are typical indications of B deficiency, particularly in B deficiency-sensitive plant species, such as *Helianthus annuus* (sunflower). Enhanced oxidation of phenols is responsible for generation of reactive quinones which subsequently produce extremely toxic O_2 species, thus resulting in the increased risk of a peroxidative damage to vital cell components such as membrane lipids and proteins. In B-deficient tissues, enhancement in levels of toxic O_2 species may also occur as a result of impairments in photosynthesis and antioxidative defence systems. Recent evidence shows that the levels of ascorbic acid, non-protein SH-compounds (mainly glutathione) and glutathione reductase, the major defence systems of cells against toxic O_2 species, are reduced in response to B deficiency. There is also increasing evidence that, in the heterocyst cells of cyanobacteria, B is involved in protection of nitrogenase activity against O_2 damage.

Introduction

Boron is an essential micronutrient element required for growth and development of vascular plants, diatoms and some species of marine algal flagellates, while bacteria, fungi, green algae and animals apparently do not require B (Loomis and Durst, 1992). Some species of cyanobacteria require B when dependent on N_2 fixation, but not when supplied with mineral N (Bonilla et al., 1990; Mateo et al., 1986). The essentiality of B for growth and development of higher plants was first demonstrated in 1923 by Warington. As reviewed by Dugger (1983), Goldbach et al. (1991), Parr and Loughman (1983), Pilbeam and Kirkby (1983), Römheld and Marschner (1991) and recently by Marschner (1995) and Shelp (1993), considerable research activities have been directed toward characterising the physiological and biochemical roles of B in growth and development of plants. However, the exact role of B in growth is still a matter of discussion.

Several physiological impairments as a result of B deficiency were reported by Parr and Loughman (1983), such as sugar transport; cell wall synthesis; lignification; cell wall structure; carbohydrate metabolism; RNA metabolism; respiration; indole acetic acid (IAA) metabolism; phenol metabolism, and membrane integrity. In addition, it has recently been reported that B deficiency may impair ascorbate metabolism (Lukaszewski and Blevins, 1996) and induce oxygen activation (Marschner, 1995). However, despite such numerous effects of B, no evidence has yet been presented that B is an enzyme constituent or

* E-mail: icak@pamuk.cc.cu.edu.tr

that it has a direct role in enzyme activities. Moreover, it is also unclear whether the changes mentioned above indicate direct functions of B or the changes are of an indirect nature. So far, the well described roles of B are its involvement in (a) cell wall intactness and synthesis, possibly by formation of B-pectin complexes (Hu et al., 1996; Loomis and Durst, 1991, 1992; Matoh, Chapter 5) and (b) plasma membrane integrity (Cakmak et al., 1995; Marschner, 1995; Shelp, 1993) It is widely accepted that the morphological and physiological changes caused by B deficiency are most probably related to formation of complexes between B and compounds having *cis*-hydroxly groups (diols) such as pectic substances in cell walls, glycoproteins or glycolipids in membranes, and *o*-diphenols (Dugger, 1983; Marschner, 1995; Römheld and Marschner, 1991; Shelp, 1993). In this review we will discuss the main impairments in cellular functions of plants in response to B deficiency. Emphasis is given to impairments in phenol metabolism, ascorbate metabolism, free radical generation and detoxification and membrane integrity.

Phenol metabolism

Cis-diol complexes and accumulation of phenolics

Boric acid has a particular ability to form stable complexes with compounds having *cis*-hydroxyl groups (*cis*-diols) (Shelp, 1993).

Several compounds are known to have such a *cis*-diol configuration required for the formation of stable borate complexes. Sugars and their derivatives (e.g. sugar alcohols and uronic acids) and some o-diphenolics (e.g. caffeic acid and hydroxyferulic acid) are the relevant compounds forming stable borate complexes. Cell walls are rich in compounds with *cis*-diol configurations. Accordingly, it has been found that B is predominantly present in cell walls, particularly under B-deficient conditions (Hu and Brown, 1994; Matoh et al., 1992). Boron present in cell walls comprises more than 96% of the total cellular B as found in *Daucus carota* (carrot) cells (Loomis and Durst, 1991). In *Nicotiana tabacum* (tobacco) cells, B comprises about 97% and 64% of total cellular B under B-deficient and B-sufficient conditions, respectively (Hu and Brown, 1994). Most B in cell walls was found to be associated with pectic substances. Monocots with lower B requirement contain less pectin in cell walls than dicots with higher B requirement (Hu et al., 1996). Under B-deficient conditions, the structural integrity and elasticity of cell walls are impaired (Findeklee and Goldbach, 1996). The roles of B in cell walls are reviewed in Chapter 5.

Increases in concentration of phenolics in B-deficient tissues are well-known, particularly in plant species having a higher B requirement such as sunflower (Marschner, 1995; Shkol'nik, 1984). Formation of *cis*-diol complexes between B and some sugars or phenolics plays a decisive role in accumulation of phenolics in B-deficient tissues. Boron forms complexes with 6-phosphogluconic acid and, thus, inhibits activity of 6-phosphogluconate dehydrogenase (Dugger, 1983), indicating existence of more free (uncomplexed) 6-phosphogluconic acid in B-deficient tissues. Consequently, under B-deficient conditions, the substrate flux is shifted from glycolysis into the pentose phosphate pathway (PPP), which leads to enhanced phenol synthesis (Dugger, 1983; Figure 1). According to Gomes-Rodriguez et al. (1987), under B-deficient conditions the activity of glucose-6-phosphate dehydrogenase, the first enzyme of PPP, is also enhanced, much more than that of 6-phosphogluconate dehydrogenase. However, the role of B in the activities of these enzymes seems not to be caused by formation of complexes with the substrates of the enzymes.

Formation of borate complexes with some phenolics (i.e. caffeic acid) can also affect the pool of phenolics in tissues. As reviewed in Lewis (1980) and Pilbeam and Kirkby (1983), binding of B to caffeic acid blocks the formation of quinones and, thus, facilitates the synthesis of phenolic alcohols which are the precursors of lignin biosynthesis. Enhancement in concentration of phenolics in B-deficient tissues as a result of either stimulation of the pentose phosphate pathway (Dugger, 1983) or restriction in biosynthesis of phenolic alcohols (Pilbeam and Kirkby, 1983) is less pronounced in graminaceous monocots as compared with the dicotyledons (Shkol'nik, 1984).

Accumulation of phenolics in B-deficient tissues is a critical step in the initiation of impairment in several cellular functions. Most phenolics are known to be phytotoxic to roots and shoots, even at low concentrations such as 1 μM (Vaughan and Ord, 1990). In B-deficient plants, caffeic acid, chlorogenic acid, ferulic acid, vanillic acid and some coumarins accumulate (Perkins and Aronoff, 1956; Rajarathnam and Lowry, 1974; Watanabe et al., 1964), and are pronounced with enhancements in light intensity (Marschner, 1995). Stimulation of phenol biosynthesis with increasing light intensity is a well-known

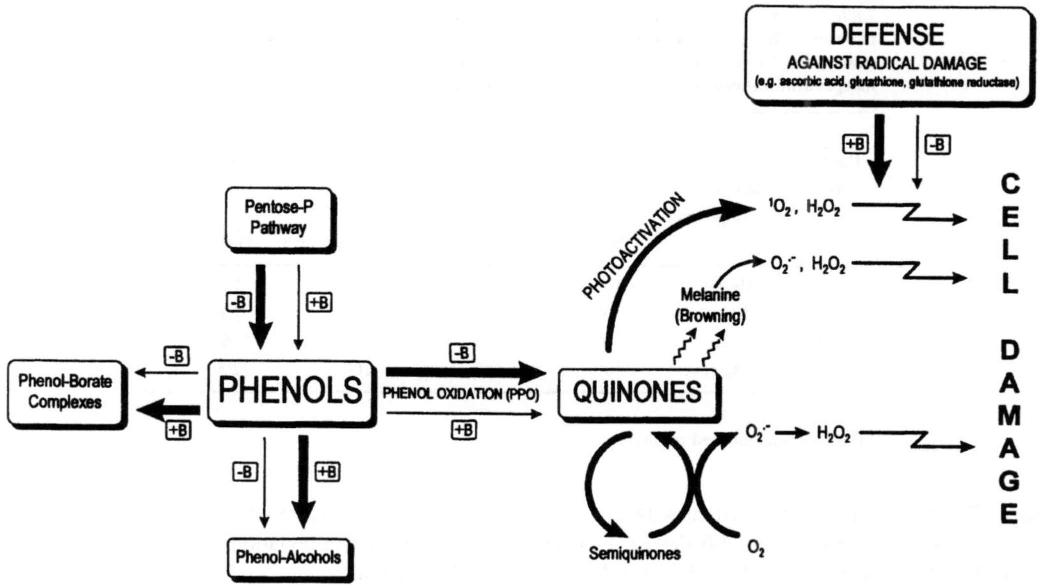

Figure 1. Model of the impairments in phenol metabolism and generation of toxic O_2 species in B-deficient plants.

phenomenon (Chattopadhyay et al., 1994; Mole et al., 1988) and attributable to induction of the enzyme phenylalanine-ammonium lyase (PAL) by light (Bolwell and Butt, 1983). Shkol'nik (1984) reported that the activity of PAL was enhanced by B deficiency, especially in B deficiency-sensitive dicotyledons.

As with phenols, auxin concentrations may be higher in B-deficient than B-sufficient plants. Higher levels of IAA in B-deficient plants were attributed to inhibition of IAA oxidase by accumulated phenolics (Coke and Whittington, 1968; Rajaratnam and Lowry, 1974), such as caffeic acid and chlorogenic acid which are inhibitors of IAA oxidase (Pilbeam and Kirkby, 1983; Shkol'nik, 1984). It has been proposed that B deficiency-induced morphological changes are a reflection of excessive accumulation of IAA in B-deficient tissues (Pilbeam and Kirkby, 1983). By contrast, it has been found that IAA concentrations did not differ among B-sufficient and B-deficient plants (Fackler et al., 1985; Hirsch et al., 1982), and the morphological symptoms caused by B deficiency and those caused by excessive accumulation of IAA are quite different (Hirsch and Torrey, 1980). Smirnov et al. (1977) suggested that deficiency symptoms of B were not related to an excess of IAA in tissues, but rather to an accumulation of oxidised phenols.

Polyphenoloxidase

Accumulation of phenolics in B-deficient tissues can activate enzymes that use phenolics as substrates. In leaf extracts of B-deficient sunflower, the activity of polyphenoloxidase (PPO) progressively increased with the severity of B deficiency symptoms, particularly under high light intensity (Marschner, 1995; Table 1). Similar to PAL, PPO activity is also inducible by increases in light intensity (Bolwell and Butt, 1983).

Plant species with different sensitivity to B deficiency differ in levels of PPO activity under B-deficient conditions. Similarly to accumulation of phenolics, PPO activity was higher in plant species with greater sensitivity to B deficiency (e.g. sunflower) than the plant species with less sensitivity (e.g. *Triticum aestivum;* wheat) (Shkol'nik, 1984; Shkol'nik et al., 1981a).

Higher activity of PPO in B-deficient leaves as compared to B-sufficient ones became more marked when PPO activity was measured by incubation of detached leaves in the test solution (Table 1). As the cell walls of the epidermal cells are one major location of PPO (Zawistowski et al., 1991), it can be suggested that higher increases in PPO activity with detached leaves over the leaf extracts are attributable to existence of an inducible PPO activity in epidermal cell

Table 1. Activity of polyphenoloxidase (relative t) in sunflower (*Helianthus annuus*) leaves in boron-sufficient (50 μM) and -deficient (0.05 μM) substrates. Enzyme activity was measured using 1mM DOPA (dihydroxyphenylalanine) as substrate. Results are means of three independent leaf extracts and four detached leaf replications

Plant age (days)	Visual deficiency symptoms	Polyphenoloxidase activity			
		Leaf extracts		Detached leaves	
		B-sufficient	B-deficient	B-sufficient	B-deficient
4	Absent	100	106	100	102
6	Mild	100	442	100	900
8	Severe	100	475	100	1466
10	Very severe	100	542	100	2200

1 Cakmak (unpublished results).

walls of B-deficient leaves. Consistent with this suggestion we also found that detached B-deficient leaves incubated in a distilled water released (exuded) PPO into test solution. Exudation of enzymes from cells is well-known. For example, a PPO-like oxidase (Tan and Kubo, 1990) and peroxidase (Vaughan et al., 1994) were released from roots in the rhizosphere. The PPO activity which was found in the test solution could be inhibited by heating (90 °C), ascorbic acid and Cu chelators such as diethyldithiocarbamate (I. Cakmak, unpublished results). Ascorbic acid and Cu chelators are well-known inhibitors of PPO (Zawistowski et al., 1991). The reason for enhanced release of PPO activity into outside is not understood and may be related to liberation of cell wall-bound PPO as a consequence of B deficiency.

Oxidation of phenolics by PPO in B-deficient leaves may lead to production of quinones (Figure 1). Quinones are known to be highly toxic and responsible for production of toxic O_2 species. It has been shown that quinones are more toxic than the phenols from which they are derived (Mason and Wasserman, 1987; Pillinger et al., 1994). Accumulation of substantial amounts of quinones in B deficiency-sensitive plant species has been considered as a major cause of cell damage and cessation of growth under B-deficient conditions (Lee and Aronoff, 1967; Shkol'nik et al., 1981b). Recently, Lukaszewski and Blevins (1996) demonstrated that root growth of *Cucurbita pepo* (squash) under B-deficient conditions was significantly improved by exogenous supply of ascorbic acid. By reducing quinones back to phenols, ascorbic acid acts as a potent inhibitor of PPO (Zawistowski et al., 1991). Thus, improvement of root growth by ascorbic acid under B-deficient conditions might partly be related to alleviation of quinone toxicity and free radical damage.

In most cases, B deficiency is associated with enhanced browning of leaf tissues. In sunflower plants, we showed that browning of leaf tissue by B deficiency was intensified by increasing light intensity (Cakmak et al., 1995). Possibly, browning of B-deficient tissues is a consequence of enhanced phenol accumulation and PPO activity. The quinones produced by PPO are subsequently polymerised to brown pigments. Also in fruits, the development of brown colour during maturation or postharvest handling was primarily associated with high levels of phenols and PPO activity (CoSeteng and Lee, 1987; Macheix et al., 1991)

Ascorbic acid, SH-groups and glutathione reductase

Concentration of ascorbic acid was substantially decreased in root tips of squash in response to B deficiency (Lukaszewski and Blevins, 1996). Recently, we also showed that in shoot tips and young leaves of sunflower, concentration of total ascorbic acid declined markedly with severity of B deficiency symptoms (Table 2). The decrease in ascorbic acid partly occurred at the expense of the reduced form of ascorbic acid, because the concentration of oxidised ascorbic acid increased, while reduced ascorbic acid decreased during B deficiency. The results indicate that biosynthesis of ascorbic acid was reduced and the consumption (oxidation) of ascorbic acid was stimulated during B deficiency. Ascorbic acid is involved in regulation

Table 2. Effect of increasing boron supply on the concentration of reduced (Asc) and oxidised (DHAsc) ascorbic acid and non-protein SH-compounds in young leaves of 10 day old sunflower (*Helianthus annuus*). Results are means from three independent replications

B supply (μM)	Visual deficiency symptoms	Asc (μmol g^{-1} fresh wt)	DHAsc (μmol g^{-1} fresh wt)	(Asc+DHAsc) (μmol g^{-1} fresh wt)	SH-compounds (μmol g^{-1} fresh wt)
0.1	Severe	0.74	1.92	2.66	0.59
1	Slight	2.30	2.22	4.52	1.13
10	Absent	3.79	1.28	5.07	1.56
100	Absent	5.16	0.89	6.04	1.61

I Cakmak (unpublished results).

of meristem activities and promotion of the growth process (Cordoba-Pedregosa et al., 1996; Liso et al., 1984). Rapid inhibitions in growth of shoot and root tips by B deficiency can, at least partly, be explained by induced consumption and restricted biosynthesis of ascorbic acid. In agreement with this, Monday and Munshi (1993) showed that B fertilisation of *Solanum tuberosum* (potato) increased concentration of ascorbic acid in tubers and reduced the discolouration of tubers.

Non-protein SH-compounds (thiols) also declined with B deficiency in sunflower leaves (Table 2). Glutathione comprises the major pool of the nonprotein SH-compounds in various plant species, e.g. 95% in *Picea abies* (spruce) needles and more than 80% in *Trifolium pratense* (red clover), *Glycine max* (soybean) and *Phaseolus vulgaris* (bean) leaves (Grill et al., 1979; Maas et al., 1987).

The reason for such significant decreases in levels of ascorbic acid and SH-compounds is not well understood. Ascorbic acid and SH-compounds (e.g. glutathione) are potent antibrowning agents used extensively during processing of foods. Ascorbic acid and SH-compounds are able to block both phenol oxidation to quinones and polymerisation of quinones to brown pigments (Cilliers and Singleton, 1990; Golan-Goldhirsh and Whitaker, 1984). Quinones formed by PPO can be reduced back to phenols by ascorbic acid and SH-compounds. Apparently, ascorbic acid and SH-compounds are consumed (oxidised) during the development of B deficiency in sunflower leaves, thus interfering with phenol oxidation and polymerisation of quinones to brown pigments or with quinone-driven free radical generation (Figure 1).

Enhanced generation of toxic O_2 species in B-deficient tissues can be expected as a result of enhanced production of quinones. Quinones produced by oxidation of phenolics can undergo one-electron reduction forming a semiquinone radical (Figure 1). Subsequent reoxidation of the radical produces the superoxide radical (O_2^-) and further toxic O_2 species such as hydrogen peroxide (H_2O_2) and hydroxyl radical (OH^-) (Appel, 1993; Kappus, 1986). In B-deficient tissues, generation of toxic O_2 species can also result directly from phenolics during their oxidation. For example, enzymatic oxidation of catechin by mushroom tyrosinase and potato phenolase produced both O_2^- and H_2O_2 (Jiang and Miles, 1993). Moreover, some polyphenols and quinones are known to be phototoxic and can be excited by light, producing toxic O_2 species such as singlett oxygen ($^1O_2$) and H_2O_2 (Bakker et al., 1983; Larson, 1986; Figure 1). Such photoactivated compounds are present in B-deficient plants (Shkol'nik, 1984; Watanabe et al., 1964).

Therefore, decreases in concentration of non-protein SH-compounds and reduced ascorbic acid by

Table 3. Relative levels of polyphenoloxidase (PPO), glutathione reductase (GRase), ascorbate peroxidase (APase), superoxide dismutase (SOD) and protein from leaf extracts of 10 day old sunflower (*Helianthus annuus*). Results are means from three independent replications

B supply (μM)	Enzyme activities (relative) PPO	GRase	APase	SOD	Protein
0.05[a]	540	28	166	102	75
50[b]	100	100	100	100	100

[a]B-deficient.
[b]B-sufficient.
I Cakmak (unpublished results).

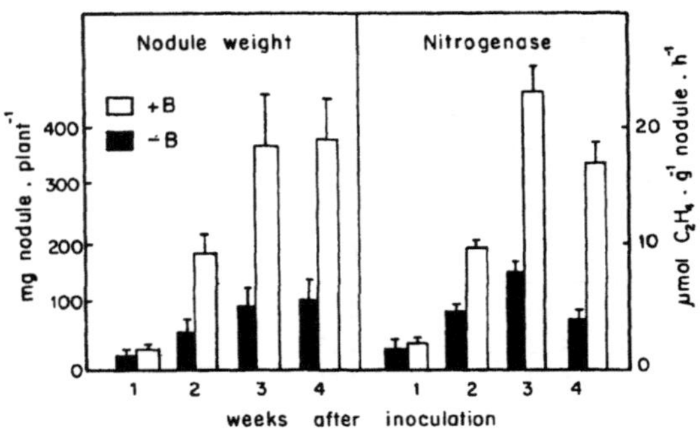

Figure 2. Nodule weight and nitrogenase activity in pea (*Pisum sativum*) roots grown with sufficient (+B) and deficient (-B) boron supply (from Bolaños et al., 1994).

B deficiency can also be a reflection of an induced oxidative stress caused by toxic O_2 radicals. Ascorbic acid and SH-compounds (i.e. glutathione) are the major antioxidants of cells and are involved in detoxification of toxic O_2 species and dissipation of excessive light energy in chloroplasts (Cakmak, 1994; Foyer et al., 1994). As with levels of ascorbic acid and SH-compounds, there was also a particular decrease in levels of glutathione reductase activity in response to B deficiency. As compared with B-sufficient plants, the decreases in activity of glutathione reductase by B deficiency were about 3.5-fold, while activity of PPO was increased by a factor of 5.5 (Table 3). The activity of other enzymes was less affected by B deficiency (Table 3). The decrease in glutathione reductase by B deficiency can be critical because it can sensitise B-deficient cells to peroxidative damage triggered by enhanced production of quinones and toxic O_2 species. Glutathione reductase is essential for the maintenance of high concentrations of glutathione and is directly involved in H_2O_2 detoxification (Foyer et al., 1994). The reason for a decrease in glutathione reductase by B deficiency is not understood at present. Possibly, this decrease in activity of glutathione reductase is related to the binding of quinones to the NADPH-binding sites of the enzyme (Bironaite et al., 1989). It seems very likely that the decreases in levels of ascorbic acid, SH-compounds and glutathione reductase during the course of B deficiency can exacerbate the development of B deficiency-induced cell damage (Figure 1).

Nitrogen fixation

Like glutathione reductase, the activity of nitrogenase is sensitive to B deficiency. Under B-deficient conditions, nodule weight and N_2 fixation capacity of legumes is usually decreased, for example in pea (*Pisum sativum*) and soybean plants (Bolaños et al., 1994; Yamagishi and Yamamoto, 1994; Figure 2). Alteration of nodule development and inhibition of nitrogenase activity in roots of B-deficient pea plants were attributed to structural degeneration of cell walls and membranes, including the peribacteroid membrane (Bolaños et al., 1994). The peribacteroid membrane surrounds the bacteroid in the nodule cell and plays an important role in establishment of nutrient transport between the symbionts. Recently, Bolaños et al. (1996) found that by stabilising membranes B might facilitate the association between bacterial cell surfaces and peribacteroid membranes and, thus, might affect the correct establishment of the symbiosis.

Role of B in maintenance of nitrogenase activity is particularly pronounced in the N_2-fixing heterocystous cyanobacteria. In the cyanobacterium *Anabaena* PPC 7119, nitrogenase activity was markedly decreased after 2 h of B-starvation, while other metabolic processes, such as photosynthesis or respiration, were not affected by B deprivation within the same period (Garcia-Gonzáles et al., 1990; Mateo et al., 1986). Interestingly, under anaerobic conditions, the B nutritional status of *Anabaena* cells had no effect on nitrogenase activity (Garcia-Gonzáles et al., 1988), indicating that B protects nitrogenase against O_2 damage. Oxygen sensitivity of nitrogenase is well-known and relates well to toxicity of O_2 species such as O_2^- and H_2O_2 (Dalton et al., 1986). Therefore, activity of O_2^-- and H_2O_2-scavenging enzymes may play an essential role in protecting nitrogenase against toxic O_2 species. Boron deficiency in *Anabaena* cells enhanced activity of O_2^-- and H_2O_2-scavenging enzymes, supporting the idea that B deficiency in *Anabaena* cells is associated with generation of toxic O_2 species (Garcia-Gonzáles et al., 1988). The reason for the increases in activity of protective enzymes in B-deficient *Anabaena* cells is not well understood. It is believed that B is required for structural integrity of the heterocyst envelope by interacting with -OH groups of glycolipid and polysaccharide layers of the envelope which are the main barriers against O_2 diffusion into the heterocyst (Garcia-Gonzáles et al., 1988, 1991). Decreases in levels of ascorbic acid caused by B deficiency (Table 2) may have negative consequences for nitrogenase activity. A

Table 4. Effect of boron pretreatment on the subsequent uptake of inorganic phosphate by root tip zones of faba bean (*Vicia faba*) and maize (*Zea mays*) grown in the presence (+B) or absence (–B) of boron

Pretreatment of root tips for 1 h	Phosphate uptake ($nmol\ g^{-1}\ h^{-1}$)			
	Faba bean		Maize	
	+B	–B	+B	–B
No boron	112	52	116	66
10^{-5} *M* B	152	108	190	171

Pollard et al. (1977).

particular role for the protection of nitrogenase activity against the toxic O_2 species is ascribed to ascorbic acid as shown in soybean nodules (Dalton et al., 1986) and *Anabaena* heterocysts (Bagchi et al., 1991). Application of ascorbic acid to N_2-fixing plants enhanced formation of nodules and concentration of N in plants (Chinoy, 1984 cited in Dalton et al., 1986). In view of the rapid decreases in ascorbic acid concentration in B-deficient plants (Lukaszewski and Blevins, 1996; Table 2), it would be quite interesting to study the relation between nitrogenase activity and the ascorbic acid pool in root nodules and heterocystous cyanobacteria.

Membrane integrity

Ion fluxes

Boron plays an important role both in structural and functional integrity of biomembranes, particularly plasma membranes (Marschner, 1995; Parr and Loughman, 1983). In many instances, it has been shown that occurrence of B deficiency was associated with substantial and rapid alterations in ion fluxes. In faba bean (*Vicia faba*) B deficiency resulted in a marked reduction in Rb^+ uptake, particularly in root tips (Robertson and Loughman, 1973). Uptake rate of phosphate was also decreased by B deficiency, and the reduction in phosphate uptake was rapidly restored by resupply of B to deficient plants for 1 h (Robertson and Loughman, 1974). More rapid effects of B on ion influx or efflux were shown by Pollard et al. (1977) in *Zea mays* (maize). They found that a resupply of B to deficient maize for 1 h caused a significant restoration of phosphate uptake (Table 4). By affecting K^+ efflux, B was also involved in enhancing stomatal opening in epidermal strips of dayflower (*Commelina communis*) (Roth-Bejerano and Itai, 1981).

As B-deficient plants accumulate high amounts of phenolic compounds (see "Phenol Metabolism"), an effect of phenolics on ion efflux or influx may be expected for B-deficient plants. Actually, supply of phenolics to plant roots at relatively low concentrations repressed uptake of ions completely, and removal of the phenols from the uptake solution rapidly restored ion uptake to the level of control plants (Glass, 1973, 1974). These changes in ion fluxes in response to treatments with phenolics were accompanied by corresponding changes in the membrane potential (Glass and Dunlop, 1974). It has been suggested that phenolics result in reversible alterations in membrane permeability, and this effect of phenolics occurs during their passage through the membranes (Glass, 1973, 1974). Increases in ion efflux by a treatment with phenolics were also found in *Escherichia coli* cells (Heipieper et al., 1991). According to Heipieper et al. (1991) cell membranes are the main target of phenolics.

H^+ extrusion and membrane potential

The influence of B on ion flux can also be mediated by direct or indirect effects of B on plasma membrane-bound proton pumping ATPases. In good agreement with this, in the experiments of Pollard et al. (1977) impairments in ion uptake by B-deficient maize roots were associated with corresponding decreases in ATPase activity. The restoration of ion uptake by resupply of B to B-deficient maize roots for 1 h was also accompanied by the restoration of the ATPase activity.

It is well established that proton efflux by membrane-bound ATPases is a main driving force for ion uptake and responsible for a gradient in electrical potential across the membranes (Poole, 1978; Serrano, 1989). There is an increasing body of evidence indicating that B nutrition has marked effects on proton secretion and creation of an electrical potential gradient across the membranes. In carrot (Blaser-Grill et al., 1989) and sunflower (Ferrol and Donaire, 1992) cells, B deficiency significantly reduced net proton release into the growing medium. The reduction in proton release caused by B deficiency was associated with a lowered activity of H^+-ATPase in isolated microsomal membranes from sunflower roots (Ferrol et al., 1993). Goldbach et al. (1990) showed that B deficiency lowered net H^+ release when proton pumping was induced by ferricyanide (Figure 3). Recently, it has been shown that B stimulated plasma membrane H^+-

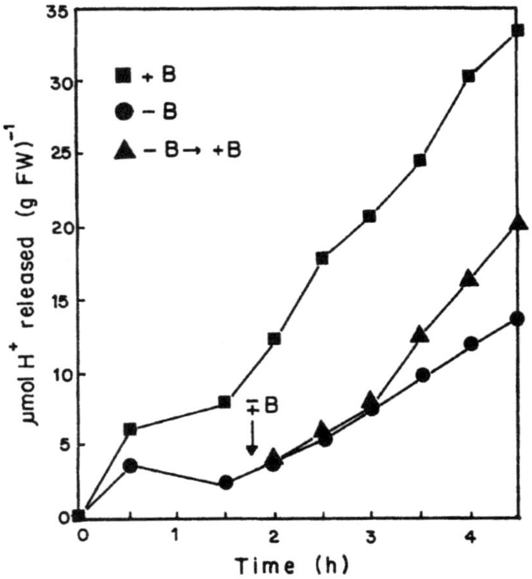

Figure 3. Proton release induced by ferricyanide in carrot (*Daucus carota*) cell cultures with sufficient (+B) and deficient (−B) boron supply. Arrow: B resupply (±B) to B-deficient carrot cell cultures (from Goldbach et al., 1990).

ATPase activity in ungerminated pollen grains of *Lilium longiflorum* (lily) (Obermeyer et al., 1996). Apparently, auxin affects B-induced H^+ extrusion. According to Goldbach et al. (1990) the stimulatory effect of B on H^+-ATPase activity requires the presence of auxin, or enhancement in proton release by auxin requires the presence of B.

In the studies with sunflower root cells and dense leaved elodea (*Elodea densa*) leaf cells, a significant depolarisation of membranes was found after transfer of cells from B-containing to B-free solution (Blaser-Grill et al., 1989). Similarly, Schon et al. (1990) showed that transfer of B-deficient sunflower root cells to B-containing solution for only 3 min caused a marked hyperpolarisation of plasma membranes. According to Schon et al. (1990), B supply enhances the activity of membrane-bound ATPase and subsequently causes hyperpolarisation of plasma membranes by stimulating ion uptake, e.g. K^+ uptake. Alternatively, B supply may stimulate proton pumping activity of membranes with a subsequent membrane hyperpolarisation, resulting in an increased driving force for K^+ influx.

Boron also stimulated activity of plasma membrane NADH oxidase and H^+ secretion in cultured carrot cells (Barr et al., 1993). These effects of B were proposed to be required for the growth of cells.

Efflux of potassium and organic solutes

The role of B in membrane integrity can also be shown by measurement of leakage of K^+ and organic solutes from cells. Recently, we showed that B deficiency in sunflower resulted in dramatic increases in membrane permeability of leaf cells (Cakmak et al., 1995). As compared with B-sufficient leaves, in severely B-deficient leaves net efflux from isolated leaves was 35-fold higher for K^+, 45-fold higher for sucrose and 7-fold higher for phenolics and amino acids (Table 5). Treatment of severely B-deficient leaves with B for 20 min restored membrane permeability to about the levels of B-sufficient leaves (Cakmak et al., 1995), indicating a particular role for B in maintaining the integrity of plasma membranes. Increases in K^+ efflux by B deficiency was also shown in hypocotyl segments of sunflower plants (Tang and Dela Fuente, 1986). Resupply of B to B-deficient sunflower plants for 3 h restored completely K^+ efflux from sunflower hypocotyls. However, based on the cell turgor measurements in root cells, Findeklee and Goldbach (1996) suggested that plasma membrane integrity was not primarily affected by B deficiency. According to these authors, increases in membrane permeability by B deficiency should be a secondary effect of B deficiency. More studies are, however, needed in this area for better understanding of the roles of B in plasma membrane integrity.

Light intensity played a major role in B deficiency-induced K^+-leakage. Permeability of leaf cell membranes to K^+ under B-deficient conditions showed a particular increase in response to a short-term exposure of leaves to a continuous high light intensity (Cakmak et al., 1995). The reason for light-induced K^+ efflux from B-deficient leaf cells is not understood. It may be proposed that light-induced generation of toxic radicals and thus peroxidative damage to membranes can make the membranes in B-deficient tissues more leaky.

Mechanisms involved in boron function in membrane integrity

Despite such clear and rapid effects of B on K^+ leakage, membrane potential and H^+ extrusion, mechanisms by which B affects structural and/or functional integrity of plasma membranes are poorly understood. Boron seems to play a critical structural role in plasma

Table 5. Effect of boron supply on B concentration, leakage of K^+, sucrose, phenolics and amino acids from leaves of 10 day old sunflower (*Helianthus annuus*)

		Leakage ($\mu g\ g^{-1}$ fresh wt $2\ h^{-1}$)			
B supply (μM)	[B] (μg B g^{-1} dry wt)	Potassium (K^+)	Sucrose (glucose equiv.)	Phenolics (caffeic acid equiv.)	Amino acids (glycine equiv.)
0.01	4.7	630	900	79	163
0.20	11.8	390	440	72	122
1.00	16.7	52	70	17	33
20.0	37.7	18	20	13	23

Cakmak et al. (1995).

membranes by its ability to bind to membrane compounds containing *cis*-diol groups, such as glycoproteins and glycolipids. The binding of B to polyhydroxy groups of membranes and, thus, maintenance of structural integrity of plasma membranes were discussed as a major reason for the stimulatory effects of B on membrane-bound ATPase activity and for maintaining integrity and controlling permeability of plasma membranes (Cakmak et al., 1995; Ferrol et al., 1993; Goldbach et al., 1990; Pollard et al., 1977; Roldan et al., 1992; Tang and Dela Fuente, 1986). Accordingly, Tanada (1983) showed that plasma membranes contain a relatively high amount of B as compared to protoplasts isolated from *Phaseolus mungo* (mungbean) hypocotyls. Nevertheless, to date the existence and nature of B complexes within and/or on plasma membranes have not been shown and need to be elucidated. To our knowledge, there is no study involving the role of B nutrition on the activity of ion channels. Based on the very rapid and drastic effects of B on ion efflux or influx (Cakmak et al., 1995), it seems quite possible that B may exert some influence on activity of ion channels in the plasma membrane.

Formation of B complexes with phenolics may be relevant to the membrane integrity for several reasons. As discussed above and shown in Figure 1, accumulation of phenolics is usual in B-deficient plants. Enzymatic or non-enzymatic oxidation of phenolics leads to generation of destructive O_2 species (Appel, 1993; Jiang and Mills, 1993). Moreover, in B-deficient tissues defence capacity of cells against toxic O_2 species was weakened due to reduced levels of ascorbic acid, SH-compounds and H_2O_2 scavenging enzymes. Therefore, it can be suggested that B deficiency-enhanced membrane leakiness and alterations in ion fluxes might, at least in part, be a consequence of peroxidative damage and structural alterations in plasma membranes caused by toxic O_2 species. Consistent with these suggestions is the enhancement of membrane permeability with increasing light intensities and with browning of leaf tissue (quinone formation) (Cakmak et al., 1995). On the same leaf, the areas showing a high discolouration (browning) had much higher K^+ leakage as well as polyphenoloxidase activity (formation of quinones) than the leaf areas without discolouration. Moreover, B deficiency reduced both total lipid and phospholipid contents in roots and leaves of *Lycopersicon esculentum* (tomato) and *Abelmoschus esculentus* (okra) plants (Desiraju et al., 1993). These decreases in phospholipids might be a result of their peroxidation by toxic O_2.

Rapid restoration of K^+ efflux from B-deficient leaves was achieved not only by a resupply of B but also by germanium (Ge) (Cakmak et al., 1995). Substitution of B by Ge is known in cell walls, and a supply of Ge to B-deficient plants greatly delayed the onset of B deficiency symptoms (Loomis and Durst, 1991; Skok, 1957). These results indicate that cellular binding sites for B in both cell walls and plasma membranes are nonspecific and can also bind Ge. Germanium and B are chemically closely related and form similar complexes with polyhydroxy phenolics. This suggests that Ge can be effective in detoxification of phenolics and repression of their oxidation to toxic quinones and O_2 radicals.

Recently, Tanada (1995) proposed that B is required for the systems which determine the generation of positive charges in the membranes. Therefore, B can influence ion fluxes across the cell membranes. According to Tanada (1995), the generation of positive charges is caused by an ability of B to capture electrons from the membrane compounds, possibly from the sulfhydryl-containing compounds.

In conclusion, together, all results indicate that by playing structural and protective roles, B primarily affects plasma membrane integrity. Further research activities are needed to better understand the distinction between the structural and the protective functions of B in cellular membranes. However, present evidence favours the conclusion that the requirement of B for the maintenance of integrity of plasma membranes is mainly related to the structural roles of B at the membrane level. Protective roles of B on membrane integrity against free radical attack might be particularly important under high light conditions and in those plant species that accumulate high amounts of phenolics in response to B deficiency, like sunflower.

Photosynthesis and photooxidation

Photosynthesis

The effect of B nutrition on photosynthesis has rarely been studied. In the cyanobacterium *Anabaena* PCC 7119 (Garcia-Gonzáles et al., 1988) and sunflower leaves (Kastori et al., 1995), B deficiency resulted in decreases in photosynthetic O_2 evolution and CO_2 assimilation. In spinach (*Spinacea oleracea*) leaves, B deficiency reduced not only photosynthetic O_2 evolution but also energy transfer from PSII to PSI (Goldbach et al., 1991). Despite the impairment of photosynthetic process, sugars and starch accumulated in B-deficient leaves (Dugger, 1983; Kastori et al., 1995; Shelp, 1993), indicating a reduced translocation of photosynthates in the sink organs. Reduced translocation of sucrose from B-deficient leaves to sink organs was attributed to the lack of sink activity in root and shoot apices (Marschner, 1995). Accordingly, an improvement in B nutritional status of plants stimulated translocation of ^{14}C-photosynthates from the source leaves to other parts of plants (Dugger, 1983).

Photooxidation

Under B-deficient conditions, increases in light intensity accelerated development of B-deficiency symptoms in duckweed (*Lemna*) cells (Tanaka, 1966) and sunflower leaves (Cakmak et al., 1995). In studies with B-deficient sunflower plants, increasing light intensity caused rapid development of leaf browning, shoot tip necrosis and decreases in chlorophyll, while B concentration in leaves was affected slightly (Cakmak et al., 1995). Similar results were also found after a partial shading of sunflower leaves. In *Vigna mungo* (black gram), partial shading of B-deficient leaves reduced development of B-deficiency symptoms (Noppakoonwong et al., 1993). In the study with black gram B requirement for leaf blade elongation was less at low light and higher at high light intensity.

These results suggest (i) that under low supply of B, high light intensity increases plant sensitivity to B deficiency, and (ii) indicate a role of photooxidation in development of visual symptoms of B deficiency. The enhancement effect of high light intensity on B deficiency symptoms can also be explained by a greater growth rate (cell wall formation) and thus a greater B requirement. In addition, increasing light intensity enhances biosynthesis of phenolics and cell wall carbohydrates, leading to immobilisation of B by formation of stable borate complexes.

The role of toxic O_2 species in development of high light-induced necrosis and chlorosis in B-deficient leaves cannot be excluded. As indicated above, free radical generation in B-deficient leaves is increased with increases in light intensity. Moreover, in B-deficient plants accumulation of carbohydrates in leaves and the impairment of photosynthesis under high light conditions may lead to a decreased utilisation of photoreductants and light energy in CO_2 assimilation. As suggested by Goldbach et al. (1991) impairments in distribution of absorbed light energy between photosystems caused by B deficiency may result in an oxidative damage to chloroplasts. Under such conditions, molecular O_2 functions as an alternative acceptor for non-utilised electrons and light energy. Transfer of electrons or light energy to O_2 leads to photogeneration of toxic O_2 species (Cakmak, 1994; Foyer et al., 1994; Robinson, 1988). Thus, a photooxidative damage to membrane and chlorophyll can be expected in B-deficient leaves as a result of impaired photosynthesis and enhanced phenol oxidation. The decreases in levels of glutathione, ascorbic acid and glutathione reductase can additionally exacerbate photooxidative damage in B-deficient leaves.

Acknowledgments

The authors would like to dedicate this work to Prof. Dr. Drs. h.c. H. Marschner who originated and drew the outline presented here. When writing this manuscript, we missed his guidance and his constructive comments. The authors are grateful to Dr. Zdenko Rengel (The University of Western Australia) for his valuable comments and correction of the English text and Miss Huriye Avsar for her excellent assistance in preparation of the manuscript. This study was supported by the Deutsche Forschungsgemeinschaft and TÜBITAK (The Scientific and Technical Research Council of Turkey).

References

Appel H M 1993 Phenolics in ecological interaction: the importance of oxidation. J. Chem. Ecol. 19, 1521–1552.

Bagchi N, Ernst A and Böger P 1991 The effect of activated oxygen species on nitrogenase of *Anabaena variabilis*. Z. Naturforsch. 66, 407–415.

Bakker J, Gommers F J, Smits L, Fuchs A and De Vries F W 1983 Photoactivation of isoflavonoid phytoalexins: Involvement of free radicals. Photochem. Photobiol. 38, 323–329.

Barr R, Böttger M and Crane F L 1993 The effect of B on plasma membrane electron transport and associated proton secretion by cultured carrot cells. Biochem. Mol. Biol. Int. 31, 31–39.

Bironaite A D, Cenas N K and Kulys J J 1989 The inhibition of glutathione reductase by quinones. Eur. J. Biochem. 178, 693–703.

Blaser-Grill J, Knoppik D, Amberger A and Goldbach H 1989 Influence of boron on the membrane potential in *Elodea densa* and*Helianthus annuus* roots and H^+ extrusion of suspension cultured *Daucus carota* cells. Plant Physiol. 90, 280–284.

Bolaños L, Estaben E, De Lorenzo C, Fernandez-Pascual M, De Felipe M R, Garate A and Bonilla I 1994 Essentiality of boron for symbiotic dinitrogen fixation in pea (*Pisum sativum*) rhizobium nodules. Plant Physiol. 104, 85–90.

Bolaños L, Brewin N J and Bonilla I 1996 Effect of boron on *rhizobium*-legume cell-surface interactions and nodule development. Plant Physiol. 110, 1249–1256.

Bolwell G P and Butt V S 1983 Photoinduced changes in *o*-dyphenol oxidase and *p*-coumarate hydroxylase activities in spinach beet seedlings and leaves. Phytochem. 22, 37–45.

Bonilla I, Garcia-Gonzàlez M and Mateo P 1990 Boron requirement in cyanobacteria. Its possible role in the early evolution of photosynthetic organisms. Plant Physiol. 94, 1554–1560.

Cakmak I 1994 Activity of ascorbate-dependent H_2O_2-scavenging enzymes and leaf chlorosis are enhanced in magnesium- and potassium-deficient leaves, but not in phosphorus-deficient leaves. J. Exp. Bot. 45, 1259–1266.

Cakmak I, Kurz H and Marschner H 1995 Short-term effects of boron, germanium and high light intensity on membrane permeability in boron deficient leaves of sunflower. Physiol. Plant. 95, 11–18.

Chattopadhyay S, Datta S K and Mahato S B 1994 Production of L-DOPA from cell suspension culture of *Mucuna pruriens f. pruriens*. Plant Cell Rep. 13, 519–522.

Cilliers J J L and Singleton V L 1990 Caffeic acid autoxidation and the effect of thiols 38, 1789–1796.

Coke L and Whittington W J 1968 The role of boron in plant growth, IV. Interrelationships between boron and indol-3-yl acetic acid in the metabolism of bean radicles. J. Exp. Bot. 19, 295–308.

Cordoba-Pedregosa M D C, Gonzalez-Reyes J A, Canadillas M D S, Navas P and Cardoba F 1996 Role of apoplastic and cell wall peroxidases on the stimulation of root elongation by ascorbate. Plant Physiol. 112, 1119–1125.

CoSeteng M Y and Lee C Y 1987 Changes in apple polyphenoloxidase and polyphenol concentrations in relation to degree of browning. J. Food Sci. 52, 985–989.

Dalton D A, Russell S A, Hanus F J, Pascoe G A and Evans H J 1986 Enzymatic reactions of ascorbate and glutathione that prevent peroxide damage in soybean root nodules. Proc. Natl. Acad. Sci. USA 83, 3811–3815.

Desiraju S, Rashmi A and Rathore V S 1993 Influence of boron deficiency on growth, protein and lipid contents in tomato and okra seedlings. Acta Physiol. Plant. 15, 25–30.

Dugger W M 1983 Boron in plant metabolism *In* Encyclopedia of Plant Physiology, new series, Vol. 15. Eds. A Lauchli and R L Bieleski. pp 626–650 Springer-Verlag, Berlin.

Fackler U, Goldbach H, Weiler E W and Amberger A 1985 Influence of boron deficiency on indol-3yl-acetic acid and abscissic acid levels in root and shoot tips. J. Plant Physiol. 119, 295–299.

Ferrol N and Donaire J P 1992 Effect of boron on plasma membrane proton extrusion and redox activity in sunflower cells. Plant Sci. 86, 41–47.

Ferrol N, Belver A, Roldan M, Rodriguez-Rosales M P and Donaire J R 1993 Effects of boron on proton transport and membrane properties of sunflower (*Helianthus annuus* L.) cell microsomes. Plant Physiol. 103, 763–769.

Findeklee P and Goldbach H E 1996 Rapid effects of boron deficiency on cell wall elasticity modulus in *Cucurbita pepo* roots. Bot. Acta 109, 463–465.

Foyer C H, Lelandais M and Kunert K J 1994 Photooxidative stress in plants. Physiol. Plant. 92, 696–717.

Garcia-Gonzáles M, Mateo P and Bonilla I 1988 Boron protection for O_2 diffusion in heterocysts of *Anabaena* sp. PCC 7119. Plant Physiol. 87, 785–789.

Garcia-Gonzáles M, Mateo P and Bonilla I 1990 Effect of boron deficiency on photosynthesis and reductant sources and their relationship with nitrogenase activity in *Anabaena* PCC 7119. Plant Physiol. 93, 560–565.

Garcia-Gonzáles M, Mateo P and Bonilla I 1991 Boron requirement for envelope structure and function in *Anabaena* PCC 7119 heterocysts. J. Exp. Bot. 42, 925–929.

Glass A D M 1973 Influence of phenolic acids on ion uptake. I. Inhibition of phosphorus uptake. Plant Physiol. 51, 1037–1041.

Glass A D M 1974 Influence of phenolic acids upon ion uptake. III. Inhibition of potassium absorption. J. Exp. Bot. 25, 1104–1113.

Glass A D M and Dunlop J 1974 Influence of phenolic acids on ion uptake. IV Depolarisation of membrane potentials. Plant Physiol. 54, 855–858.

Golan-Goldhirsh A and Whitaker J R 1984 Effect of ascorbic acid, sodium bisulphite, and thiol compounds on mushroom polyphenol oxidase. J. Agric. Food Chem. 32, 1003–1009.

Goldbach H E, Hartmann D and Rötzer T 1990 Boron is required for the ferricyanide-induced proton release by auxins in suspension-cultured cells of *Daucus carota* and *Lycopersicon esculentum*. Physiol. Plant. 80, 114–118.

Goldbach H E, Blaser-Grill J, Lindemann N, Porzelt M, Hörmann C, Lupp B and Gessner B 1991 Influence of boron on net proton release and its relation to other metabolic process. Curr. Topics Plant Biochem. Physiol. 10, 195–220.

Gomes-Rodriguez M V, Luna del Castillo de D and Alvarez-Tinaut M C 1987 The evolution of glucose-6P-dehydrogenase and 6P-gluconate-dehydrogenase activities and the ortho-diphenolic content of sunflower leaves cultivated under different boron treatments. J. Plant Nutr. 10, 2211–2229.

Grill D, Esterbauer H and Klösch U 1979 Effect of sulphur dioxide on glutathione in leaves of plants. Environ. Pollut. 19, 187–194.

Heipieper H J, Keweloh H and Rehm H J 1991 Influence of phenols on growth and membrane permeability of free and immobilised *Escherichia coli*. Appl. Environ. Microbiol. 57, 1213–1217.

Hirsch A M and Torrey J G 1980 Ultrastructural changes in sunflower root cells in relation to boron deficiency and added auxin. Can. J. Bot. 58, 856–866.

Hirsch A M, Pengelly W L and Torrey J G 1982 Endogenous IAA levels in boron-deficient and control root tips of sunflower. Bot. Gaz. 143, 15–19.

Hu H and Brown P H 1994 Localisation of boron in the cell wall and its association with pectin. Evidence of a structural role of boron in the cell wall. Plant Physiol. 105, 681–689.

Hu H, Brown P B and Labavitch J M 1996 Species variability in boron requirement is correlated with cell wall pectin. J. Exp. Bot. 47, 227–232.

Jiang Y and Miles P W 1993 Generation of H_2O_2 during enzymic oxidation of catechin. Phytochem. 33, 29–34.

Kappus H 1986 Overview of enzyme systems involved in bioreduction of drugs and in redox cycling 35, 1–6.

Kastori R, Plesnicar M, Pankovic D and Sakac Z 1995 Photosynthesis, chlorophyll fluorescence and soluble carbohydrates in sunflower leaves as affected by boron deficiency. J. Plant Nutr. 18, 1751–1763.

Larson R A 1986 Insect defences against phototoxic plant chemicals. J. Chem. Ecol. 12, 859–780.

Lee S and Aronoff S 1967 Boron in plants: a biochemical role. Science 158, 798–799.

Lewis D H 1980 Boron, lignification and the origin of vascular plants a unified hypothesis. New Phytol. 84, 209–229.

Liso R, Calabrese G, Bitonti M B and Arrigoni O 1984 Relationship between ascorbic acid and cell division. Exp. Cell Res. 150, 314–320.

Loomis W D and Durst R W 1991 Boron and cell walls. Curr. Topics Plant Biochem. Physiol. 10, 149–178.

Loomis W D and Durst R W 1992 Chemistry and biology of boron. BioFactors 4, 229–239.

Lukaszewski K M and Blevins D G 1996 Root growth inhibition in boron-deficient or aluminium-stressed squash may be a result of impaired ascorbate metabolism. Plant Physiol. 112, 1135–1140.

Maas F M, DeKok L J, Peters J L and Kuiper P J C 1987 A comparative study on the effects of H_2S and SO_2 fumigation on the growth and accumulation of sulphate and sulphydryl compounds in *Trifolium pratense* L., *Glycine max* Merr. and *Phaseolus vulgaris* L. J. Exp. Bot. 38, 1459–1469.

Macheix J-J, Sapis J-C and Fleuriet A 1991 Phenolic compounds and polyphenoloxidase in relation to browning in grapes and wines. Crit. Rev. Food Sci. Nutr. 30, 441–486.

Marschner H 1995 Mineral Nutrition of Higher Plants. pp 379–396. Academic Press, San Diego, USA.

Mason T L and Wasserman B P 1987 Inactivation of red beet β-glucan synthetase by native and oxidised phenolic compounds. Phytochem. 26, 2197–2202.

Mateo P, Bonilla I, Fernandez-Valiente E and Sanchez-Maeso E 1986 Essentiality of boron for dinitrogen fixation in *Anabaena* sp. PCC 7119. Plant Physiol. 81, 430–433.

Matoh T 1997 Boron in plant cell walls. *In* Boron in Soils and Plants. Ed. B Dell. pp. 000-000. Kluwer Academic Publishers, Dordrecht, the Netherlands.

Matoh T, Ishigaki K-I, Mizutani M, Matsunaga W and Takabe K 1992 Boron nutrition of cultured tobacco BY-2 cells. I. Requirement for and intracellular localisation of boron and selection of cells that tolerate low levels of boron. Plant Cell Physiol. 33, 1135–1141.

Mole S, Ross J A M and Waterman P G 1988 Light-induced variation in phenolic levels in foliage of rain-forest plants. I. Chemical Changes. J. Chem. Ecol. 14, 1–21.

Monday N I and Munshi C B 1993 Effect of boron on enzymatic discolouration and phenolic and ascorbic acid content of potatoes. J. Agric. Food Chem. 41, 554–556.

Noppakoonwong R N, Bell R L, Dell B and Loneragan J F 1993 An effect of shade on the boron requirement for leaf blade elongation in black *gram (Vigna mungo* L. Hepper). Plant and Soil 155/156, 317–320.

Obermeyer G, Kriechbaumer R, Strasser D, Maschessnig A and Bentrup F-W 1996 Boric acid stimulates the plasma membrane H^+-ATPase of ungerminated lily pollen grains. Physiol. Plant. 98, 281–290.

Parr A J and Loughman B C 1983 Boron and membrane function in plants. *In* Metals and Micronutrients, Uptake and Utilisation by Plants. Eds. D A Robb and W S Pierpoint. pp 87–107. Academic Press, New York.

Perkins H J and Aronoff S 1956 Identification of the blue-fluorescent compounds in boron-deficient plants. Arch. Biochem. Biophys. 64, 506–507.

Pilbeam D J and Kirkby E A 1983 The physiological role of boron in plants. J. Plant Nutr. 6, 563–582.

Pillinger J M, Cooper J A and Ridge I 1994 Role of phenolic compounds in the antialgal activity of barley straw. J. Chem. Ecol. 20, 1557–1569.

Pollard A S, Parr A J and Loughman B C 1977 Boron in relation to membrane function in higher plants. J. Exp. Bot. 28, 831–841.

Poole R J 1978 Energy coupling for membrane transport. Ann. Rev. Plant Physiol. 29, 437–460.

Rajarathnam J A and Lowry J B 1974 The role of boron in the oil-palm *(Elaeis guinensis)*. Ann. Bot. 38, 193–200.

Robertson G A and Loughman B C 1973 Rubidium uptake and boron deficiency in *Vicia faba*. J. Exp. Bot. 24, 1046–1052.

Robertson G A and Loughman B C 1974 Reversible effects of boron on the absorption and incorporation of phosphate in *Vicia faba*. New Phytol. 73, 291–298.

Robinson J M 1988 Does O_2 photoreduction occur within chloroplasts *in vivo*? Physiol. Plant. 72, 666–680.

Roldan M, Belver A, Rodriguez-Rosales P, Ferrol N and Donaire J D 1992 *In vivo* and *in vitro* effects of boron on the plasma membrane proton pump of sunflower roots. Physiol. Plant. 84, 49–54.

Römheld V and Marschner H 1991 Functions of micronutrients in plants. *In* Micronutrients in Agriculture, 2nd ed. Eds. J J Mordvedt, F R Cox, L M Shuman and R M Welch pp 297–328. SSSA Book Series, No. 4, Madison, WI, USA.

Roth-Bejerano N and Itai C 1981 Effect of boron on stomatal opening in epidermis strips of *Commelina communis*. Physiol. Plant. 52, 302–304.

Schon M K, Novacky A and Blevins D G 1990 Boron induces hyperpolarisation of sunflower root cell membranes and increases membrane permeability to K^+. Plant Physiol. 93, 566–577.

Shelp B J 1993 Physiology and biochemistry of boron in plants. *In* Boron and Its Role in Crop Protection. Ed. U C Gupta. pp 53–85. CRC Press, Boca Raton, FL, USA.

Shkol'nik M Y 1984 Trace Elements in Plants. pp. 68–109. Developments in Crop Science (6), Elsevier, New York.

Shkol'nik M Y, Krupnikova T A and Smirnov Y S 1981a Activity of polyphenol oxidase and sensitivity to boron deficiency in monocots and dicots. Fiziol. Rast. 28, 391–397.

Shkol'nik M Y, Krupnikavo T A, Timofeeva S S and Stom D I 1981b Intensification of quinone formation from exogenous polyphenols by homogenates of the leaves of sunflower plants reared under conditions of boron deficiency. Fiziol. Rast. 28, 541–546.

Serrano R 1989 Structure and function of plasma membrane ATPase. Ann. Rev. Plant Physiol. 40, 61–94.

Skok J 1957 The substitution of complexing substances for boron in plant growth. Plant Physiol. 32, 308–312.

Smirnov Y S, Krupnikova T A and Shkol'nik M Y 1977 Content of IAA in plants with different sensitivity to boron deficits. Sov. Plant Physiol. 24, 270–276.

Tan K-S and Kubo I 1990 Release of oxidases from the roots of plants. Experientia 46, 478–481.

Tanada T 1983 Localisation of boron in membranes. J. Plant Nutr. 6, 743–749.

Tanada T 1995 Boron as a transducer in some physiological process of plants. J. Plant Nutr. 18, 1743–1750.

Tanaka H 1966 Response of *Lemna paucicostata* to boron as affected by light intensity. Plant and Soil 25, 425–434.

Tang P M and Dela Fuente R K 1986 The transport of indole-3-acetic acid in boron- and calcium-deficient sunflower hypocotyl segments. Plant Physiol. 81, 646–650.

Yamagishi M and Yamamoto Y 1994 Effects of boron on nodule development and symbiotic nitrogen fixation in soybean plants. Soil Sci. Plant Nutr. 40, 265–274.

Vaughan D and Ord B G 1990 Influence of phenolic acids on morphological changes in roots of *Pisum sativum*. J. Sci. Food Agric. 52, 289–299.

Vaughan D, Cheshire M V and Ord B G 1994 Exudation of peroxidase from roots of *Festuca rubra* and its effects on exuded phenolic acids. Plant and Soil 160, 153–155.

Warington K 1923 The effect of boric acid and borax on the broad bean and certain other plants. Ann. Bot. 37, 629–672.

Watanabe R, Skok C J and Wender S H 1964 Effect of boron deficiency on polyphenol production in the sunflower. Phytochemistry 3, 391–393.

Zawistowski J, Biliaderis C G and Eskin N A M 1991 Polyphenol Oxidase. *In* Oxidative Enzymes in Foods. Eds. D S Robinson and A M Eskin. pp 217–273. Elsevier Science Publishers, England.

Plant and Soil **193:** 85–101, 1997.

Chapter 7

Boron mobility in plants

Patrick H. Brown[1] and Barry J. Shelp[2]
[1]*Department of Pomology, University of California, Davis, CA 95616, USA** [2]*Department of Horticultural Science, University of Guelph, Guelph, Ontario, N1G2Wq, Canada*

Abstract

In the majority of plant species, B distribution between plant organs and the symptoms of B deficiency and toxicity indicate that B has restricted mobility. Nevertheless, B is present in phloem and is retranslocated in phloem, often in sufficient amounts to satisfy the demands of developing sink regions that do not readily transpire. In species that produce significant amounts of polyols in source leaves, boron is readily translocated as a consequence of the formation of B-polyol complexes. Boron is thus unique among the essential plant nutrients in that it has restricted mobility in many plant species and is freely mobile in others. No other element is known to vary so greatly in mobility.

The retranslocation of B has a profound effect on the expression of B deficiency and toxicity symptoms, and the approaches needed to diagnose and correct B imbalances. Examples of the impact of B mobility on B uptake, B diagnosis, B toxicity and the breeding of species for B tolerance, are discussed here and in the relevant chapters of this volume. In the following we provide a summary of current information on the mobility of B in plants and provide insights into the physiological and agronomic consequences of these findings.

Introduction

Boron deficiency in crops is more widespread than deficiency of any other micronutrient (Gupta, 1993a). Visual symptoms of B deficiency generally become evident in dicots, maize (*Zea mays*) and wheat (*Triticum aestivum*) at tissue concentrations of less than 20–30, 10–20 and 10 mg kg^{-1} dry wt, respectively (Anon., 1991; Gupta, 1993b). Nutritional disorders attributed to B deficiency are prevalent among vegetables, fruit and nut trees. These include brown heart in storage roots of rutabaga (*Brassica napus rapifera*), turnip (*Brassica rapa*) and radish (*Raphanus sativus*), and hollow stem in cauliflower (*Brassica oleracea* var. *botrytis*) and broccoli (*Brassica oleracea* var. *italica*) (Shelp and Shattuck 1987a, 1987b; Shelp et al., 1987, 1992a). In fruit and nut trees, B deficiency often results in decreased seed set even when vegetative symptoms are absent (Nyomora et al., 1997 and references therein). The occurrence of these disorders, even when B is in ample supply in the soil, suggests that B deficiency in plants is physiological in nature (e.g. induced by rapid growth resulting from favourable environmental conditions or high fertiliser nitrogen levels) and relates to B mobility within the plant. While there is little doubt that B is translocated in xylem to sites of greatest water loss, there has been considerable controversy regarding the role that the phloem plays in providing B to sites that do not lose water readily. In this article, we review evidence for the mobility of B in phloem, highlighting the strategies used to study its translocation.

It is now clear that species differ dramatically in the extent of B mobility. Species can thus be classified into those having restricted B mobility and those in which B is highly mobile. Given the dramatic differences in the behaviour of B in these two classes of species, they are considered separately in the following text. For each, we present a review of current information and a perspective on future research directions. This is followed by a discussion of the impact of differential B mobility on the diagnosis and management of B status in plants, and is concluded with a discourse on the potential uses for species that differ in B mobility to help define the function of B in plants.

* E-mail: phbrown@ucdavis.edu

Dynamics of boron in species with restricted boron mobility

In plants, the long distance translocation of nutrient elements takes place in the vascular system consisting of the xylem and phloem; water is the translocating agent in these tissues. Upward movement from the roots to the shoots occurs in the nonliving cells of the xylem and is driven predominantly by the gradient in water potential resulting from surface water loss (transpiration) during the day. Thus, primary xylem translocation is directed mainly to the sites of highest transpiration (the large source leaves), which are not usually the sites of highest demands for nutrients (Pate, 1975). In contrast, long distance translocation in the phloem, with its living cells, occurs in both upward and downward directions. Phloem translocation is independent of transpiration and supplies the major proportion of nutrient requirements for actively growing areas, such as young leaves, fruits and seeds: organs that do not lose water readily.

During translocation, nutrients are transferred between xylem and phloem by extensive exchange processes. In the stem and leaf veins, xylem and phloem are separated by only a few cell layers, and direct exchange of unmetabolised nutrients between these two pathways may occur (DaSilva and Shelp, 1990; Pate, 1975; Shelp and DaSilva, 1990; van Bel, 1984). Indirect transfer occurs when nutrients are transformed in the leaf and immediately exported in the phloem, or stored and exported at a later stage of development. The latter process is often associated with leaf senescence during the growth of reproductive structures (flowers and fruits) and involves the translocation of reserve materials such as protein and carbohydrates. It is less associated with vegetative growth unless the plant has experienced a period of insufficient nutrient supply. Throughout this review, the term retranslocation will comprise both intraveinal exchange and the longer route involving solute movement between xylem and phloem via the mesophyll.

Nutrient distribution within plant shoots may be explained in terms of retranslocation, which has been assessed in a variety of ways including: the ratios of concentrations in younger to older parts, fruits and leaves, and in phloem exudate and xylem sap; the uptake pattern into fruit and inflorescences during growth; and the mobility of soil and foliar applied isotopes (Pate, 1975; Shelp, 1993).

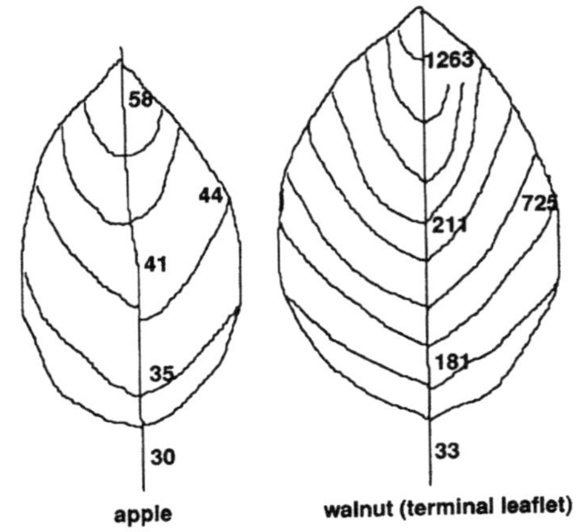

Figure 1. Leaf-B concentration (mg kg^{-1} dry wt) in field grown apple and walnut. Leaves were collected at the end of the growing season in 1995 in the Pomology orchard, Davis, California, USA. The two species were grown in close proximity and received the same irrigation.

Evidence against boron retranslocation

Boron uptake across the root plasma membrane is probably passive, involving the unassisted permeation of $B(OH)_3$ and subsequent formation of *cis*-diol complexes (Brown and Hu, 1994; Seresinhe and Oertli, 1991; Shelp, 1993). Once in the xylem, the translocation of B is related to the loss of water from shoot organs, as shown by the decreasing acropetal concentration gradient in response to increasing developmental maturity (e.g. Table 1, *Isomems* and *Ulmus*) (Bowen, 1972; Kohl and Oertli, 1961; Shelp et al., 1987; Shelp et al., 1992b; von Michael et al., 1969). Even within a particular leaf, a steep gradient in B concentration (petioles and midribs < middle of lamina < margins and tips (Oertli, 1994; Oertli and Roth, 1969; Figure 1) may result from an excessive supply of B. Furthermore, the concentration of B in leaves is higher than in phloem exudate (Marentes et al., 1997; Shelp, 1987, 1988; Shelp et al., 1996, Tammes and van Die, 1966), and this is discussed below. Symptoms of B deficiency in shoots typically occur in meristematic tissue (e.g. terminal buds and young leaves), whereas symptoms of B toxicity occur first in the margins of mature leaves, generally at the end of the transpiration stream (Marschner, 1995; Oertli, 1994; Oertli and Roth, 1969; Shelp et al., 1995). In young squash (*Cucurbita* sp.) and tomato (*Lycopersicon esculentum*) plants, B deficiency symptoms develop rapidly when the B supply is interrupted

Table 1. The relationship between sugar alcohols and B concentration (mg B kg^{-1} dry wt) in the leaf. Samples were collected from plants growing in a natural environment in Davis, California on 20 Sept., 1995. Leaves were washed for one min in deionised water. After dry ashing, B was analysed by Inductively Coupled Plasma-Mass Spectrometry (ICP-MS)

Sugar alcohol[1]	Species	B concentration		
		Expanding leaf	Mature leaf	Old leaf
None[2]	*Isomems arborea*	35	178	216
None[2]	*Ulmus parvifolia* (Chinese elm)	98	180	205
Sorbitol	*Eriobotrya deflexa*	292	285	76
Sorbitol	*Prunus elysifolia*	86	64	38
Mannitol	*Apium graveolens* (celery)	175	173	41
Mannitol	*Osmanthus fragrans* (fragrant olive)	428	144	68
Mannitol	*Phillyrea latifolia*	65	147	248

[1]Sugar alcohol information from Bourne (1958) and Bieleski (1982).
[2]No reported occurrence of sugar alcohols in this species.

(Hu and Brown, 1994; Oertli, 1993). While these data are the basis for the historical classification of B as an immobile element, they do not exclude the possibility of B translocation in phloem.

Evidence for boron retranslocation

Greenhouse grown broccoli and soybean (*Glycine max*) plants have a decreasing acropetal gradient of B concentration when grown with luxury or adequate B supply. However, in plants grown with deficient B, this gradient from old to young tissues disappears or is reversed (Al- Molla, 1986; Marentes et al., 1996; Shelp et al., 1992b). If B is mobile only in the xylem, tissue B should always be highest in old organs.

Shelp and coworkers reported that ratios of tissue B in developing sinks to old leaves of radish (Shelp et al., 1987), rutabaga (Shelp and Shattuck, 1987a), cauliflower (Shelp and Shattuck, 1987b) and broccoli (Shelp, 1988) are higher with a continuous supply of limiting B than with sufficient B. With each plant species studied, the relative element composition and, in particular, the low Ca, Mn and B contents of the sinks match what is generally known about the composition of phloem exudate (Pate, 1975; van Die and Tammes, 1975), indicating that phloem rather than xylem transport is quantitatively the most important route for nutrients.

Boron retranslocation is suggested from the decreased B concentration in mature leaves of grape (*Vitis vinifera*) (Scott and Schrader, 1947), broccoli (Benson et al., 1961; Shelp, 1988), cotton (*Gossypium hirsutum*) and turnip (McIrath, 1965), and its unchanging concentration in fruits of peanut (*Arachis hypogaea*) and subterranean clover (*Trifolium subterraneum*) (Campbell et al., 1975) when plants are subjected to B deficiency. Liu et al. (1993) investigated the B distribution in two broccoli cultivars, grown under irrigation with and without supplemental B, at three locations which differed in extractable B. A decreasing tissue B gradient up the shoot is only found in plants supplied with supplemental B. Furthermore, floret B concentration and content are rarely affected by the B treatments. Thus, in spite of the accumulation of B in transpiring organs, these studies provide circumstantial evidence for B translocation under conditions of low B supply (i.e. conditional mobility).

Recently, Marentes et al. (1997) supplied ^{10}B to the root systems of broccoli and lupin (*Lupinus albus*) plants to provide a quantitative picture of B distribution during early reproductive development. Plant B status and external B supply does not markedly influence the specific allocation to broccoli florets of ^{10}B acquired during inflorescence development; this newly acquired ^{10}B accounts for 68–99% of the floret B at 14 days after inflorescence emergence (DAIE), whereas the remainder is derived from previously acquired B (Figure 2). Boron acquired during inflorescence development is transiently accumulated in the stem, and middle and upper portions of the canopy (data not

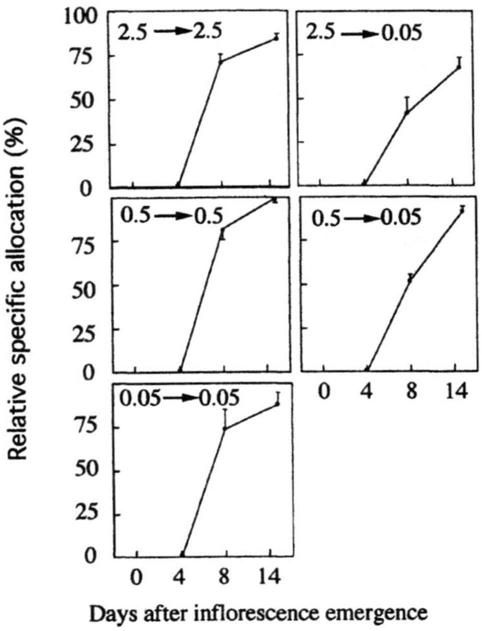

Figure 2. Relative specific allocation (proportion of B atoms that are ^{10}B labelled) in broccoli florets of ^{10}B acquired after inflorescence emergence. Arrows in the panel indicate a shift from various B concentrations supplied at natural abundance to those enriched in ^{10}B at inflorescence emergence. Data represent the mean ± SE (n = 5); where smaller than the symbol, SE is not shown. From Marentes et al. (1997).

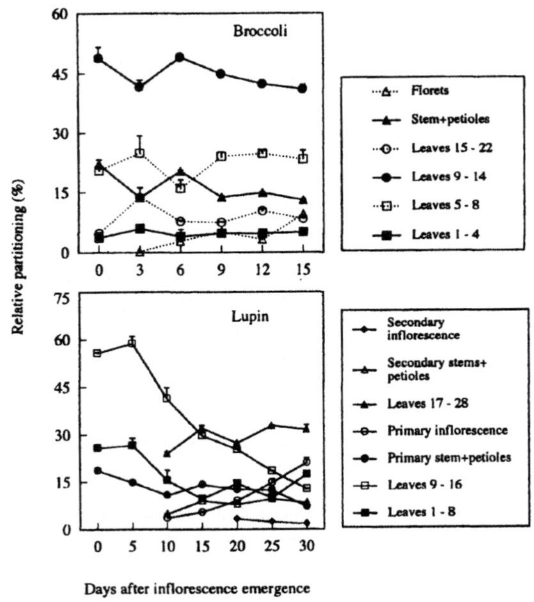

Figure 3. Relative partitioning among shoot strata during inflorescence development of ^{10}B acquired in the 10-day period before inflorescence emergence in broccoli (upper panel), and ^{10}B acquired in the 15-day period just before primary inflorescence emergence in lupin (lower panel). With the exception of the 10- or 15-day periods when broccoli and lupin plants were supplied with enriched ^{10}B at 0.5 and 0.3 mg L^{-1} respectively, they received enriched ^{11}B at the same concentration. The key indicates the various strata. Data represent the mean + SE (n = 5); where smaller than the symbol, SE is not shown. From Marentes et al. (1997).

shown). In another experiment, 8% of the ^{10}B acquired by broccoli in the 10 day period immediately prior to inflorescence emergence is recovered in the florets at 15 DAIE (Figure 3). Twenty one percent of the ^{10}B acquired by lupin in the 15 day period immediately prior to inflorescence emergence is recovered in fruits at 30 DAIE. Therefore, both newly and previously acquired B can contribute to the B requirements of reproductive structures. It is tempting to speculate that B, like N (DaSilva and Shelp, 1990; Pate, 1975; Shelp and DaSilva, 1990; van Bel, 1984), undergoes direct as well as indirect xylem to phloem transfer.

Foliar applied ^{10}B is also translocated to broccoli florets primarily via the phloem (Shelp et al., 1996), although some ^{10}B is recovered in xylem sap, presumably after transfer from phloem in roots (Pate, 1975). The florets of plants grown with a deficient, sufficient or luxury supply of B receive 25, 10 and 4%, respectively of its B from foliar fertilisation; this corresponds with 30, 29 and 16% of the ^{10}B distributed throughout the shoot. Thus, the partitioning of foliar applied B to florets is inversely related to plant B status. This translocated ^{10}B significantly enhances floret yields, regardless of plant B status, but is most effective in plants with relatively lower B status.

Schon and Blevins (1987) used a stem infusion technique to inject boric acid solutions directly into soybean stems throughout the reproductive phase of the plant. This methodology allows for a slow, but continual supply of B directly into the stem, throughout the reproductive phase of the plant. Boron treatment significantly increases the number of pods/branch and total seed yield, a result that is most likely attributed to the xylem to phloem transfer of B (Grabau et al.,1986).

Boron in phloem

From data on shoot B accumulation and water usage, Kitheka and Shelp (unpublished data) estimated that the B concentration of xylem sap from intact broccoli and lupin plants grown with sufficient B is about 5 μM;

this value is an order of magnitude lower than the B concentration found in root bleeding sap. In contrast, phloem exudate from attached inflorescences and fruits of these plants contain about 0.3 m*M* B (also Shelp, 1987, 1988). Furthermore, broccoli plants supplied with a sufficient supply of B (45 m*M*) have old leaves and florets with B concentrations of 4.6 and 0.4 m*M*, respectively (Shelp et al., 1992b), assuming 15% dry wt (Shelp, 1988). Tammes and van Die (1966) reported that in yucca (*Yucca flaccida*), B concentrations of source leaves, inflorescence and phloem exudate are 1.0 m*M* (32% dry wt), 0.3 m*M* (16% dry wt) and 0.2 m*M* (18% dry wt), respectively. Thus, the B concentration in phloem exudate is lower than that in source leaves but similar to that in young shoot tissues.

Shelp (1987) has estimated the contribution of selected nutrients from phloem to the developing sinks of broccoli grown with sufficient B. A comparison of the relative proportion by weight of all elements suggests that the phloem stream provides B to an equal or greater extent than it does N, P and K, nutrients which are generally classified as phloem mobile. The phloem supply of Zn is the only one exceeding B; the relative Ca supply is the lowest. Together, these studies suggest that phloem, rather than xylem, is the predominant source of B for developing sinks.

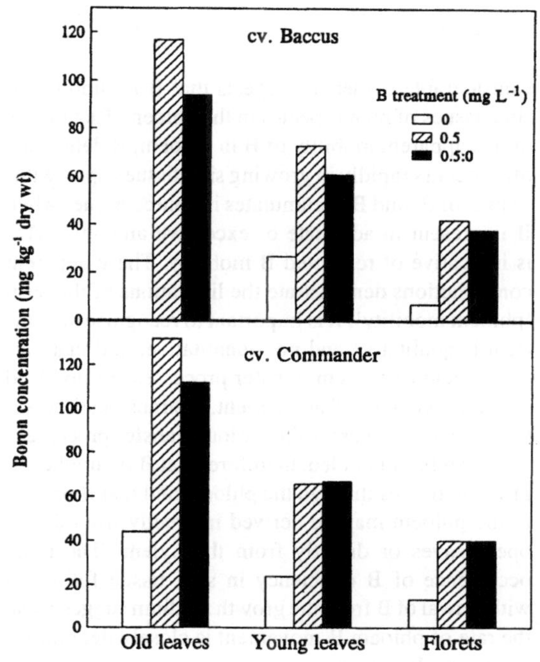

Figure 4. Boron concentrations of different parts of two broccoli cultivars grown to commercial maturity in a greenhouse. In the 0.5:0 treatment, the B supply was removed at initiation of inflorescence development. Based on Shelp et al. (1992a).

Genetic variation

Shelp and Shattuck (1987a) conducted a detailed comparison under controlled environmental conditions of two rutabaga cultivars, Laurentian (most widely grown North American cultivar) and Wihelmsberger (a European cultivar with low susceptibility to B deficiency; Shelp and Shattuck, 1987a). With sufficient B, the ratio of tissue B in storage root to old leaves is considerably less than the ratio for nitrogen and higher than that for calcium. Removal of B supply when roots are 1.0 to 1.5 cm in dia. does not markedly affect root yields, but does increase the B ratio by 110% in Laurentian and 190% in Wihelmsberger. Thus, the lower susceptibility of Wihelmsberger to B deficiency is attributed to its greater capacity for retranslocation of B to developing sinks from source leaves.

Another study determined the relative susceptibility of four broccoli cultivars including Commander, which possesses low susceptibility to the hollow stem disorder (Shelp et al., 1992a). In all cultivars, B retranslocation within the shoot is implicated from high tissue B in florets and young leaves, sinks which develop after the B supply is interrupted (Figure 4). Commander is the only cultivar that has tissue B in sinks equivalent to those in plants receiving sufficient B continuously. This is accompanied by a small but significant decline in tissue B of old leaves. Furthermore, in response to continuous B at a deficient level, the tissue B of florets from Commander is highest among the cultivars and shows the lowest percent decline relative to the controls. The stability of floret B concentration supports the view that Commander has a greater retranslocation capacity than the other cultivars examined. It is noteworthy that the authors were careful to compare the broccoli cultivars at similar developmental age. Earlier, Benson et al. (1961) provided evidence with another cultivar (Puyallup Ploycross 2) for the retranslocation of B from all except the basal leaves and youngest tissues. Together, these studies provide convincing evidence for a genetic basis for the ability of plants to retranslocate B during the reproductive growth phase and suggests that further research on the impact of plant development is required.

Summary

Considerable evidence suggests that B is supplied to sink tissues of many species in the phloem. Regardless of this apparent mobility of B in phloem, B deficiency often occurs rapidly in growing sink tissues upon withdrawal of B, and B accumulates in source tissues when B is present in adequate or excessive amounts. This is indicative of restricted B mobility. These apparent contradictions demonstrate the limitations of the term 'phloem mobility'. It is important to recognise that this term is qualitative and not quantitative, and that various xylem to phloem transfer processes are involved in the movement of an element. Retranslocation is a term that encompasses the various transfer processes.

It can be concluded, therefore, that B is supplied to growing tissues through the phloem and that B present in the phloem may be derived indirectly from developed leaves or directly from the xylem. The rapid occurrence of B deficiency in sink tissue following withdrawal of B from the growth medium suggests that the rate of phloem B movement is closely determined by current B uptake.

Dynamics of boron in species with significant boron mobility

It has recently been demonstrated that B exhibits rapid and significant phloem mobility in species for which sorbitol is a primary photosynthetic product (Brown and Hu, 1996). This conclusion was made on the basis of our experimentation, a review and reanalysis of previous work and observations drawn from plants growing in natural environments. Brown and Hu (1996) suggested that B should be phloem mobile in any species for which sorbitol, mannitol, or dulcitol are the primary photosynthates since these polyols can effectively complex B. Subsequently, this hypothesis was verified through the isolation and characterisation of B–polyol complexes from celery (*Apium graveolens*) and peach (*Prunus persica*) (Hu et al., 1997).

In the following, we review the information that led to the identification of significant species differences in phloem B mobility. This will be followed by a discussion of the physiological and biochemical mechanism of B mobility, and a description of species and environmental conditions in which this mechanism is likely to occur. The implications of B mobility for the selection and development of B efficient species and the diagnosis and correction of B disorders will then be considered.

Historic evidence for phloem boron mobility

Distribution of boron within leaves

In their classic studies, Kohl and Oertli (1961) and Oertli and Kohl (1961) demonstrated that "the patterns of B distribution in plants, and symptoms of B toxicity, are correlated with leaf venation and are consistent with the hypothesis that B moves with the transpiration stream". Further, they concluded that "once B enters a leaf it remains, with little or none moving elsewhere" (Oertli and Richardson, 1970). Thus, in a species in which B is immobile, B will accumulate at the sites of termination of leaf veins.

This principle is illustrated in Figure 1 in which the distribution of B within a mature leaf of walnut (*Juglans regia*) and apple (*Malus domestica*) is contrasted. Both species were of similar age and were growing adjacent to each other under identical management and environmental conditions. In walnut, the highest B accumulation occurred at the leaf tip and leaf margin. Boron concentration at the leaf base was less than 14% of the tip value while the petiole represented only 2% of the leaf tip value. Walnut represents a striking example of B immobility. There are many examples of species with B distributions similar to those present in walnut, while, to our knowledge there are no published exceptions to this rule (Asen and Davidsen, 1950; Eaton, 1944; Kohl and Oerli, 1961; Oertli, 1960, 1993, 1994).

Leaves of apple are distinctly different from those of walnut (Figure 1). Leaf B concentrations were significantly lower than in walnut (40 mg B kg^{-1} dry wt in contrast to 600 mg B kg^{-1} dry wt leaf average) and there was very little difference in B accumulation across the leaf. This uniform low distribution of B in apple does not correlate with leaf venation pattern and is not consistent with the hypothesis that B distribution is determined solely by transpiration. This same leaf distribution was observed in almond (*Prunus amygdalus*), peach, and plum (*Prunus salicina*) (Hu and Brown, unpublished data).

Boron distribution in plant tissues

Evidence of B mobility can also be found in the distribution of B within different organs of a species. For example, under field conditions pistachio (*Pistacia vera* cv. Kerman) had highest B concentration in

Table 2. Boron distribution (mg B kg^{-1} dry wt) in leaf and fruit organs of four tree species. Samples were collected on 23 Sept., 1994. Values are means ±SE of three replicates

Organ	Almond (*Prunus amygdalus*)	Apple (*Malus domestica*)	Pistachio (*Pistacia vera*)	Walnut (*Juglans regia*)
Leaf	42 ± 1	41 ± 2	130 ± 8	295 ± 23
Hull	170 ± 19	51 ± 2*	33 ± 1	40 ± 1
Kernel	43 ± 1	54 ± 1†	1 ± 0	4 ± 3
Shell	34 ± 2	34 ± 1#	2 ± 1	9 ± 3

Note: * = peel, † = core, # = pulp.

leaves and lowest in fruit and seed tissue (Table 2). In contrast, almond grown at the same site, had highest B concentration in hull (fruit tissue) with much lower B in the leaves (Table 2). The same contrast can be seen between apple and walnut (Table 2). Similar results were observed by Eaton et al. (1941) who reported 1300 mg B kg^{-1} in leaves of walnut, 123 mg B kg^{-1} in the hull and 14 mg B kg^{-1} in the kernel, while leaves of apricot (*Prunus armeniaca*) (grown at the same site) contained 118 mg B kg^{-1}, fruit flesh had 441 mg B kg^{-1} while the kernel contained 85 mg B kg^{-1}.

The distribution of B seen in pistachio and walnut are indicative of limited phloem B mobility in these species, whereas B is clearly phloem mobile in almond and apricot. Examples of apparent B mobility based upon the presence of relatively high B concentrations in fruit and flower tissue can be found for pear, apple, cherry (*Prunus* sp.) (Crandall et al., 1981; Eaton, 1944; Woodbridge et al., 1971), apricot (Dye et al., 1983) as well as celery (Eaton, 1944) and olive (*Olea europa*) (Delgado et al., 1994).

The concentration of B in leaves of different age within a species can also provide evidence of B mobility. The authors examined seven species growing in close proximity under similar soil and environmental conditions. The occurrence of higher B concentrations in old or mature leaves in comparison to younger leaves is evidence of B immobility while higher B concentrations in younger leaves is an indication of B mobility since these leaves have transpired less water than older leaves. Boron concentrations in fruit or apical tissues that equal or exceed leaf B concentrations also indicate phloem B mobility (van Goor and van Lune, 1980). The results presented in Table 1, suggest that B is phloem mobile in species in which sorbitol is a primary photosynthetic product (*Eriobotrya deflexa* and *Prunus elysifolia*). Species known to produce mannitol as a primary photosynthetic product [celery, fragrant olive (*Osmanthus fragrans*) and *Phillyrea latifolia*] were more variable with some clearly transporting B (celery and fragrant olive) while others did not (*P. latifolia*). The relationship between sugar alcohols and B mobility will be addressed below.

Boron toxicity symptoms

The occurrence of B toxicity symptoms in leaf margins of old leaves has long been interpreted as an indication of the immobility of B in plants (Oertli, 1993). Nevertheless, as early as 1941 evidence was available to suggest that some plant species did not express this symptom of B toxicity (Eaton et al., 1941). Eaton observed that in stone fruit trees such as prune (*Prunus domestica*) and apricot, the typical symptoms of B toxicity are absent. Unusual symptoms of B toxicity in the *Prunus*, *Malus* and *Pyrus* species have also been observed by many subsequent authors. In almond and prune (Hansen, 1948, 1955), pear, apple and cherry (Woodbridge, 1955), peach (Kamali and Childers, 1970), apricot (Dye et al., 1983), and almond, peach, plum and hybrid species (almond × peach) (El-Motaium et al., 1994), a lack of leaf symptomology in plants suffering from B toxicity has been observed. Instead of the marginal leaf burn so typical of B toxicity, these species exhibit B toxicity as tip die back in young shoots, profuse gumming in the leaf axil and the appearance of brown corky lesions along stems and petioles. In apple, apricot and pear, bud abscission and death is also observed (Crandall et al., 1981; Dye et al., 1983; Hansen, 1974). Other researchers also reported similar findings and have shown that B does not accumulate in leaves of these species to any significant extent even when trees are grown at excessive levels of B (El-Motaium et al., 1994; Hansen, 1955; Woodbridge, 1955). Further comparison of the symptoms of B toxicity is presented in Chapter 12 of this volume.

The occurrence of these 'unusual' symptoms of B toxicity are not however restricted solely to the members of the *Prunus*, *Malus* and *Pyrus* families. Francois and Clark (1979) studied B tolerance of 25 species of ornamental shrubs in sand culture. More than 20 of the species showed the typical symptoms of B toxicity, including marginal or tip leaf burn and/or leaf drop. Of the species that do not exhibit leaf burn only the growth of wax-leaf privet (*Ligustrum japonicum*) was inhibited by the B concentrations used. In this species, B toxicity is expressed as terminal stem death with no leaf burn. The symptoms described by Francois and Clark closely resemble those described for *Prunus* by El-Motaium et al. (1994). In a subsequent study, Francois (1988) contrasted the response of lettuce (*Lactuca sativa*) and celery to toxic levels of B in the medium. Whereas leaf margin and tip burn is evident in the oldest leaves of lettuce, in celery these symptoms are absent. Rather, celery responds to B toxicity by producing deformed young leaves, 'bitter' and misshapen stems.

Movement of foliar applied boron

If B is immobile in a species then application of foliar B fertilisers will result in enrichment of the treated leaf but will not result in enhanced B content of leaves formed after treatment or of tissues supplied primarily by the phloem (e.g. fruit). Evidence that foliar B is not readily transported to other tissues is readily available as has been discussed above (also Brown and Hu, 1996). Foliar B applications, however, have long been known to be an effective means of enhancing bud and flower B concentrations resulting in increased fruit set and yield in *Prunus*, *Malus* and *Pyrus* species (Batjer et al., Callan et al., 1978; Chaplin et al., 1977; Hanson, 1991b; Hanson et al., 1985; Johnson et al., 1995, and references therein).

The application of B in autumn results in enhanced B concentrations in buds and flowers in the subsequent spring. In plum, Hanson and Breen (1985) determined that dormant buds contain about 18% of the B present in senescent leaves, the remainder of this B is supplied from bark and wood of nearby branches. Boron applied to leaves of these trees in autumn apparently increases pools of B in these subtending tissues and plays a critical role in supplying the B requirements of flowers in the spring. In almond, B applied in autumn (1.7 kg B ha^{-1}) is the most effective timing of application and generally results in the highest increases in flower B concentrations in the following spring when

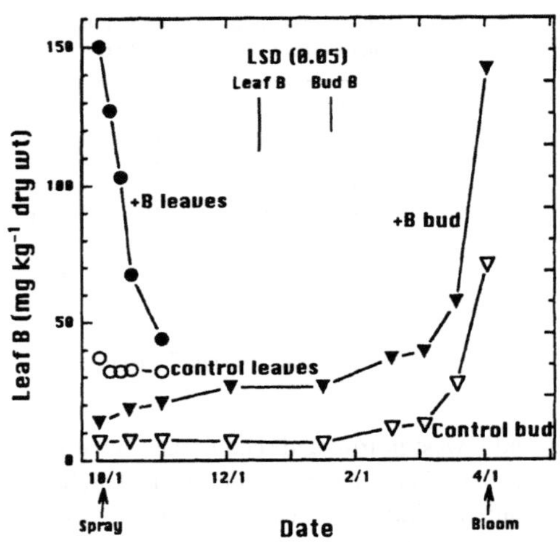

Figure 5. Changes in B levels in flower buds and leaves of 'Italian' prune following 500 mg L^{-1} B sprays applied on 3 Oct., 1978. Redrawn from Hanson et al. (1985).

compared to sprays applied at other times (Nyomora et al., 1997). Similar results have been observed for a number of *Prunus* species as well as *Malus* and *Pyrus* (Hanson, 1991a, 1991b). In most *Prunus* species, flowering precedes significant soil nutrient uptake; hence, nutrient demands for the critical periods of flowering and seed set are supplied from storage.

The mobility of autumn applied foliar B is illustrated in Figure 5 for prune. Autumn B application (October) increases leaf B concentrations 400% within 24 h of B application; over the subsequent 30 days this B concentration decreases dramatically, such that at leaf senescence (November) both treated and untreated leaves have similar B concentrations. During this period, and continuing through to early bud swell (February), buds on branches that received autumn foliar B have significantly higher B concentrations than control branches. As buds swelled during early spring, and up until flowering (April), this difference in bud B concentrations increases such that at flowering, flowers on treated branches have B concentrations 200% higher than control flowers.

The most compelling and detailed information demonstrating rapid and extensive mobility of foliar applied B followed from the introduction of isotopically enriched B isotopes into plant physiological research in the early 1990's (Brown and Hu, 1996; Brown et al., 1992; Hanson, 1991a, 1991b; Picchioni et al., 1995).

Using foliar application of ^{10}B, Hanson (1991b) reported a rapid export of isotope from the spur leaves of sour cherry (*Prunus cerasus*) 3 to 9 days after treatment, the largest concentration of isotope is subsequently found in buds subtending the treated leaves. The possibility of earlier export was not evaluated in that study. On the basis of the results of Hanson and co-workers, Picchioni et al. (1995) conducted a series of experiments on the solution retention and the kinetics of uptake and export of foliar applied, labelled B by apple, pear, and sweet cherry (*Prunus avium*) leaves. Foliar uptake of labelled B is 88–96% complete within 24 h of application. Uptake proceeds to near completion within 24 h and is limited solely by the amount of B which can be retained by the leaf surfaces. Export, in turn, is closely related to the quantity of B absorbed and is 50% complete within 6 h of B application. This rapid rate of B uptake and export is equal to or greater than the rates observed for foliar urea (Klein and Weinbaum, 1985). This is likely to be due to similarities in physiochemical properties of B (as boric acid) and urea (urea and boric acid are both non-charged and are of similar molecular weight).

In these experiments, the sink demand of nearby, mature apple fruit does not affect the rate of export of labelled B from adjacent spur leaves, but the fruit imports 16% of their total B from the solution applied during a 10 day period. Picchioni et al. (1995) also detected significant quantities of ^{10}B in the phloem sap of apple and prune 2 h after foliar ^{10}B application, further suggesting that B is transported in the phloem.

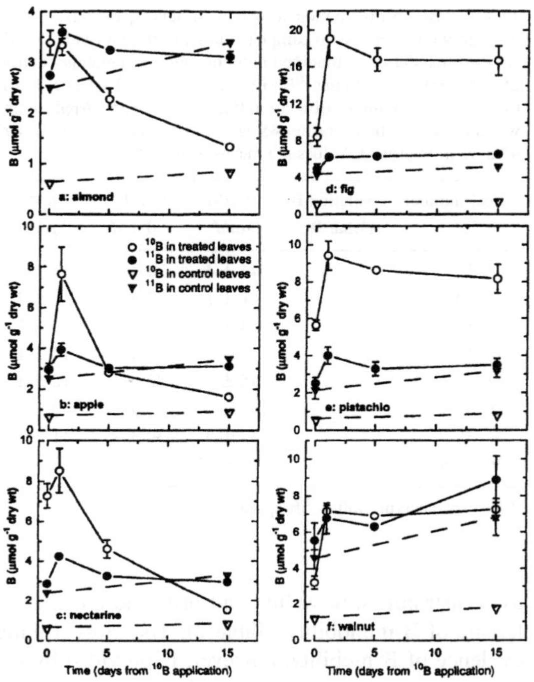

Figure 6. Changes in leaf B concentration during the 2 weeks following application of ^{10}B (50 m*M*) to leaves of six tree species. Treatments were applied on 19 May, 1995. Each point is a mean of three replicates ± SE. Where smaller than the symbol, SE is not shown. Reproduced from Brown and Hu (1996) with permission.

Phloem mobility of boron is species dependent

Despite the evidence provided by the preceding experimental results, that B is phloem mobile in *Prunus*, *Malus* and *Pyrus*, it is not evident from these studies that this is a general phenomenon or was merely a consequence of foliar B application. It was not until recently, that Brown and Hu (1996) proposed that the occurrence of significant phloem mobility of B is species dependent. Based upon the movement of isotopic B and a review of existing chemical data (Makkee et al., 1985 and references therein), they hypothesised that the apparent mobility of B in *Prunus*, *Malus* and *Pyrus* is a consequence of the use of sorbitol as a primary translocated photosynthate in these species.

Experimental verification of this hypothesis was provided in a series of experiments involving foliar and solution applied B isotopes and naturally acquired (soil) B. Three species in which sorbitol is a predominant photosynthate [apple, almond and nectarine (*Prunus persica* var. *nectarina*)] and three in which sorbitol is not known to occur [pistachio, walnut and fig (*Ficus carica*)] were contrasted. In sorbitol rich species, foliar applied ^{10}B is rapidly (0–15 days) translocated out of treated leaves to adjacent fruit. In the sorbitol poor species, by contrast, there is no evidence of ^{10}B translocation out of the leaf even though leaf B concentrations are increased by 200–500% in leaves receiving foliar B (Figure 6). Whereas foliar applied ^{10}B is rapidly and almost completely translocated out of the leaves of sorbitol rich species within 15 days, only a small percentage of the ^{11}B present in the leaf at the time of foliar application is retranslocated.

These results, though verifying the phloem mobility of foliar applied B, do not necessarily demonstrate that B, acquired from the soil, is phloem mobile. Evidence to support the notion that B is phloem mobile, irrespective of its original source, is available. The B distribution within organs, that is commonly observed in the sorbitol rich species, almond and apple, clearly suggest that soil acquired B is phloem mobile, where-

Table 3. Boron distribution in leaves of apple seedlings. Plants were grown for 36 days in complete nutrient solution (with 25 μM B), then transferred to modified nutrient solution containing either 25 μM B or 0 μM B. Foliar ^{10}B (50 mM) was applied to leaves 6–10 to half of the seedlings growing in 0 μM B treatment. Application was made every three days for 45 days. Plants were harvested after 49 days of treatment. Values are means ±SE of two replicates.

Leaf number	Solution B Foliar B	25 μM 0 mM	0 μM 50 mM	0 μM 0 μM
1–5		44 ± 4	28 ± 4	18 ± 1
6–10		41 ± 1	37 ± 2*	15 ± 2
11–15		44 ± 2	22 ± 0	12 ± 1
16–20		48 ± 1	22 ± 0	10 ± 1
21–25		45 ± 3	22 ± 1	12 ± 1
26–30		49 ± 6	25 ± 2	13 ± 2
31–35		47 ± 3	30 ± 1	11 ± 3
Meristem		73 ± 12	66 ± 31	17 ± 3

* Leaves receiving foliar ^{10}B application.

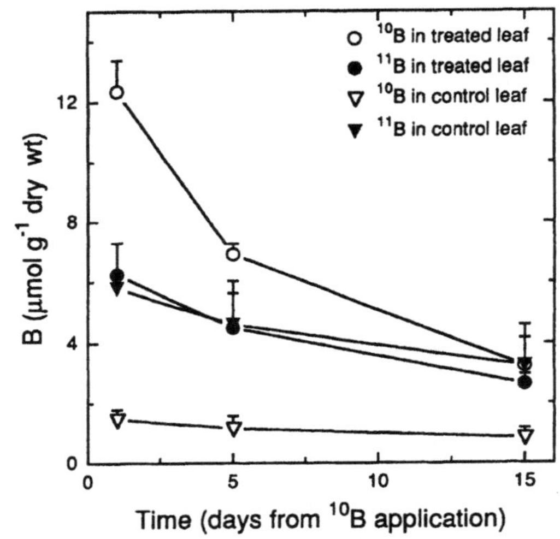

Figure 7. Changes in celery leaf B concentration following application of 50 mM ^{10}B enriched boric acid at time 0. Each point is a mean of two replicates ± SE. Reproduced from Hu et al. (1997) with permission.

as B distribution in walnut and pistachio are characteristic of B immobility (Table 2). The most striking evidence of B mobility, however, is provided by the growth of apple seedlings in solution cultures with either B supplied in the solution or as a foliar application to a limited number of mature leaves. Using this approach, Brown and Hu (1996) demonstrated that following removal of B from the rooting medium there is a gradual decline in B content of mature leaves and an apparent transfer of this B to maintain the growth of meristematic tissues. Only, when the majority of the leaves decline in B concentration to below 12 mg kg^{-1} dry wt are signs of B deficiency observed in root tissues. This does not occur until 4–6 weeks after removal of B from the root medium. Foliar B supply to leaves 6–10 (in the absence of B in the root medium), effectively maintains plant growth at control levels for the duration of the experiment (seven weeks) with no development of B deficiency symptoms in either roots or shoots (Table 3).

The ability of apple to grow normally in the absence of B in the root medium clearly demonstrates that B is freely phloem mobile in apple. Movement of B in the phloem occurs irrespective of the B source (soil, solution or foliar). Boron also appears to be phloem mobile in all species in which sorbitol is a primary translocated photosynthate and it has been suggested that B is also phloem mobile in any species in which the B binding polyols (sorbitol, mannitol or dulcitol) occur (Brown and Hu, 1996). Evidence to support phloem B mobility in other species is presented in Figure 7, which demonstrates the rapid export of foliar applied ^{10}B in celery. The distribution of B within tissues of various sorbitol and mannitol producing species further supports this hypothesis (Table 1). Boron is also freely phloem mobile in dulcitol producing species (results not shown).

Isolation and characterisation of soluble boron-polyol complexes from leaf tissue and phloem sap of celery and peach

Experimental verification of the hypothesis of Brown and Hu (1996) has recently been provided (Hu et al., 1997; Penn et al., 1997), utilising matrix assisted laser desorption Fourier transform mass spectrometry (MALDI-FTMS). This technique allows for the direct determination of mass and charge of molecules in liquid samples collected from plant tissues. This technique does not require any sample pretreatment and the vaporisation technique utilised does not disrupt the integrity of small molecular weight biological compounds. MALDI-FTMS has a mass resolution of 0.000001 atomic mass units which, in combination with known isotopic distributions, allows for the absolute identification of essentially all complexes with a molecular weight of less than 1000.

MALDI-FTMS analysis of celery phloem sap and vascular exudate as well as sap collected from phloem fed nectaries of peach, demonstrates that B forms sta-

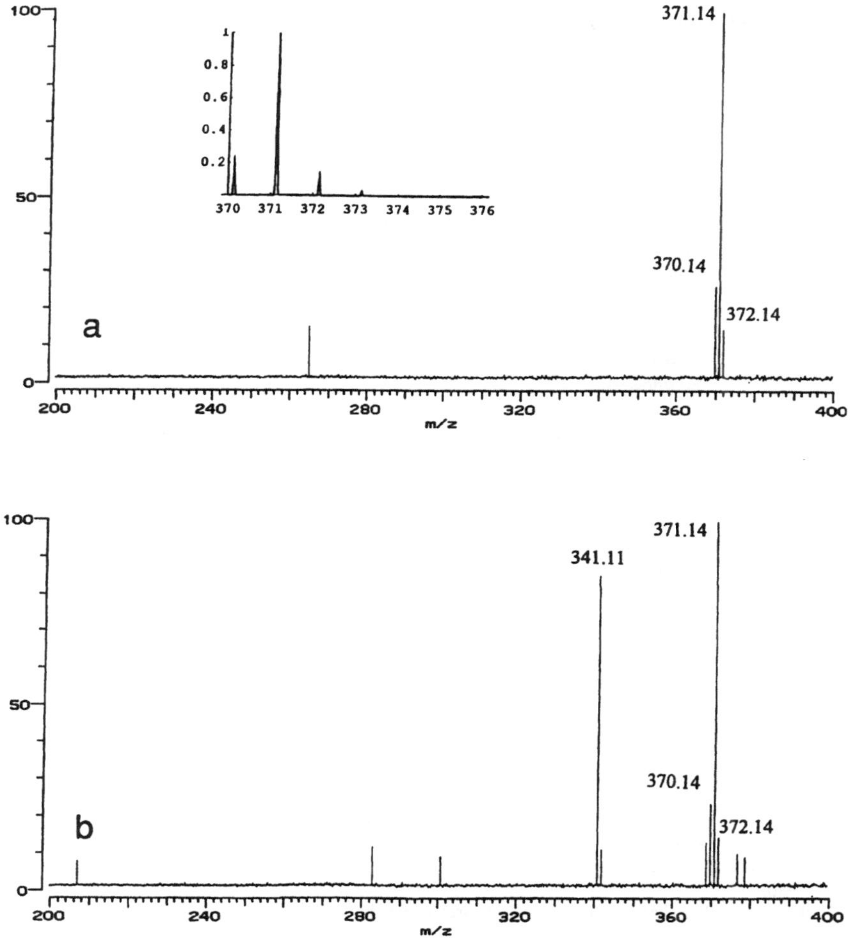

Figure 8. Spectra of matrix assisted laser desorption -Fourier transform mass spectrometry (MALDI-FTMS: (a) 10 m*M* mannitol/1m *M* boric acid. Insert: Predicted peak height of mannitol–B–mannitol complex based upon relative isotopic abundance: (b) celery phloem sap. Mass to charge ratio of B complexes is identified as following: 370.14, mannitol-^{10}borate-mannitol; 371.14, mannitol-^{11}borate-mannitol; 372.14, mannitol-^{11}borate-mannitol which contains one ^{13}C; 341.11, deprotonated disaccharide (sucrose). Reproduced from Hu et al. (1997) with permission.

ble complexes with the ligands (Lg) mannitol, sorbitol and fructose (Hu et al., 1997). A typical mass charge spectrum and peak allocation is presented in Figure 8. Presuming normal concentrations of both B (10 $\mu M - 1$ m*M*) and Lg (10–500 m*M*), and based upon the theoretical calculations of Makkee et al. (1985) it is predicted that essentially all B is present as the BLg_2 complex. This is verified by MALDI-FTMS. Additional peaks are also present at higher masses corresponding to various polymeric combinations of B with the predominant ligand i.e. B_2Lg_3, B_3Lg_4 etc. (Penn et al., 1997). No free B is observed in any collected sample, and this is predicted given the ratio of B to ligand in these samples (Makkee et al., 1985).

The distribution of polyols in higher plants

The demonstration of the occurrence of phloem mobile B–polyol complexes clearly provides a mechanistic explanation for the phloem mobility of B in celery and peach. Polyols are also known to be significant in other *Prunus*, *Malus* and *Pyrus* species as well as members of the Apiaceae [celery, carrot (*Daucus carota*)],

Table 4. List of economically important plants reported to produce polyols

Sugar alcohol[1]	Genus/species		
Dulcitol	*Catha*	*Euonymus*	*Maytenus*
	Celastrus		
Mannitol	*Allium* (onion)	*Coffea arabica* (coffee)	*Olea europaea* (olive)
	Apium graveoleus (celery)	*Daucus carota* (carrot)	*Phaseolus vulgaris* (bean)
	Asparagus (asparagus)	*Foeniculum vulgare* (fennel)	*Pisum* (pea)
	Brassica (cabbage, cauliflower)	*Fraxinus* (ash)	*Punica granatum* (pomegranate)
Sorbitol	*Crataegus monogyna* (English hawthorn)	*Prunus amygdalus* (almond)[2]	*Prunus salisina* (plum)[3]
	Cydonia oblonga (quince)	*Prunus armeniaca* (apricot)	*Prunus serotina* (black cherry)
	Eriobotrya deflexa	*Prunus avium* (cherry)[2]	*Pyrus communis* (pear)[2]
	Eriobotrya japonica (loquat)	*Prunus domestica* (prune)[2]	*Pyrus pyrifolia* (Asian pear) pear)[3]
	Heteromyles arbutifolia (Christmas berry)	*Prunus elysiforlia*	*Pyracantha buxifolia*
	Malus domestica (apple)[2]	*Prunus glandulosa* (flowering almond)	*Vauquelinia californica*
	Osteomeles enthyllidifolia	*Prunus persica* (peach)	
	Osteomeles schwerinae	*Prunus persica* var. *nectarina* (nectarine)[2]	

[1]Information on sugar alcohols from Bieleski (1982), Bourne (1958), Plouvier (1963), Wallaart (1980).
[2]Boron mobility already demonstrated by Brown and Hu (1996), Hanson (1991a), Picchioni et al. (1995).
[3]Predicted to contain sorbitol from Bieleski (1982).

Oleaceae [olive and ash (*Fraxinus*)] and Celestraceae (*Celastrus* and *Euonymus*).

Sorbitol is characteristic of the woody Rosaceae, including members of the economically important genera, *Malus*, *Pyrus* and *Prunus* (peach, nectarine, cherry, plum and apricot) where it accounts for 60–90% of the carbon exported from the leaf (Loescher, 1987; Zimmermann and Ziegler, 1975). Mannitol is found in many important members of the family Scrophulariaceae [snapdragon (*Antirrhinum majus*)], Apiaceae [carrot, celery, caraway (*Carum* sp.), fennel (*Foeniculum vulgare*) and parsley (*Petroselinum crispum*)], Rubiaceae [coffee (*Coffea arabica*) and gardenia (*Gardenia augusta*)] and Oleaceae (olive and ash) (Bieleski, 1982; Bourne, 1958). Dulcitol is found in some species of the Lauraceae and Celastraceae (Bourne, 1958). Low and variable quantities of polyols, however, may be present in many species (Bieleski, 1982). A list of economically important species thought to produce polyols is presented in Table 4.

Indications that B is mobile in these species and families was presented earlier (Table 2). Currently, our knowledge of the occurrence and physiology of polyols is insufficient to predict the extent of B mobility in agricultural species. It is also likely that there are significant environmental and phenological effects on polyol production which might also influence B mobility. For example, Delgado et al. (1994) provided evidence to suggest that foliar applied B is retranslocated out of leaves of olive at only certain times of the year. This may be interpreted as evidence that mannitol is not always the primary translocated photosynthetic

Table 5. Comparison of B uptake (mg kg^{-1}) between wild type tobacco (*Nicotiana tobacum*) 'Strain SR1' and transgenic tobacco 'Strain S11' (genetically engineered with sorbitol dehydrogenase gene). Plants grown for three weeks with 3/4 strength Hoagland solution. ^{10}B was supplied as 95.91% enriched (atom%) boric acid to the roots. Values are means of 4 observations

Treatment ^{10}B	^{10}B in leaf		^{10}B in root	
(m*M*)	Wild type	Transgenic	Wild type	Transgenic
0.01	25	28	39	45
0.1	27	32	37	70
1	27	69	41	204

product of olive or that mannitol is converted to another sugar form prior to loading of the carbohydrate into the phloem. Recently it was demonstrated that stachyose, not mannitol, is the primary transported photosynthate in olive (Flora and Madore, 1993). If a significant fraction of the mannitol is converted to stachyose in the phloem companion cells, then B transport would be limited since stachyose does not complex B and is not expected to facilitate B transport. The extent of this interconversion may be environmentally or phenologically determined.

Summary

The research described above demonstrates that accepted dogma regarding the mobility of B in plants is incorrect for a broad range of important agricultural species. The evidence provided demonstrates that B is transported as a complex with polyols in species in which polyols are the primary photosynthetic product. Species that produce polyols include many important crop and forest species; however, we have insufficient information to fully determine the extent of polyol distribution in higher plants. Further, it is likely that environmental and phenological factors would influence the production and distribution of polyols and hence B mobility.

Impact of boron mobility on breeding and selection of crops for improved tolerance of boron toxicity and deficiency

Adequate B supply is essential to reproductive processes and the requirement for B in reproductive growth may be higher than the B requirement for vegetative growth (Nyomora et al., 1997). Thus, B deficiency may reduce fruit and seed set even when B concentrations in the tissue are adequate for optimal vegetative growth. As a result of the general immobility of B that is characteristic of most species, plants require a constant supply of B during all phases of plant growth. Even a short term decrease in B availability (as a result of drought, etc.) can influence plant growth. If B deficiency occurs during critical reproductive stages very significant crop losses can occur.

The short duration and variable nature of many B deficiencies makes prediction and correction of B deficiency difficult. Breeding and selection for tolerance to B deficiency (specifically short term deficiencies) has not been effective since (1) we did not understand the physiology of B movement and (2) the variable occurrence of B deficiency greatly complicates field screening techniques. The polyol transporting species we have identified are immune to short term deficiencies of B by virtue of their ability to retranslocate B within the plant. The selection or genetic manipulation of plants for improved ability to retranslocate internal B by increasing the production of polyols may, therefore, result in cultivars that are resistant to short term B deficiency.

Recently, Bellaloui and Brown (1997) demonstrated that tobacco plants genetically engineered to produce sorbitol have a significantly greater rate of B uptake than control plants. Increases in root and shoot B concentration of greater than 200% are observed (Table 5) suggesting that the expression of sorbitol is sufficient to affect B metabolism. It is significant that sorbitol concentrations in these early transformants are only relatively low (100–700 nmole g^{-1} fresh wt) which is approximately 0.1% of that present in species such as *Prunus*. Nevertheless, this amount of sorbitol is sufficient to influence B uptake. Interestingly, increasing B concentrations in the medium increased the production of sorbitol by these transgenic plants. One may speculate that the formation of the sorbitol–B complex releases end product feedback facilitating additional sorbitol production. This has not been verified.

It may not, however, be necessary to genetically engineer species for enhanced polyol production to enhance B mobility. Many families of plants, including the economically important Brassicaceae are reported to have low but variable mannitol production (Bourne, 1958). It is plausible that *Brassica* genotypes with relatively high mannitol concentration may be selected from within wild or improved genotypes. Increased mannitol production may, in turn, be associated with B mobility and hence tolerance of short term B deficiency. This has not been experimentally verified.

Impact of boron mobility on management of boron in crops

Diagnosis of plant boron status

The mobility of B in polyol transporting species influences B diagnosis and correction. Currently practiced sampling techniques and symptom descriptions are based on the premise that B is not mobile. Selection of tissue samples and determination of critical nutrient concentrations are all fundamentally dependent on the phloem mobility of B. Table 1 illustrates that B does not accumulate in the older leaves of species in which B is mobile. Thus, old leaves are an inadequate tissue for determination of B toxicity as has been observed by El Motaium et al. (1994) and others (discussed previously). The B concentration of fruit tissues may be a superior indicator of B status in these species since B accumulates readily in these tissue (Table 2). This observation has led to the widespread use of hull B as a determinant of B status in almond in California (Nyomora et al., 1997).

Even within species that do not exhibit a high degree of B retranslocation the use of a recently matured leaf is inappropriate because the xylem has provided a significant portion of the B present in these leaves. The B concentration of a developed leaf may not, therefore, reflect the B status of growing tissues for which a constant B supply is most critical. Since tissue expansion is one of the first processes influenced by B deficiency it is clear that B diagnosis must reflect current B availability, this can only be achieved by sampling growing tissues. This is an inherently difficult and inconsistent process. By contrast, mature leaves are appropriate for assessing B deficiency in B mobile species (both annual and perennial) especially if remedial action is required. This issue is discussed in greater detail in Chapter 10 of this book.

Improved fertilisation strategies

The management of B fertilisation in plants is greatly influenced by patterns of B mobility. In general, foliar application of B has not been widely used in agriculture as it was believed that the immobility of B would limit its effectiveness. The mobility of foliar applied B in polyol transporters suggests that foliar application of B in those species can be used effectively at any time functional leaves are present to correct B deficiency, and to supply B to flower and fruit tissue (Brown and Hu, 1996; Nyomora et al., 1977; Picchioni et al., 1995). Since B deficiency dramatically affects fruit set and quality in many species this is a finding of significant importance.

The significant response to foliar B application on fruit set that has been observed in numerous tree species may be in part due to the limited mobility of B to the opening flowers. This may even occur when plants have adequate B for vegetative growth (Hanson, 1991b; Nyomora et al., 1997). In species in which B is phloem mobile, the enrichment of bud B concentrations can be ensured by supplying foliar B at a time when leaves are still photosynthetically active and buds represent the primary carbohydrate sink. This occurs after fruit harvest and prior to significant leaf senescence in early autumn. In species in which B is immobile, foliar B applications should be made directly to the swelling bud in the spring time. Earlier B applications may not be absorbed and may be leached in winter rains; soil B applications though potentially effective, may not be available to the plant at the critical stage of bud formation or expansion.

The demonstration that foliar B sprays can supply B to broccoli florets also suggests the utility of foliar B sprays (discussed in "Dynamics of boron in species with restricted boron mobility"). The amount of B supplied in this manner is, however, limited and the effectiveness of the spray is short lived. This approach is of most use in high value crops in which the reproductive structure is the organ of economic value.

Summary

Boron is unique amongst all essential mineral elements in that species differ dramatically in their ability to retranslocate B within the plant. This phenomenon fundamentally alters the manner in which B nutrition is managed and provides a direction for the selection of genotypes resistant to B deficiency. The great diversity that exists in the expression of B toxicity and the distribution of B between plant parts provides a striking example of the role of phloem mobility in nutrient management.

Acknowledgments

Patrick H Brown would like to thank the United States Department of Agriculture (#9601359) and U S Borax for financial support. Research on B in the laboratory of BJS has been supported by grants from US Borax, the Ontario Ministry of Agriculture and Food, and Agriculture Canada/Natural Sciences and Engineering Research Council of Canada Partnerships Program.

References

Al-Molla R M M 1986 Some physiological aspects of soybean development and yield as affected by boron fertilisation. Diss. Abstr. Int. B 46, 3268.

Anon. 1991 Boron in Agriculture. *In* Fertilizer Borate Guide. Ed. V M Shorrocks. pp 1–16. US Borax, Rosemont, IL, USA.

Asen S and Davidson O W 1950 The boron distribution in greenhouse rose plants. Proc. Amer. Soc. Hort. Sci. 56, 433–437.

Batjer L P, Rogers B L and Thompson A H 1953 "Blossom blast" of pears: an incipient boron deficiency. Proc. Amer. Soc. Hort. Sci. 62, 119–122.

Bellaloui N and Brown P H 1997 The relationship between sugar alcohols and boron uptake in tobacco and celery. Plant Physiol. (submitted).

Benson N R, Degman E S and Chmelir I C 1961 Translocation and re-use of boron in broccoli. Plant Physiol. 36, 296–301.

Bieleski R L 1982 Sugar alcohols. *In* Encyclopedia of Plant Physiology, new series. V13A Plant Carbohydrates. I. Intercellular Carbohydrates. Eds F A Loewus and W Tanner. pp 158–192. Springer-Verlag, New York.

Bourne E J 1958 The polygydric alcohols. Acyclic polyhydric alcohols. Encycl. Plant Physiol. VI. Springer-Verlag, New York, USA pp. 345–361.

Bowen J E 1972 Effects of environmental factors on water utilization and boron accumulation and translocation in sugarcane. Plant Cell Physiol. 13, 703–714.

Brown P H and Hu H 1994 Boron uptake by sunflower, squash and cultured tobacco cells. Physiol. Plant. 91, 435–441.

Brown P H and Hu H 1996 Phloem mobility of boron is species dependent. Evidence for phloem mobility in sorbitol rich species. Ann. Bot. 77, 497–505.

Brown P H, Picchioni G, Jenkin M and Hu H 1992 Use of ICP-MS and ^{10}B to trace the movement of boron in plants and soil. Comm. Soil Sci. Plant Anal. 23, 2781–2807.

Callan N W, Thompson M M, Chaplin M H, Stebbins R L and Westwood M N 1978 Fruit set of 'Italian' prune following fall foliar and spring boron sprays. J. Amer. Soc. Hort. Sci. 103, 253–257.

Campbell L C, Miller M H and Loneragan J F 1975 Translocation of boron to plant fruits. Aust. J. Plant Physiol. 2, 481–487.

Chaplin M H, Stebbins R L and Westwood M N 1977 Effect of fall-applied boron sprays on fruit set and yield of 'Italian' prune. HortScience 12, 500–501.

Crandall P C, Chamberlain J D and Garth J K L 1981 Toxicity symptoms and tissue levels associated with excess boron in pear. Comm. Soil Sci. Plant Anal. 12, 1047–1057.

DaSilva M C and Shelp B J 1990 Xylem-to-phloem transfer of organic nitrogen in young soybean plants. Plant Physiol. 92, 797–801.

Delgado A, Benlloch M and Fernandez-Escobar R 1994 Mobilisation of boron in olive trees during flowering and fruit development. HortScience 29, 616–618.

Dye M H, Buchanan L, Dorofaeff F D and Beecroft F G 1983 Die-back of apricot trees following soil application of boron. New Zealand J. Exp. Agric. 11, 331–342.

Eaton F M 1944 Deficiency, toxicity, and accumulation of boron in plants. J. Agric. Res. 69, 237–277.

Eaton F M, McCallum R D and Mayhugh M S 1941 Quality of irrigation waters of the Hollister area of California. US Dept. Agr. Tech. Bull. 746, Washington, D.C., USA

El-Motaium R, Hu H and Brown P H 1994 The relative tolerance of six *Prunus* rootstocks to boron and salinity. J. Amer. Soc. Hort. Sci. 119, 1169–1175.

Flora L F and Madore M A 1993 Stachyose and mannitol transport in olive (*Olea europaea* L.) Planta 189, 484–490.

Francois L E 1988 Yield and quality responses of celery and crisp-head lettuce to excess boron. J. Amer. Soc. Hort. Sci. 113, 538–542.

Francois L E and Clark P H 1979 Boron tolerance of twenty-five ornamental shrub species. J. Amer. Soc. Hort. Sci. 104, 319–322.

Grabau L J, Blevins D G and Minor H C 1986 Stem infusions enhance methionine content of soybean storage protein. Plant Physiol. 82, 1013–1018.

Gupta U C 1993a Introduction. *In* Boron and its Role in Crop Production. Ed. U C Gupta. p 1. CRC Press, Boca Raton, FL, USA.

Gupta U C 1993b Deficiency, sufficiency, and toxicity levels of boron in crops. *In* Boron and its Role in Crop Production. Ed. U C Gupta. pp 137–145. CRC Press, Boca Raton, FL, USA.

Hansen C J 1948 Influence of rootstock on injury from excess boron in French (*Agen*) prune and President plum. Proc. Amer. Soc. Hort. Sci. 51, 239–244.

Hansen C J 1955 Influence of rootstock on injury from excess boron in Nonpareil almond and Elberta peach. Proc. Amer. Soc. Hort. Sci. 65, 128–132.

Hansen P 1974 The effect of boron upon leaf development and growth of the apple cultivar 'Cox's Orange Pippin'. J. Hort. Sci. 49, 211–216.

Hanson E J 1991a Movement of boron out of tree fruit leaves. HortScience 26, 271–273.

Hanson E J 1991b Sour cherry trees respond to foliar boron applications. HortScience 26, 1142–1145.

Hanson E J and Breen P J 1985 Effects of fall boron sprays and environmental factors on fruit set and boron accumulation in 'Italian' prune flowers. J. Amer. Soc. Hort. Sci. 110, 389–392.

Hanson E J, Chaplin M H and Breen P J 1985 Movement of foliar applied boron out of leaves and accumulation in flower buds and flower parts of 'Italian' prune. HortScience 20, 747–748.

Hu H and Brown P H 1994 Localisation of boron in cell walls of squash and tobacco and its association with pectin. Plant Physiol. 105, 681–689.

Hu H, Penn S G, Lebrilla C B and Brown P H 1997 Isolation and characterization of soluble B-complexes in higher plants. Plant Physiol. 113, 649–655.

Johnson F, Allmendinger D F, Miller V L and Dolley D 1955 Fall application of boron spray as a control for blossom blast and twig dieback of pears. Phytopath. 45, 110–114.

Kamali A R and Childers N F 1970 Growth and fruiting of peach in sand culture as affected by boron and a fritted form of trace elements. J. Amer. Soc. Hort. Sci. 95, 652–656.

Klein I and Weinbaum S A 1985 Foliar application of urea to almond and olive: leaf retention and kinetics of uptake. J. Plant Nutr. 8, 117–129.

Kohl H C Jr and Oertli J 1961 Distribution of boron in leaves. Plant Physiol. 36, 420–424.

Liu L, Shelp B J and Spiers G A 1993 Boron distribution and retranslocation in field-grown broccoli (*Brassica oleracea* var. *italica*). Can. J. Plant Sci. 73, 587–600.

Loescher W H 1987 Physiology and metabolism of sugar alcohols in higher plants. Physiol. Plant. 70, 553–557.

Makkee M, Kieboom A P G and van Bekkum H 1985 Studies on borate esters III. Borate esters of D-mannitol, D-glucitol, D-fructose and D-glucose in water. Recl. Trav. Chim. Pays-Bas. 104, 230–235.

Marentes E, Shelp B J, Vanderpool R A and Spiers G A 1997 Retranslocation of boron in broccoli and lupin during early reproductive growth. Physiol. Plant. (*In press*).

Marschner H 1995 Mineral Nutrition of Higher Plants, 2nd ed. Academic Press, New York. pp 379–396.

McIrath W J 1965 Mobility of boron in several dicotyledonous species. Bot. Gaz. 126, 27–30.

Nyomora A M S, Brown P H and Freeman M. 1997 Foliar applied boron increases tissue boron concentration and nut set of almond. J. Amer. Soc. Hort. Sci. (*In press*)

Oertli J J 1960 The distribution of normal and toxic amount of boron in leaves of rough lemon. Agron. J. 52, 530–532.

Oertli J J 1993 The mobility of boron in plants. Plant and Soil 155/156, 301–304.

Oertli J J 1994 Non-homogeneity of boron distribution in plants and consequence for foliar diagnosis. Comm. Soil Sci. Plant Anal. 25, 1133–1147.

Oertli J J and Kohl H C 1961 Some considerations about the tolerance of various plant species to excessive supplies of boron. Soil Sci. 92, 243–247

Oertli J J and Richardson W F 1970 The mechanism of boron immobility in plants. Physiol. Plant. 23, 108–116.

Oertli J J and Roth J A 1969 Boron supply of sugar beet, cotton and soybean. Agron. J. 61, 191–195.

Pate J S 1975 Exchange of solutes between phloem and xylem and circulation in the whole plant. *In* Encyclopedia of Plant Physiology, new series. Vol. 1, Transport in Plants. I. Phloem Transport. Eds. M H Zimmermann and J A Miburn. pp 451–473. Springer-Verlag, New York.

Penn S G, Hu H, Brown P H and Lebrilla C B 1997 Direct analysis of sugar alcohol borate complexes in plant extracts, by matrix assisted laser desorption fourier transform mass spectrometry (MALDI-FTMS). Anal. Chem. (*In press*).

Picchioni G A, Weinbaum S A and Brown P H 1995 Retention and the kinetics of uptake and export of foliage-applied, labelled boron by apple, pear, prune, and sweet cherry leaves. J. Amer. Soc. Hort. Sci. 120, 28–35.

Plouvier V 1963 Distribution of aliphatic polyols and cyclitols. *In* Chemical plant taxonomy. Ed T Swain. pp 313–336. Academic Press, New York.

Schon M K and Blevins D G 1987 Boron stem infusions stimulate soybean yield by increasing pods on lateral branches. Plant Physiol. 84, 969–971.

Scott L E and Schrader A L 1947 Effect of alternating conditions of boron nutrition upon growth and boron content of grape vines in sand culture. Plant Physiol. 22, 526–537.

Seresinhe P S J W and Oertli J J 1991 Effects of boron on growth of tomato cell suspensions. Physiol. Plant. 81, 31–36.

Shelp B J 1987 The composition of phloem exudate and xylem sap from broccoli (*Brassica oleracea* var. *italica*) supplied with NH_4^+, NO_3^- or NH_4NO_3. J. Exp. Bot. 38, 1619–1636.

Shelp B J 1988 Boron mobility and nutrition in broccoli (*Brassica oleracea* var. *Italica)*. Ann. Bot. 61, 83–91.

Shelp B J 1993 Physiology and biochemistry of boron in plants. *In* Boron and its Role in Crop Production. Ed. U C Gupta. pp 53–85. CRC Press, Boca Raton, FL, USA.

Shelp B J and DaSilva M C 1990 Distribution and metabolism of xylem-borne ureido and amino compounds in developing soybean shoots. Plant Physiol. 94, 1501–1511.

Shelp B J and Shattuck V I 1987a Boron nutrition and mobility, and its relation to elemental composition of greenhouse grown root crops. I. Rutabega. Comm. Soil Sci. Plant Anal. 18, 187–201.

Shelp B J and Shattuck V I 1987b Boron nutrition and mobility, and its relation to hollow stem and the elemental composition of greenhouse grown cauliflower. J. Plant Nutr. 10, 143–162.

Shelp B J, Penner R and Zhu Z 1992a Broccoli (*Brassica oleracea* var. *italica*) cultivar response to bóron deficiency. Can. J. Plant Sci. 72, 883–888.

Shelp B J, Shattuck V I and Proctor J T A 1987 Boron nutrition and mobility, and its relation to elemental composition of greenhouse grown root crops. II. Radish. Comm. Soil Sci. Plant Anal. 18, 203–219.

Shelp B J, Marentes E, Kitheka A M and Vivekanandan P 1995 Boron mobility in plants. Physiol. Plant. 94, 356–361.

Shelp B J, Stattuck V I, McLellan D and Liu L 1992b Boron nutrition and composition of glucosinolates and soluble nitrogen compounds in two broccoli (*Brassica oleracea* var. *italica*) cultivars. Can J. Plant Sci. 72, 889–899.

Shelp B J, Vivekanandan P, Vanderpool R A and Kitheka A M 1996 Translocation and effectiveness of foliar-applied boron in broccoli plants of varying boron status. Plant and Soil 183, 309–313.

Tammes P M L and van Die J 1966 Studies on phloem exudation from *Yucca flaccida* Haw. IV. Translocation of macro and micro nutrients by the phloem stream. K. Ned. Akad. Wet. Ser. C Biol. Med. Sci. 65, 677–679.

van Bel A J E 1984 Quantification of the xylem to phloem transfer of amino acids by use of inulin[^{14}C]carboxylic acid as xylem transport marker. Plant Sci. Lett. 35, 81–85.

van Die J and Tammes P M L 1975 Phloem exudation from monocotyledonous axes. *In* Encyclopedia of Plant Physiology, new series. Eds M H Zimmermann and J A Milburn. Vol. 1, Transport in Plants. I. Phloem Transport. pp 196–222. Springer-Verlag, New York.

van Goor B J and van Lune P 1980 Redistribution of potassium, boron, iron, magnesium and calcium in apple trees determined by an indirect method. Physiol. Plant. 48, 21–26.

von Michael G, Wilberg E and Kouhsiahi-Tork K 1969 Durch hohe Luftfeuchtigkeit induzieter Bormangel. Z. Pflanzen. Bodenk. 122, 1–3.

Wallaart A M 1980 Distribution of sorbitol in Rosaceae. Phytochem. 19, 2603–2610.

Woodbridge C G 1955 The boron requirements of stone fruit trees. Can. J. Agric.Sci. 35, 282–286.

Woodbridge C G, Venegas A and Crandall P C 1971 The boron content of developing pear, apple and cherry flower buds. J. Amer. Soc. Hort. Sci. 96, 613–615.

Zimmermann M H and Ziegler H 1975 List of sugars and sugar alcohols in sieve-tube exudates. *In* Encyclopedia of Plant Physiology. V1. Transport in Plants I. Phloem Transport. Eds. M H Zimmermann and J A Milburn. pp 480–503. Springer-Verlag, New York.

Plant and Soil **193:** 103–120, 1997.

Chapter 8

Physiological response of plants to low boron

Bernie Dell and Longbin Huang
*School of Biological and Environmental Sciences, Murdoch University, Perth 6150, Australia**

Abstract

This review focuses on physiological responses in higher plants to B deficiency at the whole plant and organ level. Plants respond to decreasing B supply in soil solutions by slowing down or ceasing growth. Boron deficiency inhibits root elongation through limiting cell enlargement and cell division in the growing zone of root tips. In the case of severe B deficiency, the root cap, quiescent centre and protoderm of root tips disappear and root growth ceases, leading to the death of root tips. Although vascular bundles are weakly developed in B-deficient roots, early effects of B deficiency on their initiation and differentiation is poorly understood. Inhibited leaf expansion by low B indirectly decreases the photosynthetic capacity of plants, though exact roles of B in photosynthesis remain to be explored. The early inhibition of root growth, compared to shoot growth, increases the shoot:root ratio. It is hypothesised that this may enhance the susceptibility of plants to environmental stresses such as marginally deficient supplies of other nutrients and water deficit in soil.

In the field, sexual reproduction is often more sensitive to low soil B than vegetative growth, and marked seed yield reductions can occur without symptoms being expressed during prior vegetative growth. In flowers, low B reduces male fertility primarily by impairing microsporogenesis and pollen tube growth. Post-fertilisation effects include impaired embryogenesis, resulting in seed abortion or the formation of incomplete or damaged embryos, and malformed fruit. However, there is a great diversity of effects of low B on reproductive growth among species, and within the same species between sites and seasons. Much of this diversity is not explained by the current literature. Key processes in reproductive development which may be impaired under B deficiency are proposed and discussed. These include the formation of a diverse array of cell wall types, the supply of carbohydrates for growth and storage reserves, and the production of flavonols. Inflorescence architecture, floral morphology, canopy structure and prevailing weather conditions are suggested as being important for xylem B delivery into flowers because of their impact on transpiration. The extent of phloem translocation of B into reproductive organs has yet to be fully assessed. The timing of B sensitive stages in reproduction of most crop plants need defining in order to facilitate appropriate timing of corrective B treatments.

As most container studies have imposed B deficiency by withholding B, much of the data on severely B-deficient plants requires re-evaluation. Further studies are warranted to understand the effects of realistically low levels of B in solution on the growth of meristematic tissues and floral organs. A B-buffered solution culture system is recommended for some of this work.

Introduction

Boron deficiency in higher plants causes adverse effects on cellular functions (Cakmak and Römheld, Chapter 6) and physiological processes (Dugger, 1983; Marschner, 1995). It decreases or inhibits the growth of vegetative and reproductive plant parts, depending on the timing and extent of B deficiency. During vegetative growth, stunted root and shoot tips are commonly observed in severely B-deficient plants. These growth defects at apical meristems are the consequences of B having some function in cell walls (Matoh, Chapter 5) and cellular activities (Cakmak and Römheld, Chapter 6). The present review will focus on physiological responses to B deficiency at the whole plant and organ level. For convenience, it is divided into sections which deal separately with vegetative and reproductive growth. The vegetative section considers effects of B

* E-mail: dell@central.murdoch.edu.au

deficiency on leaf and root growth, vascular differentiation and assimilate partitioning.

This review will not describe in any detail the range of visible symptoms in B-deficient plants, nor will it catalogue quantitative changes in the components of reproductive yield. These areas have been well illustrated (e.g. Bergmann, 1992; Shorrocks, 1989) or reviewed in recent publications (Bell, Chapter 10; Gupta, 1993). Given the need for more detailed work on the role of B in reproduction in flowering plants, more emphasis is given in this chapter to this area than vegetative growth. Although B deficiency often affects reproductive growth more acutely than vegetative growth, there is a very large diversity of effects of B deficiency on reproductive growth. Much of this variety of effects is not explained by the current literature, although recent discoveries concerning phloem transport of B (Brown and Shelp, Chapter 7) and the chemistry of B in cell walls (Matoh, Chapter 5) are likely to be contributing factors. Accordingly, we have chosen to examine those stages and processes in reproductive growth where B may be essential for development, and suggest areas where research effort can be placed in the future.

Boron in vegetative growth

Root growth

In higher plants, the most rapid response to B depletion or deficiency is the inhibition or cessation of root elongation (Figure 1a–c) in both the main and lateral roots and this has been described in a number of reviews (Dugger, 1983; Marschner, 1995; Shelp, 1993). Inhibition of root tip elongation occurs within hours after the removal of B (-B) from nutrient solution (Cohen and Albert, 1974; Cohen and Lepper, 1977; Hirsch and Torrey, 1980). Bohnsack and Albert (1977) observed that root tip elongation of –B squash (*Cucurbita pepo*) slowed down after 3 h, was severely inhibited within 6 h and completely stopped after 24 h of –B treatment in the nutrient solution. Inhibition of root elongation in tomato (*Lycopersicum esculentum*) was observed within 3 h after withdrawal of B from the nutrient solution (Dugger, 1983). Repressed roots can resume elongation if B is resupplied before permanent damage occurs in the meristematic region of root tips. For example, squash roots transferred to –B solution for 12 h resumed rapid elongation as soon as the B supply was restored (Bohnsack and Albert, 1977).

Boron deficiency appears to impair cell enlargement (including changes in length and width) and division in the root meristematic region, resulting in the inhibition of total root elongation. This inhibition may be primarily caused by adverse effects of B deficiency on wall structure and enlargement of newly divided cells. Abnormal radial cell wall thickening is a common phenomenon observed in B-deficient root tips (Loomis and Durst, 1992). Kouchi and Kumazawa (1975) examined the effects of B deficiency on the cellular growth and development of root tips of tomato grown in nutrient solution without B for 3–6 days. In –B root tips, the apical cells had thickened walls and had lost their ability to elongate and divide. Hirsch and Torrey (1980) also observed, using an electron microscope, that cell wall thickness in –B root tips of sunflower (*Helianthus annuus*) increased as early as 6 h after withholding B supply, and the increase in thickness was unevenly distributed around the cell wall.

Plant cell walls consist of rigid microfibrils of cellulose molecules embedded in a gel-like matrix composed of hemicellulose, pectin, lignin and protein (Becker and Deamer, 1991; Brett and Waldron, 1996). The primary cell wall has properties of elasticity, which is reversible, and extensibility which allows the wall to expand in a plastic manner in response to cell growth. When a plant cell reaches its final size and shape, it may deposit a secondary cell wall on the inner surface of the primary wall. The secondary wall contains more cellulose and lignin than the primary wall, giving rise to its inelasticity and inextensibility, thus fixing the size and shape of the cell (Becker and Deamer, 1991). The abnormal structural development in cell walls caused by B deficiency may be related to the dramatic decrease in cell wall plasticity, as observed by Hu and Brown (1994), leading to the failure of newly divided cells to enlarge. The malformation of cell walls in B-deficient roots may also lead to abnormality in the final shape and size of the newly divided cells. Cohen and Lepper (1977) showed that as –B squash roots ceased elongation, the length of cells in the region of 0–320 μm from the meristem continued to increase, but their width was substantially decreased. The cortical and endodermal cells in tomato root tips enlarged abnormally after 3 days without B in the nutrient solution (Kouchi and Kumazawa, 1975).

Inhibition of cell division is probably the secondary response of the root meristematic region to B deficiency. The inhibition or cessation of cell division has only been observed in root tips subjected to pro-

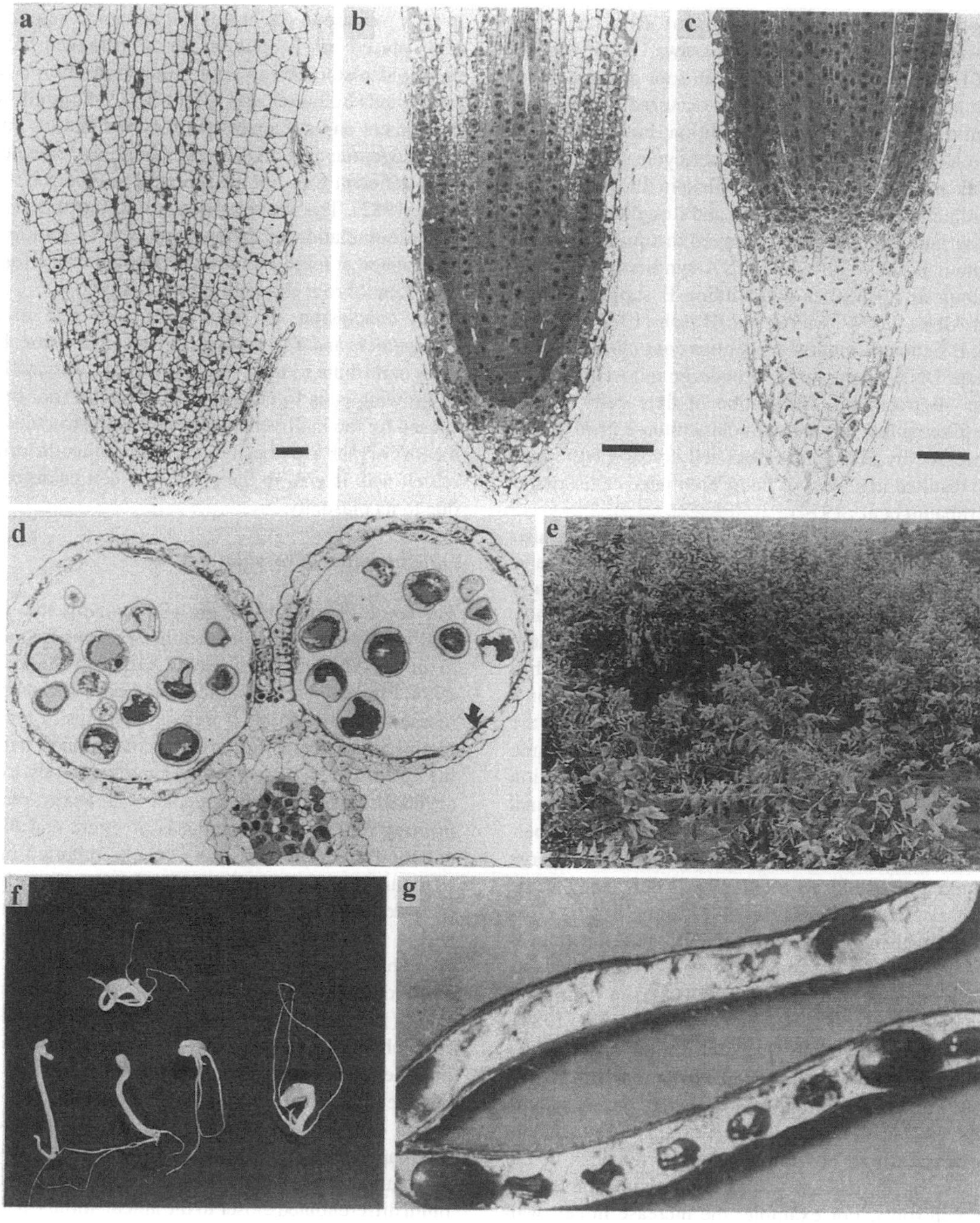

Figure 1. Effects of low boron on vegetative and reproductive growth of flowering plants (a–c) LS root tips of sunflower (*Helianthus annuus* cv. Hysun 42) seedlings grown for 32 h in nutrient solution in which B was buffered using the B specific resin Amberlite IRA 743; treatments were (a) 0.02 μM B, (b) 0.04 μM B, (c) 10 μM B. In (a) cell growth is abnormal and the apical meristem is absent. (b) shows a reduced root cap and increased cell maturation close to the meristem. (c) is a normal root. Bar = 50 μM. (d) TS anther of wheat (*Triticum aestivum* cv. Senora) at anthesis showing a range of pollen abnormalities and normal endothecium (arrow). (e) prostrate blue-gum (*Eucalyptus globulus*) trees (foreground) in Yunnan Province, P.R. China. The wood in these trees had low lignin content. (f) black gram (*Vigna mungo*) seedlings from plants grown with low B showing missing primary leaves, roots and terminal buds. (g) black gram pod showing a high percentage of aborted seeds.

longed B deficiency (more than 6 h of -B) (Cohen and Albert, 1974; Kouchi and Kumazawa, 1975; Moore and Hirsch, 1983). A decrease in root elongation of faba bean (*Vicia faba*) seedlings occurred within 24 h from transfer to –B nutrient solution, but cells in the regions of the endodermis and pericycle continued to divide in a radial manner rather than in the normal longitudinal direction (Robertson and Loughman, 1974). Cessation of mitosis was observed in squash root tips without B for 6.5 h while DNA synthesis continued as long as 20 h after withholding B supply (Cohen and Albert, 1974). Moore and Hirsch (1983) showed that B-deficient sunflower root tips continued to incorporate DNA precursors and undergo mitosis up to 72 h of –B treatment. Elongation of fibre cells in cotton (*Gossypium hirsutum*) ovules cultured *in vitro* was depressed by B deficiency but cell division was rapid and resulted in a mass of undifferentiated callus tissue (Birnbaum et al., 1974).

A primary role of B in cell enlargement is also supported by the fact that lateral root tips continue to emerge even when the elongation of main root tip is severely inhibited. In B-deficient plants, it is usually observed that numerous lateral root tips are initiated behind the zone of the root tip in which elongation has prematurely ceased. In tomato roots without B for 3 days, lateral root primordia formed normally from the pericycle opposite the xylem poles, but the tips became necrotic (Kouchi and Kumazawa, 1975). The initiation of lateral roots is the consequence of the longitudinal and unequal cell division of primordia within the pericycle. As a result, lateral root tips would not emerge behind a main root tip stunted by B deficiency, if the primary effect of B was on cell division.

It has been proposed that inhibition of root elongation is caused by the accumulation of excessive levels of endogenous indoleacetic acid (IAA) in the meristematic regions of B-deficient roots (Dugger, 1983). This hypothesis is largely based on the superficially similar morphological appearance of roots with B deficiency and supraoptimal levels of IAA. These include reduced root elongation and abnormal cell expansion. However, there are fundamental differences in the effects of B deficiency and supraoptimal levels of IAA in roots. For example, the increase in cell wall thickness and loss of membrane integrity observed in roots within 6 h of –B treatment are not observed in auxin-treated roots (Hirsch and Torrey, 1980). In B-deficient squash, root elongation was inhibited (within 6 h) before an increase in IAA oxidase activity could be detected (Bohnsack and Albert, 1977). The decrease in IAA oxidase in the apical and subapical regions of roots placed in –B solution for 12 h lagged behind the rapid restoration in root elongation after B was resupplied (Bohnsack and Albert, 1977). In sunflower, B-deficient and sufficient root tips had similar levels of endogenous free IAA, even though the elongation of B-deficient root tips was greatly inhibited (Hirsch et al., 1982). The above evidence suggests that excessive accumulation of endogenous IAA in root apical and subapical regions is not the primary cause for the inhibition of root elongation by B deficiency.

In conclusion, B plays a primary role in cell enlargement and a secondary role in cell division; both roles contribute to total root elongation. The inability of growing cells to enlarge in B-deficient root tips is caused by the malformation of cell wall structure and the loss of plastic extensibility. In particular, the primary cell wall is vital in the process of cell enlargement due to its plasticity.

Differentiation of vascular tissues

The wood of some trees, growing on soils low in B, may be so poorly lignified that the branches are unable to support the weight of leaves (Dell and Malajczuk, 1994; Figure 1e). So far, there has been no systematic research on what roles B deficiency plays in the differentiation and maturation of vascular tissues, namely the xylem and phloem in roots, stems and leaves. Lewis (1980a) suggests that B, together with auxin, plays a primary role in the biosynthesis of lignin and differentiation of xylem. However, results reported in the literature are often conflicting with regards to effects of B deficiency on the differentiation of vascular tissues.

Some evidence suggests that B deficiency causes abnormal development of xylem and phloem cells. Spurr (1957) observed that the cell walls of the phloem parenchyma in celery (*Apium graveolens*) were thicker in B-deficient plants than B-sufficient plants. In faba bean seedlings treated with –B for more than 24 h, new xylem cells became lignified even though they were misshapen (Robertson and Loughman, 1974). Kouchi and Kumazawa (1975) observed that the vascular cambium differentiated closer to the apical initials in tomato root tips without B for 6 days. However, it is believed that enhancement of cell division in cambial stem tissue, and impaired xylem differentiation, are not direct effects of boron deficiency (Marschner, 1995).

On the basis of the limited available information, it is difficult to conclude what role B plays in the process of vascular initiation and differentiation in the plant. Existing data were mostly obtained in plant tissues deficient in B for many hours and even days. This makes it difficult to distinguish primary events from secondary events that are exacerbated by tissue damage caused by extreme B deficiency. Further studies are needed to examine the time course of events that occur in the whole process of xylem and phloem formation from the onset to later stages of B deficiency. Recent advances in microscopy provide a means for investigating in greater detail the early structural changes in vascular tissues of B-deficient plants.

Leaf growth and photosynthesis

Since B plays an important role in cell wall structure and plasticity (Loomis and Durst, 1992; Hu and Brown, 1994), B deficiency limits the expansion or elongation of newly initiated leaves. As a result, B deficiency symptoms in the shoot first appear at the shoot apex and in actively growing leaves. When plants are marginally B-deficient, the expansion of immature leaves is limited, resulting in small leaves, dark-green in colour. Leaf symptoms develop progressively with the severity of B deficiency, from deformed in shape with unequal leaf expansion and downward curvature, to purple in colour, followed by necrotic spots and then senescence. The shoot apex can still produce young leaf initials in the early stage of B deficiency, but become necrotic if severe B deficiency is prolonged.

Numerous studies have shown that the elongation rate of the most actively growing leaf decreases in response to a decrease or interruption in the external B supply (e.g. Kirk and Loneragan, 1988; Bell et al., 1990; Hu and Brown, 1994; Huang et al., 1996). In soybean (*Glycine max*), B concentrations in young trifoliolate leaves declined after 11 days of 1 μm B treatment and leaf elongation became depressed three days later (Kirk and Loneragan, 1988). Furthermore, cell size is generally smaller in B-deficient than B-adequate leaves (Hu and Brown, 1994). This inhibition in leaf expansion or elongation is the consequence of a reduction in the plasticity of B-deficient cells as described earlier (Hu and Brown, 1994).

There is no evidence of a direct role for B in photosynthesis (Dugger, 1983; Shelp, 1993). However, B deficiency may indirectly affect photosynthesis by decreasing photosynthetic area and altering leaf constituents. Leaves developed under B deficiency can also have lower stomatal frequencies and smaller stomatal apertures than those with sufficient B, resulting therefore, in a decrease in stomatal conductance to CO_2 (Sharma and Ramchandra, 1990). Boron deficiency reduces chlorophyll and soluble protein (largely photosynthetic enzymes) contents of leaves, which affects the Hill reaction activity and net photosynthesis (Sharma and Ramchandra, 1990). In cyanobacteria (*Anabaena* PCC 7119) under N_2-fixation conditions, chlorophyll and phycobiliprotein contents declined after 24 h of culture in –B medium; and photosynthetic O_2 evolution and CO_2 fixation rates fell after 24 and 48 h, respectively (Garcia-González et al., 1990).

Several studies have shown that the extent of B deficiency in plants can be modulated by altering photon flux density (Cakmak, et al., 1995; MacInnes and Albert, 1969; Tanaka, 1966). The mechanisms involved in the interaction between B deficiency and photon flux density are yet to be unravelled. Changing photon flux density may alter the external or/and internal B requirements for growth. In black gram (*Vigna mungo*), decreasing sunlight from 70 to 35% of full sunlight, by shading in a glasshouse trial, lowered the B concentration for leaf elongation from 15 to 10 mg B kg^{-1} dry wt (Noppakoonwong etal., 1993). Cakmak et al. (1995) demonstrated that increasing the photon flux density or photoperiod enhanced the membrane permeability of leaf discs of B-deficient sunflower. On the other hand, the interaction between photon flux density and B deficiency in plants may be related to the indirect effects of B deficiency on leaf photosynthetic capacity – especially contents of chlorophyll and photosynthetic enzymes. Therefore, B deficiency could increase the susceptibility of leaves to photoinhibition by lowering the photosynthesis saturation point. The accumulation of soluble sugars in B-deficient leaves (Dugger, 1983; Marschner, 1995) may also produce feedback inhibition to net photosynthesis. Nevertheless, these suggestions remain speculative pending definitive experimental data.

Assimilate partitioning between shoot and root

Root growth is more sensitive to B deficiency than shoot growth, leading to an increase in shoot to root ratios. In general, root growth responds to the onset of low or deficient B supply in plants 2–3 days earlier than leaf growth. In soybean grown in nutrient solution with 1 μm B, root elongation rate decreased by day 12, but the elongation rate of the most actively expanding leaf did not decline until day 14 (Kirk and Loneragan, 1988). In green gram (*Vigna radiata*), leaf elongation was inhibited by interruption of the B supply (–B) 5

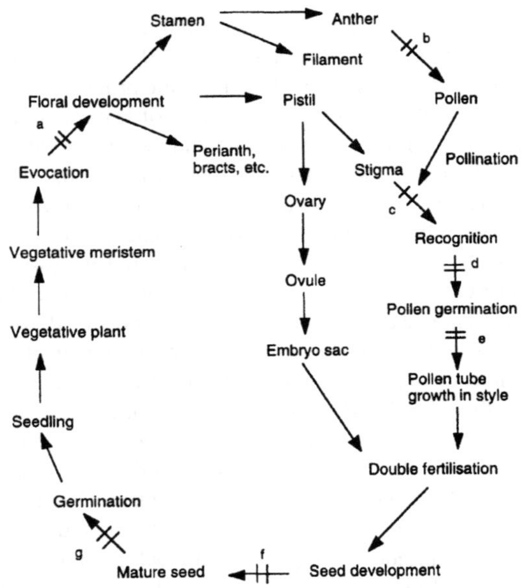

Figure 2. Life-cycle of an angiosperm emphasising those stages when an inadequate boron supply may directly or indirectly impact on reproductive development. Consequences of B deficiency shown at (a)–(f) are: (a) impaired inflorescence/flower formation; (b) infertile or aborted pollen; (c) reduced recognition of pollen by the stigmatic surface; (d) impaired pollen germination; (e) impaired pollen tube growth in stylar tissue leading to reduced seed and fruit set; (f) impaired seed development, e.g. hollow heart, shrivelled seed; (g) abnormal seedlings, reduced seedling vigour.

days later than the appearance of B deficiency symptoms in the roots (Bell et al., 1990). In oilseed rape (*Brassica napus*) subject to B concentrations from 0.4 to 100 μg B L^{-1} in solution culture, shoot/root ratios significantly increased with decreasing B concentrations (Huang et al., 1996).

The impact of B deficiency on assimilate partitioning may greatly influence the ability of plants to cope with other unfavourable environmental conditions such as soil water deficit and low supply of other nutrients. In the field, B-deficient plants with higher shoot/root ratios would be less resilient to a decline in available water and nutrients (particularly macronutrients) in soil, as there would be a higher shoot demand for water and nutrients per unit root length in B-deficient plants.

Boron in reproductive growth

Reductions in crop reproductive yield or seed/fruit quality in low B soils can be due to impaired reproductive development early or late in the flowering/fruiting cycle (Figure 2). It has often been observed that reproductive growth, especially flowering, fruit and seed set and seed yield, is more sensitive to B deficiency than vegetative growth (e.g. Dear and Lipsett, 1987; Noppakoonwong et al., 1997; Woodbridge et al., 1971). Furthermore, significant yield reductions can occur without expression of symptoms during prior vegetative growth. That reproductive development is sensitive to low B should not be surprising given that: flower development often occurs over a very narrow time window in the life of a plant; the flowering process is generally sensitive to a range of environmentally imposed stresses; some reproductive structures (e.g. pollen chamber, embryo sac) have poorer access to the vascular system than any vegetative organ; and sexual reproduction involves a large number of specialised cell types, many of which have distinctive cell walls.

Factors affecting the impact of a low external B supply on sexual reproduction in flowering plants are likely to include: the capacity of roots to acquire B from soil (Hu and Brown, Chapter 4); the mobility of B in the phloem (Brown and Shelp, Chapter 7); the relative sink size in floral parts for photosynthate; the capacity to redistribute B from vegetative tissues to reproductive organs; the rate of transpiration by floral organs; the functional requirement for B in reproductive tissues; and the distribution and abundance of B-binding compounds in the apoplast pathway between the vein endings and the most distal floral tissue. Further, responses can be expected to differ between species or cultivars and this aspect is discussed later in this book (Rerkasem and Jamjod, Chapter 11).

That B is required for flowering is indicated by the sensitivity of pollen development to low B (Figure 1d) and the generally high concentrations of B that occur in reproductive parts of the flower. Under conditions of low external B, concentrations in the anthers and pistils do not decline to the low levels measured in leaves (Table 1). Table 1 also shows that B concentrations are often higher in the stamen than in the pistil.

The physiological roles for B in sexual reproduction have yet to be fully defined and there is a great need for experimentation in this area. Many of the experiments that have been undertaken do not give definitive information because plants were critically deficient in B at the time of flowering, or the B status of floral tissues was not determined at the time of impairment in cellular development/function, or cell structure and metabolism were examined long after the primary effect of B took place. An example is the observation

Table 1. Boron concentrations (mg B g^{-1} dry wt) in young reproductive parts and leaves in oilseed rape (*Brassica napus*), wheat (*Triticum aestivum*) and green gram (*Vigna radiata*) in relation to B supply

	B supply	Flower	Stamen	Pistil	Leaf
Oilseed rape[1]	Low		19	19	10
	Adequate		38	40	17
Green gram[2]	Low	30		19[3]	11
	Adequate	61		50[3]	34
Wheat[4]	Adequate		16[5]	8	3

[1]Pot study in soil by Zhang et al. (1994).
[2]Pot study in solution culture by Bell et al. (1990). The data for the low B plants is after transfer to solution without B for 4 days.
[3]Data for young pods.
[4]Field study by Rerkasem (1996). The data are means of five cultivars.
[5]Data for anthers.

that B deficiency results in male sterility, a condition that can be induced by deficiencies in other nutrients (e.g. Mn, Cu) or unfavourable environmental conditions (e.g. water deficit, high/low temperature). These observations reveal nothing about the process that is being affected by low B supply and do not enable us to conclude whether the effect of low B supply on male sterility is a direct or indirect event. Because root function is greatly impaired in severely B-deficient plants in containers and this can impact on whole plant physiology, the requirement for B for pollen development should be studied under conditions of controlled external supply. In the following sections of this chapter, we examine the reproductive response to B deficiency and identify stages in reproductive growth which may be especially sensitive to constraints in B supply. Much of the discussion is, by necessity, highly speculative at this time.

Flowering response

There is no evidence that B deficiency prevents evocation or delays floral development. When plants are grown without B or are transferred into B-free nutrient solutions in pot experiments, the apical meristems may abort and flowers therefore do not develop, as shown for peach (*Prunus persica*) (Kamali and Childers, 1970). However, under field conditions, where B-deficiency stresses occur more gradually, plants may have the time to adapt to deficiency. For example, in peanut (*Arachis hypogaea*), the flowering period was extended in B-deficient plants (Harris and Brolman, 1966a), resulting in low-B plants producing as many flowers as B-adequate plants. In species where the flowers occur in compact inflorescences, and these are terminal on the stem (e.g. sunflower, wheat: *Triticum aestivum*), low B has a greater impact on reproduction because the plant has less ability to modulate reproductive growth than species with axillary inflorescences and indeterminate growth. Our observations indicate that this first group of plants is more prone to pollen sterility under low B than the latter group of plants. In wheat, we often observe a shift in floret (to higher orders) and spikelet position where seed fill occurs, as well as a greater role for the tillers, in low-B plants, presumably as a consequence of the timing of B supply to flowers. Finally, low B can result in plants being functionally male sterile [e.g. wheat (Li et al., 1978); rice (*Oryza sativa*) (Garg et al., 1979); barley (*Hordeum vulgare*) (Simojoki, 1972)], although cases of female sterility have been reported [e.g. in maize (*Zea mays*) (Vaughan, 1977); avocado (*Persia americana*) (Coetzer and Robbertse, 1987)]. The external B supply does not appear to alter the frequency of unisexual flowers, although some workers have attempted unsuccessfully to do this by applying B sprays (Singh and Choudhury, 1988).

Anther development

The process of pollen development
Compared to ovule growth, male gametogenesis is particularly sensitive to low B supply especially in cereals (Rerkasem and Loneragan, 1994). Before examining

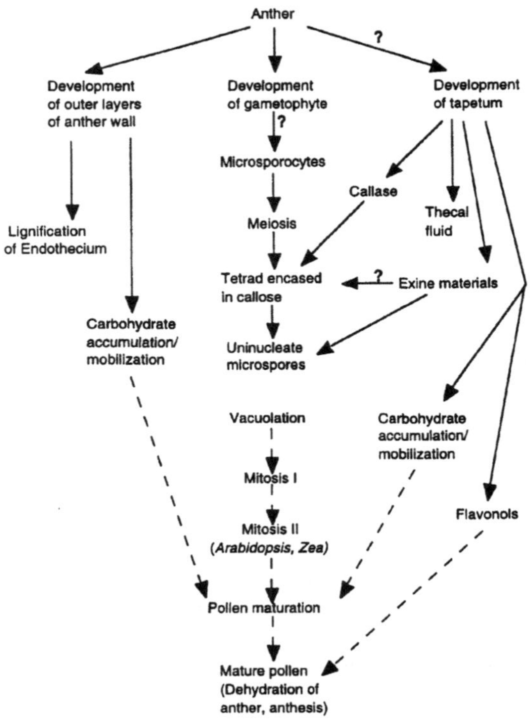

Figure 3. Pollen development and maturation. The broken arrows indicate the processes that B deficiency is most likely to disturb and the arrows with ? indicate events where there are unconfirmed observations of low B impairing development.

whether there is a specific requirement for B in the formation of the male gametophyte, the development of pollen is briefly reviewed in order to familiarise the reader with the chain of events preceding anthesis (Figure 3).

Microsporogenesis has been described in detail in some crop plants, such as maize (Chang and Neuffer 1989), and *Arabidopsis thaliana* (Owen and Makaroff, 1995). Generally, before the anther has reached 25% of its final length, a large number of centrally located cells (microspore mother cells or microsporocytes) undergo synchronous meiotic division to form tetrads. At the end of this stage, the haploid cells are encased as tetrads by a massive common wall rich in ß-1, 3-glucan (callose). Exine (future outer pollen wall) patterns are initiated and appear within the callose (Takahasi, 1993). The tetrads lie within a central locule surrounded by the nutritive tapetum, a cytoplasmically rich tissue lining the locule. The cellulosic intine wall layer, which is similar to the primary walls of somatic cells, is laid down after the exine forms. Once the uninucleate microspores have been released from the tetrads (a few species retain the microspores in the tetrad e.g. *Drosera* spp.), they rapidly grow, take up water, expand and become vacuolated (Bedinger, 1992). The nucleus in the vacuolate microspore divides by mitosis giving rise to a binucleate microspore. Cytokinesis partitions the young pollen grain asymmetrically into a large transcriptionally active vegetative cell and a small quiescent generative cell (Mascarenhas, 1990). In some plants, such as members of the Crucifereae and Gramineae, the generative cell divides again and the mature pollen contains three nuclei. In species with binucleate pollen, the second mitotic division occurs in the pollen tube after germination. During maturation of the pollen in the anther, the large vacuole disappears, the cytoplasm becomes rich in organelles and membranes, reserve substances (lipid, carbohydrate, protein) accumulate and the deposition of tapetum-derived surface wall deposits in the exine is completed. During these latter stages, the anther enlarges, the filament extends, the tapetum degenerates and the endothecium becomes fully differentiated. In mature pollen of tobacco (*Nicotiana tabacum*) (Goldberg et al., 1993), maize and *Tradescantia paludosa* (Mascarenhas, 1989) the pollen mRNAs are more prevalent than in shoot tissue, suggesting a requirement for the rapid synthesis of a large quantity of protein either during the final stages of pollen maturation or during germination. Prior to anther dehiscence and pollen release (anthesis), the anther dehydrates losing about 50% of its water (Keijzer, 1983).

When considering effects of low B supply on pollen development it is also important to study the tapetal tissue. The tapetum is pivotal to pollen formation and maturation because:

- it provides enzymes to degrade the callose-rich walls of the pollen tetrads (Steiglitz, 1977),
- it provides the thecal fluid which bathes the pollen initials as they undergo cell division,
- it secretes, or releases by autolysis, the sporopollenin, lipid and possibly protein constituents of the exine wall of the pollen grain (Mascarenhas, 1975),
- it is the supply route through which water and nutrients (inorganic and organic) pass to reach the pollen,
- it accumulates and mobilises the carbohydrate reserves needed for starch and lipid deposition in the maturing pollen grain, and

- it is the source of flavonols and other compounds which accumulate in the exine just prior to pollen release.

It is worth noting that the main cause of physiological and genetic male sterility has been attributed to persistence or premature breakdown of the tapetum (Kaul 1988).

Pollen development in boron deficient plants

As mentioned earlier, male sterility is induced by B deficiency in many species of monocots and dicots. At anthesis, the time when most workers examine pollen morphology, the pollen grains in well-formed anthers appear empty, misshapen, shrivelled, or may be normal in shape but lack reserves of storage materials such as starch. Clearly, these are the consequences rather than the causes of B-imposed male sterility. In severely deficient wheat, some anthers have their development arrested early resulting in small arrow-shaped structures largely devoid of cells in the anther locules. Small stamens have also been observed in B-deficient oilseed rape (Zhang et al., 1994). There is a tendency for pollen from plants with a deficient B supply to aggregate, possibly because the cells have leaked part of their contents.

From an experimental point of view, care is required in sampling large inflorescences for anatomical studies as the degree of male sterility can vary considerably with position in the inflorescence and between inflorescences. This is particularly evident in wheat where male infertility is often higher in the more distal spikelets and fertility in late tillers can be different from that in the ear of the main shoot. Differences may also occur with floret position within a spikelet.

Anatomical investigation of pollen development in a B-sensitive (SW41) and a B-insensitive (Sonora) wheat failed to detect any abnormality in pollen until after the uninucleate vacuolate stage had been passed (Rerkasem et al., 1997a). This is in sharp contrast to copper deficiency which has been shown to impair microsporogenesis in wheat early in meiosis (Graham, 1975). In wheat, reduced starch accumulation in anthers close to anthesis is an indicator of a potential problem but this is a general stress response and is therefore not specific to B deficiency.

In short-term transfer experiments with wheat in our laboratory, transferring plants into nutrient solution without B for a few days induced abnormal pollen to develop when the treatment was imposed just prior to meiosis (Pant et al., unpublished data). However, visible impairment in pollen development did not appear until after the microspores were released from the tetrads. This means that meiosis proceeded satisfactorily but it does not rule out a direct early effect of B deficiency that is expressed structurally later in pollen development. In follow up experiments, Rawson (1996b) confirmed these observations and defined the critical period as extending from after emergence of the flag leaf tip to shortly after the flag leaf has become fully expanded. This period, of about a week, occurs before ear emergence and occurs before any external symptoms can be seen.

Since we do not know from these transfer experiments how long it takes before there is run-down in the B content of reproductive tissues, the timing of the developmental process that is impaired by the low B supply remains to be determined. More detailed studies are required to ascertain whether a deficiency in B can impair nuclear and cell division later in pollen development. Future studies should also look at anther development in general as, in barley anthers, copper deficiency causes amoeboid development of the tapetum (normally a glandular type of tapetum is present in this species) and this has been implicated in male sterility (Jewell et al., 1988). So far, we have not observed any unusual features in tapetal development in wheat. However, Zhang et al. (1994) reported abnormal development in the tapetum of oilseed rape, but this requires confirmation. In wheat, low B did not alter the structural appearance of the mature pollen wall under the transmission electron microscope (Rerkasem et al., 1997a). By contrast, Smith et al. (1997) noted that the exine was less extensively developed in pollen from B-deficient avocado.

There have been too few studies of microsporogenesis and microgametogenesis in B-deficient plants to draw definitive conclusions to the following questions.

- Is there a stage in pollen development which is highly sensitive to B deficiency?
- Does pollen development in graminaceous monocots have the same anther tissue B requirement as the non-graminaceous monocots and dicots?
- Are the observed abnormalities in pollen development a consequence of a direct or an indirect requirement for B?
- Does B deficiency impair tapetal function and how does this affect pollen development?
- Does the extent of flower packing in inflorescences alter the B requirement for microsporogenesis?

Supply of boron to the anther

Delivery of B to the anther is likely to occur in vascular tissue and the amount delivered is governed by sink demand in the case of phloem-mediated delivery and transpiration rate for xylem delivery. In the few species where anatomical studies have been undertaken (Esau, 1977), vascular bundles from the stem pass through the filament (as for leaf petioles) and either end at the base of the anther or extend to the apex of the anther connective. Since there are no vascular connections to the sporogenous tissue or the tapetum, inorganic and organic nutrients must traverse short distances without the assistance afforded by specialised transport cells. This is analogous to the transport pathway that occurs at all vegetative apical meristems. Anatomical and morphological structures that may influence water loss include the degree of cutinisation of the anther, the degree of stomatal control, the degree of enclosure of anthers by floral parts, and inflorescence size, position and structure. Anecdotal evidence suggests that the stomata may be permanently open in some plants (Esau, 1977).

Perhaps some of the diversity of effects of B deficiency on reproductive growth among species, and within the same cultivar between sites and seasons, is a consequence of plant canopy structure and environmental control over transpiration. For example, the severe male sterility discussed above in B-deficient wheat has not been recorded in maize (Agrawala et al., 1981; Vaughan, 1977): here low B in the pistil appears to be the limitation to sexual reproduction. The stamens in maize, unlike those in wheat, are borne in a separate inflorescence which is exposed at the top of the plant above the canopy of leaves in an advantageous position for transpiration. By contrast, the female inflorescence is embedded in the canopy and is enclosed by leafy bracts. Even though the panicle is largely developed before emergence from the leaves, pollen shedding may extend over a week indicating that some pollen maturation is occurring after inflorescence release. Opening of the anther pore at anthesis is due to desiccation at the anther tip and this is possibly caused by a sudden increase in evaporation as the flower opens (Keijzer et al. 1996). It would be interesting to explore the distribution and timing of B translocation into these two types of inflorescence to determine if the terminal panicle has better access to internal B than the axillary cobs.

Vapour pressure deficit (VPD) is the environmental variable with most influence on transpiration rate (Rawson, 1996a); VPD is determined by temperature and relative humidity. Rawson argues that the interactive effects of weather on male sterility might be a consequence of altered movement of micronutrients, such as B, from the soil to the flowers around the time of meiosis in the anther. This assumes that B moves into developing flowers via the transpiration stream. Evidence that B delivery into the flower might be dependent on transpiration comes from a glasshouse trial with wheat in sand culture (Rawson 1996b) where pollen sterility resulted from enclosing plants in plastic bags during the critical period even though ample B was present in the root medium. This experiment appears to mimic the observation that grain set in cereal crops in the subtropics can be low even where vegetative growth is normal (Galrao and Sousa, 1988; Ganguly, 1979). That is, if flower development is dependent on a continuous supply of xylem-fed B from the soil, then shutting down transpiration, even for a few days, may reduce male fertility and ultimately seed yield. Given the tiny amounts of B that each anther accumulates, it is surprising that other internal sources of B can not the meet the anther B demand for normal pollen development in short-term transfer experiments.

Although there appears to be a strong link between weather, pollen development and B, we should not ignore the possibility that B movement into the anther in some species may be phloem driven. For example, broccoli (*Brassica oleracea*) flowers acquire some of their B from phloem transport (Liu et al., 1993; Shelp et al., 1995). Boron can move in the phloem with carbohydrates in those species containing sugar-alcohols, especially sorbitol and mannitol (Brown and Hu, 1996; Brown and Shelp, Chapter 7). Isotopic studies are required to determine which of these two supply routes are operational for B delivery, the degree of exchange between these pathways and the impact of weather and external B on B supply to the stamen.

Possible roles for boron in anther and pollen development

Since boric acid binds to *cis*-diols, the composition of cell walls in the anther will directly (functional requirement for B) or indirectly (B capture by the cell wall) dominate the anther B sink. The formation of borate-rhamnogalacturonan II (RG-II) complexes (Matoh et al. 1996) and the relationship between tissue B requirement and cell wall chemistry (Hu et al. 1996) are discussed in detail in an earlier chapter (Matoh, Chapter

5). The primary walls of grasses contain as little as 6% of the RG-II present in walls of broad-leafed plants (Darvill et al. 1980), and this is low compared to other flowering plants. These findings are based on vegetative cells, so it remains to be determined whether they apply to anthers. It is possible that the pectic polysaccharide composition of the cell walls of pollen has been conserved during evolution of the grasses.

The largest binding surface for B is likely to occur within the primary cell walls of the anther and the intine of the pollen grains. No comparative studies have so far been undertaken to compare the composition of the intine walls of the Gramineae to determine if these walls are distinctive from those in pollen of dicotyledonous plants. In the few broad-leafed species that have been examined, the intine is largely pectocellulosic (Li et al. 1995; Van Aelst and Van Went 1992). Monoclonal antibody analysis has established that, following secretion of acidic pectins in their esterified form, de-esterification occurs with increasing distance from the cell membrane, presumably due to enzyme activity within the pollen wall (Geitmann et al., 1995). It is interesting that, under the electron microscope, no differences can be observed in pollen from B-deficient wheat until intine synthesis is well advanced (Rerkasem et al., 1997a), that is, until the matrix of hemicellulose, pectin and protein have presumably been laid down in the cellulosic wall.

How boric acid reacts with callosic walls has not been studied. The structure of the ß 1,3-linked glucans (Brett and Waldron, 1996) is unlikely to provide binding sites for boric acid. However, as few callosic walls have been fully analysed, the possibility remains that other carbohydrates may be present which could bind B. Were this the case, then the callose wall, which acts as a filter to prevent large molecules from the surrounding sporophytic tissue from entering into the developing male gametophyte, could also prevent the entry of B by complexing most of the available B under conditions of limited supply. This is purely speculative and in any case would only be important where a cytoplasmic or cell membrane requirement for B exists and so far this has not been established. Again, anatomical studies by the authors on wheat have not shown any physical impairment in the amount of callosic wall present in tetrads in anthers from B-deficient plants. It is interesting to note that any interference in either deposition or enzymatic removal of the callose can seriously affect later pollen development. For example, premature callose degradation is sufficient to cause male sterility in tobacco (Worrall et al., 1992; Tsuchiya et al., 1995). Small reserves of cytoplasmic callose have been detected in some pollen (Pacini, 1996) and there are unconfirmed early reports of a callose envelope forming temporarily in the generative cell wall of some species (Mascarenhas, 1975).

Earlier in this chapter we discussed the view of Lewis (1980a) and others that B may be required for lignification in higher plants. There is no evidence that B deficiency impairs lignification within the anthers, as has been observed in severely Cu-deficient plants (Dell, 1981). Anthers from B-deficient wheat plants, containing fully aborted pollen, had a well developed and lignified endothecium and vascular tissue (Dell, unpublished). We can conclude from this that, at least in wheat, the pollen maturation phase is far more sensitive to B deficiency that the lignification process in anthers and that vascular differentiation is therefore not a limiting factor for B delivery into the filament.

As mentioned earlier, normal tapetal function is essential for pollen maturation. Effects on carbohydrate supply and metabolism of phenolic compounds are two ways in which low B could impair tapetal function. In Chapter 6 of this book, Cakmak and Römheld discuss the theory that B directly regulates the phenylpropanoid pathway. Key enzymes in this pathway, phenylalanine ammonia lyase (PAL), chalcone synthase and cinnamic acid 4-hydrolase, are active in the locule or exine cavities of developing pollen (Mascarenhas, 1993). These enzymes are probably produced by the tapetal cells (Kehrel and Wiermann, 1985; Mascarenhas, 1993). In broccoli, PAL activity peaked in anthers with mononucleate microspores and the level of enzyme activity was positively correlated with the number of fertile pollen grains that formed (Kishitani et al., 1993). Interrupting the phenyl propanoid pathway may therefore have consequences for the maturation of pollen. As yet, there is no evidence that low B supply affects pollen development through this mechanism.

Can low B affect the accumulation and mobilisation of carbohydrate reserves required for pollen growth and the formation of storage reserves in maturing pollen? Mature pollen in mildly B-deficient plants often contains reduced amounts of starch, but this does not mean that low B causes carbohydrate starvation. Many stress factors can inhibit starch accumulation in pollen. For example, applying several days of water deficit to rice resulted in pollen sterility even though sugars continued to accumulate in the anther during the stress period (Sheoran and Saini, 1996). Given that the phloem mobility of B differs with plant taxa (Brown and Shelp, Chapter 7), and that the tapetum was recorded as abnormal in oilseed rape grown under low B (discussed in "Pollen development in boron deficient plants"), it would be worthwhile exploring whether the B supply can modulate carbohydrate delivery to the pollen. In those species which contain sugar-alcohols, it would be interesting to know whether these forms pass into the thecal fluid. More studies are required along the lines of those undertaken on *Lilium* (Clément and Audran, 1995; Clément et al., 1994, 1996) in which the anther is the main sink for floral sugars during early microsporogenesis (up to vacuolate microspores). Recently, Pacini (1996) has postulated that cytoplasmic carbohydrates and sucrose are involved in protecting pollen viability during exposure and dispersal.

Pistil development

The pistil differentiates into regions for pollen germination (stigma), pollen tube growth (style) and embryo sac formation (ovule within the protective ovary). The pistil is generally well supplied with vascular bundles which cross the placenta into the chalaza and extend through the ovary wall and into the style. In many plants, ovary development lags behind growth of the stamen. In contrast to microsporogenesis, only one cell in each ovule undergoes meiosis. Competence for the pistil to support pollen germination and tube growth can extend over a broad developmental window. These factors may make the pistil less sensitive than the anther to variation in B supply during organogenesis. However, detailed studies on embryo sac development and developmental physiology are lacking under conditions of a low but continuous supply of B. Selected examples of low B effects on development and function of the pistil are given below.

In oilseed rape, Xu et al. (1993) observed that some of the ovules and embryo sacs showed arrested development in flowers from low B plants. They also noted unusual development of stigmatic papillae. Perhaps these plants were so severely deficient in B that all growth was impaired. This is suggested by the abortion of pollen mother cells in the same plants. At a cellular level, boron deficiency resulted in abnormal fibre development on ovules of cotton grown *in vitro* (Birnbaum et al., 1974) and impaired metabolism (Dugger and Palmer, 1980). Evidence from cross pollination trials, using donor pollen from plants supplied with adequate B and pistils of plants under low B, have yielded contrasting results for the limited number of species examined so far. Rerkasem has shown in field and pot trials that grain set in ears of B-deficient wheat can be increased by manual pollination, or natural cross-pollination, with fertile pollen (Rerkasem et al., 1993b; Rerkasem and Jamjod, Chapter 10). In maize, low B appears to be more limiting for pistil than anther function (Vaughan, 1977) even though the style contains a lower B concentration (4 mg B kg^{-1}) than the pollen (11 mg B kg^{-1}) (Agrawala et al., 1981). Two experiments that have been undertaken with avocado give conflicting results. In South Africa, application of pollen from low B trees to B-sufficient pistils resulted in higher pollen germination than when B-sufficient pollen was applied to low B pistils (Coetzer and Robbertse, 1987). In Australia, *in vitro* germination of pollen suggests that pollen from low B trees may be functionally impaired as is the case in mango (*Mangifera indica*) (de Wet et al., 1989). Further study is required to determine if climate, cultivar or B status are factors which could explain the difference.

Pollination to fertilisation

Lewis (1980b) questioned whether normally compatible pollen-stigma interactions are deranged in B-deficient plants. It is of interest that the exine of petunia and tobacco pollen contains flavonols which are suggested to have a role in recognition and pollen tube growth (Ylstra et al., 1992). These compounds are formed in the chalcone pathway in the tapetum referred to earlier. In maize mutants lacking chalcone synthase, pollen tubes grow abnormally in the style (Schrauwen and Wullems, 1997). However, flavonols do not seem to be required for functional pollen tube growth in *Arabidopsis thaliana* (Ylstra et al., 1996).

What is clear, is that boron is required for pollen germination and pollen tube growth *in vitro* (Cheng and Rerkasem, 1993; Vasil, 1964) and presumably *in vivo* in stigmatic tissue. Concentrations of B in solution that maximise pollen germination are typically around 1.5 m*M*, but a wide range has been reported (Shorrocks, 1996). There have been too few measurements to substantiate the claim made by some workers (e.g. Vasil, 1964) that pollen grains are generally deficient in B and that this is overcome by comparatively high levels of B in stigmatic and stylar tissues. As indicated earlier, the capacity of pollen and pistils from B-deficient plants to effect fertilisation varies between species. It has been proposed that B deficiency can alter the accumulation of phenolic compounds (Cakmak and Römheld, Chapter 6). Phenolic compounds which accumulated on the stigma in B-deficient *Campsis grandiflora* appeared to block the germination of viable pollen (Dhakre et al. (1994).

The pollen tube has apical growth (Steer and Steer, 1989), similar to that of root hairs and fungal hyphae, and the wall is chemically distinct from somatic cells. The tube wall consists of an inner layer rich in callose and an outer layer rich in acidic pectins (Anderson et al., 1987; Geitmann et al., 1995; Hasegawa et al., 1996; Meikle et al., 1991; Schlupmann et al., 1993). The tip of the pollen tube is free of callose (Rae et al., 1985). Partial analysis of *in vitro* grown pollen tube pectin reveals a high content of galacturonic acid in camellia (*Camellia japonica*) (Nakamura and Yoshida, 1980). In lily (*Lilium longiflorum*), there is a matrix of unesterified pectins that extends from the pollen tube wall to the surface of stylar transmitting tract cells (Jauh and Lord, 1996). Presumably these pectins contain constituents such as RG-II to which boric acid is linked.

Pollen tube growth has been reviewed recently (Cresti and Tiezzi, 1990; Derksen, 1996; Mascarenhas, 1993). Whilst there is strong evidence that tip growth is influenced by Ca (Cai et al., 1997; Malho and Trewavas, 1996), the data on B is less convincing. This is because the cellular distribution of B in the pistil and pollen tube has not been investigated and most studies have been undertaken *in vitro* which does not reflect the chemical environment in the pistil where pollen-induced signals are known to initiate a cascade of developmental events in the carpel (Stead, 1992). There are, however, two studies which reveal effects of B on pollen tube growth in the style. Firstly, in oilseed rape, pollen tube growth was slowed by B deficiency (Shen et al. 1994). Other effects of low B included a reduction in pollen germination and a reduction in the number of pollen tubes in the style (Table 2). Secondly, petunia (*Petunia* sp.) pollen tubes grew out from the cut ends of styles only when the bathing fluid contained boric acid (Robbertse et al., 1990b).

A number of workers have suggested that low B alters membrane function in rapidly expanding pollen tubes (e.g. Jackson, 1991; Rodriguez-Rosales et al., 1989) but these *in vitro* experiments may not apply to pollen tube growth in the style. Further, as there is no convincing evidence that low B consistently modulates auxin synthesis or oxidation through changes in the PAL pathway, auxin effects on pollen tube growth are not considered here.

Clearly, as for anther development, there is a need for more detailed studies using intact pistils to address such questions as:

- What is the distribution of B at a tissue and cell level in the pistil?

- How much B is required in the stigma and style for pollen germination and growth? Presently, suggested values range from 4 mg B kg^{-1} in maize (Agrawala et al., 1981) to 75 mg B kg^{-1} in avocado (Robbertse et al., 1990a).

- Do species with fast pollen tube growth (those with tricellular pollen e.g. maize) have the same B requirement in the pistil as those with slow pollen growth (bicellular pollen)? In maize, rates of pollen tube growth have been reported to be as high as 10 mm h^{-1} (Bedinger, 1992) and 35 mm h^{-1} (Mascarenhas, 1989).

- Does the acidic pectin composition of pollen tube walls vary greatly between the major groups of flowering plants?

- Could boric acid stimulation of the plasma membrane H^+-ATPase (Obermeyer et al., 1996) be a factor in pollen tube growth?

Seed and fruit development

As for pollen development and pollen tube growth, low B responses can vary widely between species. For example, decrease in pod number in response to B deficiency occurs in black gram (Rerkasem et al., 1988) whereas in soybean (Rerkasem et al., 1993a) and green gram (Bell et al., 1990) the number of seeds in each pod declines. Furthermore, in wheat, seed size increases in response to low B (Rerkasem and Loneragan, 1994) and subterranean clover (*Trifolium subterraneum*) (Dear and Lipsett, 1987) whereas it decreases in peanut (Keerati-Kasikorn et al., 1991). Further examples are illustrated by Mozafar (1993).

It is unclear whether these different responses are related to the capacity of the plant to mobilise B for seed fill (Brown and Shelp, Chapter 7) and growth of the ovary wall, to differences between experiments such as in weather and soil conditions, or to the size of the B sink in fruits. Differences in symptom expression could also be related to the timing of the onset of B deficiency. The most tenable hypothesis is that the sink for B uptake into the seed and fruit is created by the synthesis of cell wall pectins, though cell wall storage polysaccharides in some seeds could also be involved. Where the demand can not be met, fruits and seeds

Table 2. Effect of boron supply on pollen germination and pollen tube growth in the stigma of oilseed rape (*Brassica napus* Ningyou No. 8) and seed yield. Data from Shen et al. (1994)

Parameter measured	Low B	Adequate B
Number of pollen grains on stigma	31	83
% of germinated pollen	29	82
Number of pollen tubes penetrating the style	2	37
Seed yield (g $plant^{-1}$)	0.5	3.0

may abort or abscise (Figure 1g). Those seeds that can acquire sufficient B for growth may benefit from the reduced competition for assimilates by forming larger than normal seeds. Where the B supply becomes limiting later in seed and fruit development, seeds and fruits may exhibit symptoms of abnormal growth. Examples include embryo defect or hollow heart in peanut (Harris and Brolmann 1996b), premature staining of the testa in avocado (Harkness et al., 1959) and misshapen, cracked or bleeding fruit (e.g. mango: Ram et al., 1989). The latter effects can all be interpreted in terms of malformation of cell walls or premature cell wall senescence.

Fruits that are dependent on xylem-delivered B, are most at risk of B deficiency when xylem water flux, relative to their growth rate, is limited due to constraints on transpiration, as discussed earlier for anthers. Some of the rapidly growing cucurbit fruits would be ideal organs to demonstrate if there are quantitative relationships between water loss and B delivery in the xylem. In subterranean fruits, transpiration is obviously restricted. Campbell et al. (1975) concluded that B must be transported in the phloem into peanut fruit, but this has not yet been confirmed. On the other hand, the accumulation of B in many commercial fruits in the family Rosaceae can be explained by phloem transport of B-sorbitol complexes (Brown and Hu, 1996).

Seed viability and germination

Several studies have revealed that in some species, apparently normal-looking seeds, produced by plants under low B, may have altered physiological and growth properties. In black gram, germination and seedling morphology was shown to be more sensitive to low B supply during seed production than was plant growth and seed yield (Bell et al., 1989). In this study, low B had a further effect in that it enhanced germination due to a reduction in the percentage of hard seeds. Examples of seedling defects are given in Figure 1f. In an earlier study, Leggatt (1948) describes how the addition of B to the germination medium prevented the development of abnormalities due to B deficiency in pea (*Pisum sativum*) seedlings. Lastly, the relationship of seed B content or concentration to subsequent plant vigour, has been investigated for several legumes (Rerkasem et al., 1990, 1997b), but further study is required particularly for other grains.

Conclusions and future directions

Although it is known that B deficiency inhibits root and leaf growth, existing research evidence is mostly based on cases of extreme B-deficient plants after a lengthy depletion of B in culture solution. The critical B concentration in soil solution, below which early physiological effects on root and leaf growth may develop, remains to be determined. Boron-buffered solution cultures provide useful alternatives to soil for elucidating the early events of B deficiency in the process of root and shoot tip growth. Early effects of B deficiency on the differentiation of vascular bundles remain to be understood. It is of practical interest to understand these effects because poor development of roots and vascular bundles in roots would diminish the ability of plants to absorb and transport other nutrients. Further, reduced xylem lignification may have profound implications to crop survival and wood quality in trees.

Although field and pot trials continue to illustrate the essentiality of B for sexual reproduction in flowering plants, the requirement of B at the organ level in reproductive tissues remains enigmatic. The recent availability of sensitive isotopic procedures (Sah and Brown, Chapter 2) provides the opportunity to reexamine some of the puzzling aspects of plant reproductive responses to B deficiency. In particular, it will allow work to proceed on the timing and pathway of B delivery into reproductive parts and the source of B for reproductive growth. In the past, most emphasis has been given to quantifying whole reproductive responses to low B or particular structures have been described (e.g. abnormal pollen) long after B has impacted on developmental physiology. In future studies, therefore, there is a need to focus research on early stages in reproductive growth. The exploration of physiological, biochemical or structural responses to changes in B supply in these tissues must be able to be related to tissue B levels. Suggested model plants for this work are maize (monoecious, large inflorescences), lily (large accessible stamens and pistils) and oilseed rape (large inflorescence). Though wheat flowers are more difficult to handle during early stages in microsporogenesis, there is a range of B-efficient and B-inefficient genotypes available for experimentation (Rerkasem and Jamjod, Chapter 11). Care will have to be taken to ensure that environmental conditions do not mask or override primary effects of B, especially on pollen development.

Other approaches which could be taken in conjunction with the above studies include organ culture (Astwood and Hill, 1995) and the use of molecular techniques. Cultured wheat spikelets set seeds and produce viable seedlings (Trione and Stockwell, 1989). An increasing number of genes expressed during flower development are being identified (Huang et al., 1997; McCormick, 1993; Utsugi et al. 1996) and some of these may be useful markers which can assist in more closely identifying the primary function of B in microsporogenesis and pollen tube growth.

From the practical viewpoint, identification of the stages in reproduction which are sensitive to reduced B supply, such as the period just prior to ear emergence in wheat, may allow corrective sprays of B to be timed for field crops to prevent loss in reproductive yield. This is especially important where there are no early warning symptoms of B deficiency in vegetative organs or where foliar analysis does not adequately predict B requirement for reproductive growth (Bell, Chapter 10).

References

Agrawala S C, Sharma P N, Chatterjee C and Sharma C P 1981 Development and enzymatic changes during pollen development in boron deficient maize plants. J. Plant Nutr. 3, 329–336.

Anderson M A, Harris P J, Bonig I and Clarke A E 1987 Immunogold localization of α-L-arabinofuranosyl residue in pollen tubes of *Nicotiana alata* Link et Otto. Planta 171, 438–442.

Astwood J D and Hill R D 1995 Barley spikelet culture: An effective tool for the analysis of biochemical events in flower development. Plant Cell, Tissue and Organ Culture 40, 93–96.

Becker W M and Deamer D W 1991 The World of the Cell, 2nd ed. The Benjamin/Cummings Publ. Company Inc., Sydney.

Bedinger P 1992 The remarkable biology of pollen. Plant Cell 4, 879–887.

Bell RW, McLay L, Plaskett D, Dell B and Loneragan J F 1990 Internal boron requirements of green gram (*Vigna radiata*). *In* Plant Nutrition – Physiology and Application. Ed. M L van Beusichem. pp 275–280. Kluwer Academic Publishers, Dordrecht, The Netherlands.

Bell, RW, McLay L, Plaskett D, Dell B and Loneragan J F 1989 Germination and vigour of black gram (*Vigna mungo* (L.) Hepper) seed from plants grown with and without boron. Aust. J. Agric. Res. 40, 273–279.

Bergmann W 1992 Nutritional Disorders of Plants – Development, Visual and Analytical Diagnosis. Gustav Fischer Verlag, Jena, Stuttgart.

Birnbaum E H, Beasley C A and Dugger W M 1974 Boron deficiency in underfertilised cotton (*Gossypium hirsutum*) ovules grown *in vitro*. Plant Physiol. 54, 931–935.

Bohnsack C W and Albert L S 1977 Early effects of boron deficiency on indoleacetic acid oxidase levels of squash root tips. Plant Physiol. 59, 1047–1050.

Brett C T and Waldron K W 1996 Physiology and Biochemistry of Plant Cell Walls, 2nd ed. Chapman and Hall, London.

Brown P H and Hu H 1996 Phloem mobility of boron is species dependent: Evidence for phloem mobility in sorbitol-rich species. Ann. Bot. 77, 497–505.

Cai G, Moscatelli A and Cresti M 1997 Cytoskeletal organization and pollen tube growth. Trends in Plant Sci. 2, 86–91.

Cakmak I, Kurz H and Marschner H 1995 Short-term effects of boron, germanium and high light intensity on membrane permeability in boron deficient leaves of sunflower. Physiol. Plant. 95, 11–18.

Campbell L C, Miller M H and Loneragan J F 1975 Translocation of boron to plant fruits. Aust. J. Plant Physiol. 2, 481–487.

Chang M T and Neuffer M G 1989 Maize microsporogenesis. Genome 32, 232–244.

Cheung A Y 1996 Pollen-pistil interactions during pollen-tube growth. Trends in Plant Sci. 1, 45–51.

Cheng C and Rerkasem B 1993 Effects of Boron on pollen viability in wheat. Plant and Soil 155/156, 313–315.

Clément C and Audran J-C 1995 Anther wall layers control pollen sugar nutrition in *Lilium*. Protoplasma 187, 172–181.

Clément C, Chavant L, Burrus M and Audran J-C 1994. Anther starch variations in *Lilium* during pollen development. Sex Plant Reprod. 7, 347–356.

Clément C, Burrus M and Audran J-C 1996 Floral organ growth and carbohydrate content during pollen development in *Lilium*. Amer. J. Bot. 83, 459–469.

Coetzer L A and Robbertse P J 1987 Pollen biology of *Persea americana* Fuerte. S. Afr. Avocado Grow. Assoc. Yearbook 10, 43–45.

Cohen M S and Lepper R Jr 1977 Effects of boron on cell elongation and division in squash roots. Plant Physiol. 59, 884–887.

Cohen M S and Albert L S 1974 Autoradiographic examination of meristems of intact boron-deficient squash roots treated with tritiated thymidine. Plant Physiol. 54, 766–768.

Cresti M and Tiezzi A 1990 Germination and pollen tube formation. *In* Microspores: Evolution and Ontogeny. Eds. S Blackmore and R B Knox. pp 239–263. Academic Press, London.

Darvill A, McNeil M, Albersheim P and Delmer D P 1980. The primary cell wall of flowering plants. *In* The biochemistry of plants. A comprehensive treatise. Ed. N E Tolbert. pp 91–161. Academic Press, New York.

de Wet E, Robbertse P J and Groeneveld H T 1989 The influence of temperature and boron on pollen germination in *Mangifera indica* L. S Afr. J. Plant Soil 6, 228–234.

Dear B S and Lipsett J 1987 The effect of boron supply on the growth and seed production of subterranean clover (*Trifolium subterraneum* L.). Aust. J. Agric. Res. 38, 537–546.

Dell B 1981 Male sterility and anther wall structure in copper deficient plants. Ann. Bot. 48, 599–608.

Dell B and Malajczuk N 1994 Boron deficiency in eucalypt plantations in China. Can. J. For. Res. 24, 2409–2416.

Delvallée I and Dumas C 1988. Anther development in *Zea mays*: changes in proteins, peroxidase, and esterase patterns. J. Plant Physiol. 132, 210–217.

Derksen J 1996 Pollen tubes – a model system for plant cell growth. Bot. Acta 109, 341–345.

Dhakre G, Singh, J and Chauhan S V S 1994 Analysis of phenolics and boron in the stigma of seedless *Campsis grandiflora* K. Schum. Indian J. Exp. Biol. 32, 816–818.

Dugger W M 1983 Boron in plant metabolism. *In* Encyclopedia of Plant Physiology, New Series, vol. 15B. Eds. A Läuchli and R L Bieleski. pp 626–650. Springer-Verlag, Berlin.

Dugger W M and Palmer R L 1980 Effect of boron on the incorporation of glucose into cotton fibers grown *in vitro*. Plant Physiol. 65, 266–273.

Esau K 1977 Anatomy of Seed Plants, 2nd. ed. John Wiley & Sons, New York.

Galrao E Z and Sousa D M G 1988 Effect of boron on male sterility of wheat in an organic paddy soil. Revista Brasileira de Ciencia do Solo 12, 147–152.

Ganguly B 1979 Note on seedlessness in some wheat varieties caused by boron deficiency. Indian J. Agric. Sci. 49, 384–386.

Garcia-González M, Mateo P and Bonilla I 1990 Effect of boron deficiency on photosynthesis and reductant sources and their relationship with nitrogenase activity in *Anabaena* PCC 7119. Plant Physiol. 93, 560–565.

Garg O K, Sharma A N and Kona G R S S 1979 Effect of boron on the pollen vitality and yield of rice plants (*Oryza sativa* L. var. Jaya). Plant and Soil 52, 591–594.

Geitmann A, Hudák J, Vennigerholz F and Walles B 1995 Immunogold localization of pectin and callose in pollen grains and pollen tubes of *Brugmansia suaveolens* – Implications for the self-incompatability reaction. J. Plant Physiol. 147, 225–235.

Goldberg R B, Beals T P and Sanders P M 1993 Anther development: basic principles and practical applications. Plant Cell 5, 1217–1229.

Graham R D 1975 Male sterility in wheat plants deficient in copper. Nature, Lond. 254, 514–515.

Gupta U C (Ed.) 1993 Boron and its Role in Crop Production. CRC Press, Boca Raton, FL, USA.

Harkness R W 1959 Boron deficiency and alternate bearing in avocados. Proc. Florida State Hort. Soc. 72, 311–317.

Harris H C and Brolman J B 1966a Comparison of calcium and boron deficiencies of the peanut. I. Physiological and yield differences. Agron. J. 58, 575–578.

Harris H C and Brolman J B 1966b Comparison of calcium and boron deficiencies of the peanut. II. Seed quality in relation to histology and viability. Agron. J. 58, 578–582.

Hasegawa Y, Nakamura S, and Nakamura N 1996 Immunocytochemical localization of callose in germinated pollen of *Camellia japonica*. Protoplasma 194, 133–139.

Hirsch A M and Torrey J G 1980 Ultrastructural changes in sunflower root cells in relation to boron deficiency and added auxin. Can. J. Bot. 58, 856–866.

Hirsch A M, Pengelly W L and Torrey J G 1982 Endogenous IAA levels in boron-deficient and control root tips of sunflower. Bot. Gaz. 143, 15–19.

Hu H, Brown P H and Labavitch J M 1996. Species variability in boron requirement is correlated with cell wall pectin. J. Exp. Bot. 47, 227–232.

Hu, H and Brown, P H 1994 Localization of boron in cell walls of squash and tobacco and its association with pectin. Plant Physiol. 105, 681–689.

Huang L, Ye Z and Bell R 1996a The importance of sampling immature leaves for the diagnosis of boron deficiency in oilseed rape (*Brassica napus* cv. Eureka). Plant and Soil 183, 187–198.

Huang S R, An Y Q, McDowell, J M, McKinney E C and Meagher R B 1997 The *Arabidopsis* ACT11 actin gene is strongly expressed in tissues of the emerging inflorescence, pollen, and developing ovules. Plant Mol. Biol. 33, 125–139.

Jackson J F 1991 Borate control of energy-driven protein secretion from pollen and interaction of borate with auxin or herbicide – a possible role for boron in membrane events. Curr. Top. Plant Biochem. Physiol. 10, 221–229.

Jauh G Y and Lord E M 1996 Localization of pectins and arabinogalactan-proteins in lily (*Lilium longiflorum* L.) pollen tube and style, and their possible roles in pollination. Planta 199, 251–261.

Jewell A W, Murray B G and Alloway B J 1988 Light and electron microscope studies on pollen development in barley (*Hordeum vulgare* L.) grown under copper-sufficient and deficient conditions. Plant Cell Env. 11, 273–281.

Kamali A R and Childers N F 1970 Growth and fruiting of peach in sand culture as affected by boron and fritted form of trace elements. J. Amer. Soc. Hort. Sci. 95, 652–656.

Kaul M L H 1988 Male Sterility in Higher Plants. Springer-Verlag, Berlin.

Keerati-Kasikorn P, Bell R W and Loneragan J F 1991 Response of two peanut cultivars (*Arachis hypogaea*) to boron and calcium. Plant and Soil 138, 61–66.

Kehrel B and Wiermann R 1985 Immunochemical localization of phenylalanine ammonia-lyase and chalcone synthase in anthers. Planta 163, 183–190.

Keijzer C J, Leferink-ten Klooster H B and Reinders M C 1996 The mechanics of the grass flower: anther dehiscence and pollen shedding in maize. Ann. Bot. 78, 15–21.

Keijzer C J 1983 Hydration changes during anther development. *In* Pollen, Biology and Implications for Plant Breeding. Eds. D L Mulcahy and E Ottaviano. pp 197–201. Elsevier Biomedical, New York.

Kirk G J and Loneragan J F 1988 Functional boron requirement for leaf expansion and its use as a critical value for diagnosis of boron deficiency in soybean. Agron. J. 80, 758–762.

Kishitani S, Yomoda A, Konno, N and Tanak Y 1993. Involvement of phenylalanine ammonia-lyase in the development of pollen in broccoli (*Brassica oleracea* L.). Sex Plant Reprod. 6, 244–248.

Kouchi H and Kumazawa K 1975 Anatomical responses of root tips to boron deficiency II. Effect of boron deficiency on the cellular growth and development in root tips. Soil Sci. Plant Nutr. 21, 137–150.

Leggatt C W 1948 Germination of boron-deficient peas. Sci. Agric. 28, 131–139.

Lewis D H 1980a Boron, lignification and the origin of vascular plants – a unified hypothesis. New Phytol. 84, 209–229.

Lewis D H 1980b Are there inter-relations between the metabolic role of boron, synthesis of phenolic phytoalexins and the germination of pollen? New Phytol. 84, 261–270.

Li B H, Li W H, Kui M C, Chao W S, Jern H P, Li C R, Chu W J and Wang C L 1978 Studies on the cause of sterility of wheat. J. Northeast. Agric. Coll. 3, 1–19 (*In Chinese*).

Li Y, Faleri C, Geitmann A, Zhang H and Cresti M 1995. Immunogold localization of arabinogalactan proteins, unesterified and esterified pectins in pollen grains and pollen tubes of *Nicotiana tabacum* L. Protoplasma 189, 26–36.

Liu L, Shelp B J and Spiers G A 1993. Boron distribution and retranslocation in field-grown broccoli (*Brassica oleracea* var. *italica*). Can. J. Plant Sci. 73, 587–600.

Loomis W D and Durst R W 1992 Chemistry and biology of boron. Biofactors 3, 229–239.

MacInnes C B and Albert L S 1969 Effect of light intensity and plant size on rate of development of early boron deficiency symptoms in tomato root tips. Plant Physiol. 44, 965–967.

Malho R and Trewavas A J 1996 Localized apical increases of cytosolic free calcium control pollen tube orientation. Plant Cell 8, 1935–1949.

Marschner H 1995 Mineral Nutrition of Higher Plants. 2nd ed. Academic Press, London.

Mascarenhas J P 1975 The biochemistry of Angiosperm pollen development. Bot. Rev. 41, 259–314.

Mascarenhas J P 1989 The male gametophyte of flowering plants. Plant Cell 1, 657–664.

Mascarenhas J P 1990 Gene activity during pollen development. Ann. Rev. Plant Physiol. Plant Mol. Biol. 41, 317–338.

Mascarenhas J P 1993 Molecular mechanisms of pollen tube growth and differentiation. Plant Cell 5, 1303–1314.

Matoh T, Kawaguchi S and Kobayashi M 1996 Ubiquity of a borate-rhamnogalacturonan II complex in the cell walls of higher plants. Plant Cell Physiol. 37, 636–640.

McCormick S 1993 Male gametophyte development. Plant Cell 5, 1265–1275.

Meikle P J, Bonig I, Hoogenraad N J, Clarke A E and Stone B A 1991 The location of $(1\rightarrow3)$-β-glucans in the walls of pollen tubes of *Nicotiana alata* using a $(1\rightarrow3)$-β-glucan-specific monoclonal antibody. Planta 185, 1–8.

Moore H M and Hirsch A M 1983 Effects of boron deficiency on mitosis and incorporation of tritiated thymidine into nuclei of sunflower root tips. Amer. J. Bot. 70, 165–172.

Mozafar A 1993 Role of boron in seed production. *In* Boron and its Role in Crop Production. Ed. UC Gupta, pp 185–206. CRC Press, Boca Raton, FL, USA.

Nakamura N and Yoshida K 1980 A pectic substance extracted from the pollen tube wall of *Camellia japonica*. Japan J. Palynol. 25, 11–16.

Noppakoonwong R N, Bell R W, Dell B and Loneragan J F 1993 An effect of shade on the boron requirement for leaf blade elongation in black gram (*Vigna mungo*). Plant and Soil 155/156, 317–320.

Noppakoonwong R N, Rerkasem B, Bell R W, Dell B and Loneragan J F 1997 Prognosis and diagnosis of boron deficiency in black gram (*Vigna mungo* L. Hepper) in the field by using plant analysis. *In* Proceeding of Boron in Soils and Plants. Eds. R W Bell and B Rerkasem. pp 00–00. Kluwer Academic Publishers, Dordrecht, The Netherlands.

Obermeyer G, Kriechbaumer R, Strasser D, Maschessnig A and Bentrup F W 1996 Boric acid stimulates the plasma membrane H^+- ATPase of ungerminated lily pollen grains. Physiol. Plant. 98, 281–290.

Owen H A and Makaroff C A 1995 Ultrastructure of microsporogenesis and microgametogenesis in *Arabidopsis thaliana* (L.) Heynh. ecotype Wassilewskija (Brassicaceae). Protoplasma 185, 7–21.

Pacini E 1996 Types and meaning of pollen carbohydrate reserves. Sex. Plant Reprod. 9, 362–366.

Pollak P E, Hansen K, Astwood J D and Taylor L P 1995 Conditional male fertility in maize. Sex. Plant Reprod. 8, 231–241.

Rae A L, Harris P J, Bacic A and Clarke A E 1985 Composition of the cell walls of *Nicotiana alata* Link et Otto pollen tubes. Planta 166, 128–133.

Ram S, Bist L D and Sirohi S C 1989 Internal fruit necrosis of mango and its control. Acta Hort. 231, 805–813.

Rawson H M 1996a Parameters likely to be associated with sterility. *In* Sterility in Wheat in Sub-tropical Asia: Extent, Causes and Solutions. Eds. H M Rawson and K D Subedi. pp 13–31. ACIAR Proc. No. 72.

Rawson H M 1996b The developmental stage during which boron limitation causes sterility in wheat genotypes and the recovery of fertility. Aust. J. Plant Physiol. 23, 709–717.

Rerkasem B 1996 Boron and plant reproductive development. *In* Sterility in Wheat in Sub-tropical Asia: Extent, Causes and Solutions. Eds. H M Rawson and K D Subedi. pp 32–35. ACIAR Proc. No. 72.

Rerkasem B and Loneragan J F 1994 Boron deficiency in two wheat genotypes in a warm, subtropical region. Agron. J. 86, 887–890.

Rerkasem B, Bell R W and Longeragan J F 1990 Effect of seed and soil boron on early seedling growth of black and green gram. *In* Plant Nutrition – Physiology and Application. Ed. M L Van Beusichem. pp 281-285. Kluwer Academic Publishers, Dordrecht, The Netherlands.

Rerkasem B, Lordkaew S and Dell B 1997a Boron requirement for reproductive development in wheat. Proc. XIII Int. Plant Nutr. Colloq., Tokyo (*In press*).

Rerkasem B, Bell R W, Lordkaew S and Loneragan J F 1993a Boron deficiency in soybean (*Glycine max* L. Merr.), peanut (*Arachis hypogaea* L.), and black gram (*Vigna mungo* L. Hepper): symptoms in seeds and differences among soybean cultivars in susceptibility to boron deficiency. Plant and Soil 150, 289–294.

Rerkasem B, Bell R W, Lordkaew S, and Loneragan J F 1997b Relationship of seed boron concentration to germination and growth of soybean (*Glycine max* L. Merr.). Nutr. Cycling in Agroecosystems. (*In press*).

Rerkasem B, Netsangtip R, Bell R W and Loneragan J F 1988 Comparative species responses to boron on a Typic Tropaqualf in Northern Thailand. Plant and Soil 106, 15–21.

Rerkasem B, Netsangtip R, Lordkaew S and Cheng C 1993b Grain set failure in boron deficient wheat. Plant and Soil 155/156, 309–312.

Robbertse P J, Lock J J, Stoffberg E and Coetzer L A 1990a Effect of boron on directionality of pollen tube growth in *Petunia* and *Agapanthus*. S. Afr. J. Bot. 56, 487–492.

Robbertse P J, Coetzer L A, Bezuidenhou J J, Vorster L and Swart N G N 1990b The influence of boron on fruit set in avocado. Acta Hort. 275, 587–594.

Robertson G A and Loughman B C 1974 Response to boron deficiency: a comparison with responses produced by chemical methods of retarding root elongation. New Phytol. 73, 821–832.

Rodriguez-Rosales M P, Roldan M, Belver A and Donaire J P 1989 Correlation between *in vitro* germination capacity and proton extrusion in olive pollen. Plant Physiol. Biochem. 27, 723–728.

Schrauwen J A M and Wullems G J 1997 Regulation of flavonol biosynthesis during anther and pistil development, and during pollen tube growth in *Solanum tuberosum*. Plant J. 11, 105–113.

Schlupmann H, Bacic A and Read S M 1993 A novel callose synthase from pollen tubes of *Nicotiana*. Planta 191, 470–481.

Sharma P N and Ramchandra T 1990 Water relations and photosynthesis in mustard plants subjected to boron deficiency. Indian J. Plant Physiol. 33, 150–154.

Shelp B J 1993 Physiology and biochemistry of boron in plants. *In* Boron and Its Role in Crop Production. Ed. U C Gupta. pp 53–85. CRC Press, Boca Raton, FL, USA.

Shelp B J, Marentes E, Kitheka A M and Vivekanandan P 1995 Boron mobility in plants. Physiol. Plant. 94, 356–361.

Shen Z, Zhang X, Wang Z and Shen K 1994 On the relationship between boron nutrition and development of anther (pollen) in rapeseed plant. Scientia Agric. Sinica 27, 51–56.

Sheoran I S and Saini H S 1996 Drought-induced male sterility in rice: changes in carbohydrate levels and enzyme activities associated with the inhibition of starch accumulation in pollen. Sex. Plant Reprod. 9, 161–169.

Sherrell C G 1983 Effect of boron application on seed production of New Zealand herbage legumes. New Zealand J. Exp. Agric. 11, 113–117.

Shorrocks V M 1989 Boron Deficiency – its Prevention and Cure. Borax Consolidated, London.

Shorrocks V M 1996 Boron – important for pollen, fertilisation, seed and fruit set. Boron in Agriculture 16, 1–2.

Simojoki P 1972 Boron deficiency, pollen sterility and ergot disease of barley. Ann, Agric. Fenniae 11, 333–341 (*In Finnish*).

Singh R K and Choudhury B 1988 Differential response of chemicals on sex modification of three genera of cucurbits. Indian J. Hort. 45, 89–99.

Smith T E, Hetherington S E, Asher C J and Stephenson R A 1997 Boron deficiency of avocado. 1. Effects on pollen viability and fruit set. *In* Proc. Int. Symp. on Boron in Soils and Plants. Eds. R W Bell and B Rerkasem. pp 00–00. Kluwer Academic Publishers, Dordrecht, The Netherlands.

Spurr A R 1957 The effect of boron on cell-wall structure in celery. Amer. J. Bot. 44, 565–636.

Stead A D 1992 Pollination-induced flower senescence – a review. Plant Growth Regul. 11, 13–20.

Steer M W and Steer J M 1989 Tansley Review No.16. Pollen tube growth. New Phytol. 111, 323–358.

Steiglitz H 1977 Role of β-1,3-glucanase in postmeiotic microspore release. Dev. Biol. 57, 87–97.

Takahasi M 1993 Exine initiation and substructure in pollen of *Caesalpinia japonica* (Leguminosae: Caesalpinoideae). Amer. J. Bot. 80, 192–197.

Tanaka H 1966 Response of *Lemna paucicostata* to boron as affected by light intensity. Plant and Soil 25, 425–434.

Trione E J and Stockwell V O 1989 Development of detached wheat spikelets in culture. Plant Cell Tiss. Org. Cult. 17, 161–170.

Tsuchiya T, Toriyama K, Yoshikawa M, Ejiri S, and Hinata K 1995 Tapetum-specific expression of the gene for an endo-beta-1,3-glucanase causes male sterility in transgenic tobacco. Plant Cell Physiol. 36, 487–494.

Utsugi S, Sakamoto W, Ogura Y, Murata M and Motoyoshi F 1996 Isolation and characterization of cDNA clones corresponding to the genes expressed preferentially in floral organs of *Arabidopsis thaliana*. Plant Mol. Biol. 32, 759–765.

Van Aelst A C and Van Went J L 1992 Ultrastructural immunolocalization of pectins and glycoproteins in *Arabidopsis thaliana* pollen grains. Protoplasma 168, 14–19.

Vasil I K 1964 Effect of boron on pollen germination and pollen tube growth. *In* Pollen physiology and fertilisation. Ed. H F Liskens. pp 107–119. North Holland Publishing Company, Amsterdam.

Vaughan A K F 1977 The relation between the concentration of boron in the reproductive and vegetative organs of maize plants and their development. Rhod. J. Agric. Res. 15, 163–170.

Woodbridge C G, Venegas A and Crandall P C 1971. The boron content of developing pear, apple, and cherry flower buds. J. Amer. Soc. Hort. Sci. 96, 613–615.

Worrall D, Hird D L, Hodge R, Paul W, Draper J and Scott R 1992 Premature dissolution of the microsporocyte callose wall causes male sterility in transgenic tobacco. Plant Cell 4, 759–771.

Xu H, Huang Q, Shen K and Shen Z 1993 Anatomical studies on the effects of boron on the development of stamen and pistil of rape (*Brassica napus*). Acta Bot. Sinica 35, 453–459.

Ylstra B, Muskens M and van Tunen A J 1996 Flavonols are not essential for fertilisation in *Arabidopsis thaliana*. Plant Mol. Biol. 32, 1155–1158.

Ylstra B, Toyraev A, Moreno R M B, Stoger E, van Tunen A J, Vicente O, Mol J N M and Heberle-Bors E 1992 Flavonols stimulate development, germination, and tube growth of tobacco pollen. Plant Physiol. 100, 902–907.

Zhang X, Shen Z, and Shen K 1994 The effect of boron on the development of floral organs and seed yield of rape. Acta Pedologica Sinica 31, 146–151. (*In Chinese*).

Plant and Soil **193:** 121–148, 1997.

Chapter 9

The occurrence and correction of boron deficiency

Victor M. Shorrocks
*Micronutrient Bureau, Wigginton, Tring, Hertfordshire, HP23 6ED, UK**

Abstract

Positive responses to B application, which provide clear evidence of B deficiency, have been reported in over 80 countries and on 132 crops over the last 60 years. It is estimated that about 15 million ha are annually treated with B. In a few regions in the world, B deficient soils exist over a wide area, as for example in eastern and southern China. Elsewhere B deficiency is restricted to particular soil types and crops. Acrisols and Podzols and to a much lesser extent Andosols, Luvisols and Ferralsols, appear to be the soil groups most likely to produce B deficient crops. Soil parent material and texture are considered to be the major soil factors associated with the occurrence of B deficiency. For many crops it is the B requirement for successful fertilisation that is of critical importance; even crops with a small B requirement, such as the cereals, can suffer impaired seed set due to B shortage at a critical time. Climate, particularly high light intensity and low temperature are factors that need to be considered in relation to the occurrence of B deficiency. Boron deficiency can be readily prevented and corrected by both soil and foliar applications. Most reliance is placed on refined sodium borates, but crushed ores are used both in the manufacture of boronated fertilisers and on their own. Depending on the ore, its particle size and method of application, B supply may be extended by using an ore which dissolves slowly.

Introduction

Unequivocal proof that B is an essential micronutrient was provided by Warington (1923) who showed not only that field bean (*Vicia faba*) died when B was not supplied in the nutrient solution but that B was needed for revival. However before Warington, there had been several studies – notably by Agulhon (1910a, 1910b) and Maze (1915) in France showing beneficial effects of B on plant growth. Agulhon (1910a, 1910b) provided the first documented evidence of the occurrence of B deficiency by demonstrating that B stimulated growth of oilseed rape (*Brassica napus* var. *oleifera*), turnip (*B. rapa*) and maize (*Zea mays*) in a field trial and of oat (*Avena sativa*), radish (*Raphanus sativus*), and wheat (*Triticum aestivum*) in a sand culture experiment but he did not prove that normal growth was impossible in the absence of B.

Maze (1915) presented data to show that B (together with aluminium, fluorine and iodine) was needed for the best growth of maize in nutrient solution but he did not prove essentiality. Indeed, none of the reports before Warington (1923), satisfied the conditions of essentiality that were subsequently laid down by Arnon and Stout (1939); even the Warington study did not show that the role of B could not be replaced by another element or that B was directly involved in plant metabolism.

Occurrence of boron deficiency

Boron deficiency in crops may be suspected on the basis of symptoms and by plant analysis, and anticipated by soil analysis (Bell, Chapter 10). In this review neither symptoms nor soil or plant analyses have been used as a basis in compiling the lists of crop/country situations in which B deficiency occurs. All the crops/countries listed in Tables 1, 2, 3 and 4 are based on reported positive responses to B application in the field over the last 50 years. References for the majority of the approximately 500 listed responses can be found in the abstract bulletins, Boron in Agriculture (first series 1948–1979, second series 1980-1989) and Micronutrient News and Information (1990–1996). The use of soil B analysis to assess soil B status

* FAX No: +44-1442-822720.

Table 1. Europe: occurrence of boron deficiency based on responses

		Country*
Allium ampeloprasum	Leek	De
A. cepa	Onion	Fr
Apium graveolens var. *dulce*	Celery	Be, Ei, Fr, Gd, Ho, Uk
A. graveolens var. *rapaceum*	Celeriac	Ho
Avena sativa	Oats	Fi
Beta vulgaris	Sugar beet	Ab, As, Be, Bg, Cz, De, Ei, Fi, Fr, Gd, Gr, Ho, Hu, It, Po, Ru, Se, Sp, Si, Tk, Uk, Ur, Yu
B. vulgaris	Red beet	Gd, Fr, Uk, Ur
Betula spp.	Birch	No
Brassica napus	Oilseed rape	Cz, De, Ei, Fi, Fr, Gd, Hu, Po, Se, Uk
B. oleracea var. *botrytis*	Cauliflower	Be, Ei, Fi, Gd, No, Uk
B. oleracea var. *capitata*	Cabbage	Be, Cz, Fi, Gd, Ur, Uk
B. rapa	Turnip	Be, Ei, Gd, Uk
B. rutabaga	Swede	De, Ei, Fi, Gd, Ho, Se, Uk, Ur
Camellia sinensis	Tea	Ur
Chrysanthemum spp.	Chrysanthemum	Uk
Citrus spp.	Citrus	Sp, Ur
Cucumis sativus	Cucumber	Bg, Fi, Uk
Daucus carota var. *sativa*	Carrot	Be, De, Gd, Fi, Ho, Po, Uk, Ur
Dianthus caryophyllus	Carnation	Uk
Fragaria spp.	Strawberry	Be, Fr, Sp, Ur
Glycine max	Soybean	Ur
Gossypium hirsutum	Cotton	Bg, Sp, Ur
Helianthus annuus	Sunflower	Bg, Cz, Fr, Hu, Ru, Sp, Yu
Hordeum vulgare	Barley	Fi, Ur
Humulus lupulus	Hop	Be, Cz, It, Ur
Lactuca sativa	Lettuce	Fr, Uk
Lavandula spp.	Lavender	Bg
Linum usitatissimum	Flax	Ur
Lycopersicon esculentum	Tomato	Ei, Fr, Ho, No, Uk, Ur
Malus domestica	Apple	As, Be, Bg, Fi, Fr, Gd, Ho, Hu, It, No, Po, Pr, Sp, Uk, Ur, Yu
Medicago sativa	Lucerne	Be, Bg, Cz, Fr, Gd, Ho, Hu, Sa, Ur, Uk
Nicotiana tabaccum	Tobacco	Bg, Sp, Ur
Olea europaea	Olive	Gr, It, Pr, Sp, Tk, Yu
Onobrychis viciifolia	Sainfoin	Bg
Oryza sativa	Rice	Ur
Panicum miliaceum	Millet	Ur
Papava somniferum	Poppy	Cz, Gd, Po
Phaseolus vulgaris	Bean, french	Fr
Picea abies	Norway spruce	Fi, Gd, No, Se
Pinus spp.	Pines	Fi, No, Se, Tk, Ur
Pisum sativum	Pea	Ur
Prunus armeniaca	Apricot	Cz, Ur
P. cerasus	Cherry	Fr, Ur
P. domestica	Plum	Hu
Pyrus communis	Pear	As, Bg, Fr, Gd, Pr, Sp, Yu
Raphanus sativus	Radish	Fr, Po, Uk
Solanum tuberosum	Potato	Be, Cz, Fi, Gd, Hu, Se, Ur
Trifolium spp.	Clovers	Cz, Fi, No, Ru, Se, Ur
Triticum spp.	Wheat	Bg, Fi, Se, Ur, Yu
Tulipa spp.	Tulip	Ho
Vicia faba	Bean, field	Po, Sp, Uk
Vitis vinifera	Grape	As, Bg, Cz, Fr, Gd, Hu, It, Pr, Ru, Sp, Si, Ur, Yu
Zea mays	Maize	Bg, Fr, Ho, Po, Si, Ur, Yu

*Countries: Ab, Albania. As, Austria. Be, Belgium. Bg, Bulgaria. Cz, Czechoslovakia. De, Denmark. Ei, Ireland. Fi, Finland. Fr, France. Gd, Germany. Gr, Greece. Ho, Netherlands. Hu, Hungary. It, Italy. No, Norway. Po, Poland. Pr, Portugal. Ru, Romania. Sa, Saudi Arabia. Se, Sweden. Si, Switzerland. Sp, Spain. Tk, Turkey. Ur, USSR. Uk, United Kingdom. Yu, Yugoslavia.

Table 2. Australasia: occurrence of boron deficiency based on responses

		Country*
Allium cepa	Onion	In
A. sativum	Garlic	Id, In, Th
Ananas comosus	Pineapple	Au, Ch, Ta
Apium graveolens var. *dulce*	Celery	Ch, Jp, Nz
Arachis hypogaea	Groundnut	Ch, In, Ko, Th
Beta vulgaris	Sugar beet	Ch, Ir, Jp, Ko, In
B. vulgaris	Red beet	Au, Nz
Boehmeria nivea	Ramie	Ch
Brassica chinensis	Chinese cabbage	Ch, Jp, Ko, Ta
B. juncea	Mustard	In, Pk
B. napus	Oilseed rape	Au, Ch, In, Jp, Pk
B. oleracea var. *acephala*	Kale	Nz
B. oleracea var. *botrytis*	Cauliflower	In, Jp, Ko, Nz, Ta
B. oleracea var. *capitata*	Cabbage	Ch, In, Ko, Nz, Pg
B. rapa	Turnip	Au, Nz
B. rutabaga	Swede	Au, Nz
Cajanus cajan	Pigeon pea	In
Camellia sinensis	Tea	Ch, In, Pg, Sr
Carica papaya	Pawpaw	Au, In, Ms, Pg, Ta
Carthamus tinctorius	Safflower	In
Casuarina spp.	Casuarinas	Ch, Pg
Castanea spp.	Chestnut	Ko
Cicer arietinum	Chick pea	In
Citrullus lanatus	Water melon	In
Citrus spp.	Citrus	Ch, In, Jp, Ta
Cocos nucifera	Coconut	Id, In, Ph, Sr
Coffea spp.	Coffee	Ms, Pg
Cucumis melo	Melon	In
C. sativus	Cucumber	In
Dimocarpus longan	Longan	Ch
Elaeis guineensis	Oil palm	Id, Ms, Th
Eleusine coracana	Finger millet	In
Eucalyptus spp.	Eucalypt	Ch, Pg
Glycine max	Soybean	Ch, In, Ko, Th
Gossypium hirsutum	Cotton	Ch, In, Pk
Grevillea robusta	Grevillea	Sr
Helianthus annuus	Sunflower	Ch, In, Th
Hibiscus cannabinus	Kenaf	Ch
Hordeum vulgare	Barley	Ch, Ko
Humulus lupulus	Hop	Nz
Ipomoea batatas	Sweet potato	In, Pg
Lactuca sativa	Lettuce	Ko
Lens culinaris	Lentil	In
Linum usitatissimum	Flax	In
Litchi chinensis	Litchi	Ch, In
Lycopersicon esculentum	Tomato	Ch, In, Pg
Macadamia spp.	Macadamia	Au
Malus domestica	Apple	Au, Ch, In, Jp, Ko, Nz, Th, Ta
Mangifera indica	Mango	Au, Ch, In, Ta
Manihot esculenta	Cassava	In
Medicago sativa	Lucerne	Au, In, Ko, Nz
Mentha arvensis	Mint	In
Morus nigra	Mulberry	Ch, Ko
Musa spp.	Banana	Au, Ch
Nicotiana tabaccum	Tobacco	Ch, Id, Jp, Ko, Nz, Sr, Th

Table 2. Continued.

		Country*
Olea europaea	Olive	Ch
Oryza sativa	Rice	Ch, In, Jp
Pachyrhizus erosus	Yam bean	In
Papaver somniferum	Poppy	Au, In
Persea americana	Avocado	Au
Phaseolus vulgaris	Bean, french	In, Jp
Physalis peruviana	Cape gooseberry	Pg
Pinus spp.	Pines	Au, Ch, Nz, Pg, Ph, Vt
Pisum sativum	Pea	In
Prunus armeniaca	Apricot	Ch
P. cerasus	Cherry	Nz
P. domestica	Plum	Nz
P. persica	Peach	Ch, In, Ko, Nz
Punica granatum	Pomegranate	In
Pyrus communis	Pear	Ch, In, Ko, Nz
Psidium spp.	Guava	In, Jp, Sr
Raphanus sativus	Radish	In
R. sativus longipinnatus	Japanese radish	In, Jp
Ricinus communis	Castor oil	In
Rubus idaeus	Raspberry	Nz
Saccharum officinarum	Sugar cane	Ch, Id, In
Sesame indicum	Sesame	Ch, In
Sinapis alba	Mustard	In
Solanum melongena	Aubergine	In
S. tuberosum	Potato	Au, Ch, In
Sorghum vulgare	Sorghum	In, Pk
Trifolium spp.	Clovers	Au, Ch, In, Nz
T. subterraneum	Sub. clover	Au
Triticum spp.	Wheat	Bn, Ch, In, Ne, Pk, Th
Tulipa spp.	Tulip	Jp
Vicia faba	Bean, field	Ch
Vigna mungo	Gram, black	In, Th
V. radiata	Gram, green	In, Th, Ch
Vitis vinifera	Grape	Au, Ch, In, Jp, Ko, Nz
Zea mays	Maize	Ch, In, Ko, Pk
Zingiber officinale	Ginger	In
Ziziphus mauritiana	Ber	In

* Countries: Au, Australia. Bn, Bangladesh. Ch, China. Id, Indonesia. In, India. Ir, Iran. Jp, Japan. Ko, Korea. Ms, Malaysia. Np, Nepal. Nz, New Zealand. Pk, Pakistan. Ph, Philippines. Pg, Papua New Guinea. Sr, Sri Lanka. Ta, Taiwan. Th, Thailand. Vt, Vietnam.

and the linking of soil mapping units to the regional occurrence of B deficiency will be considered separately.

History of responses to boron application

The crops and countries where B deficiency was first identified by field experiments in the period 1920 to 1946 are listed in Table 5. Further details of the early work can be found in Anon. (1935, 1936), Dennis (1937, 1947/1948), Dennis and Dennis (1939,

Table 3. North and South America: occurrence of boron deficiency based on response

		Country*
Acer spp.	Maple	Cn
Aleurites spp.	Tung	Us
Allium ampeloprasum	Leek	Us
A. cepa	Onion	Bz
A. sativum	Garlic	Bz, Co
Ananas comosus	Pineapple	Us
Apium graveolens var. dulce	Celery	Us
Arachis hypogaea	Groundnut	Us
Asparagus officinalis	Aparagus	Cn, Us
Beta vulgaris	Sugar beet	Ci, Us
B. vulgaris	Red beet	Cn, Us
Brassica napus	Oilseed rape	Cn, Us
B. oleracea var. *botrytis*	Cauliflower	Bz, Cn, Us
B. oleracea var. *capitata*	Cabbage	Bz, Us
B. rutabaga	Swede	Cn, Us
Carica papaya	Pawpaw	Bz, Pe, Pu
Citrus spp.	Citrus	Bz, Co, Us
Citrullus lanatus	Water melon	Us
Cocos nucifera	Coconut	Mx
Coffea spp.	Coffee	Bz, Co, Cr, Ec, Pu
Coryllus spp.	Hazel	Us
Cucumis melo	Melon	Us
C. sativus	Cucumber	Bz, Us
Cynodon dactylon	Bermuda grass	Us
Daucus carota var. *sativa*	Carrot	Cn, Us
Elaeis guineensis	Oil palm	Co, Ec
Eucalyptus spp.	Eucalypt	Bz, Co, Ci
Fragaria spp.	Strawberry	Cn, Us
Ficus carica	Fig	Us
Glycine max	Soybean	Bz, Us, Ve
Gossypium hirsutum	Cotton	Bz, Us
Helianthus annuus	Sunflower	Ar, Bz
Hordeum vulgare	Barley	Cn
Ipomoea batatas	Sweet potato	Us
Juglans regia	Walnut	Us
Lactuca sativa	Lettuce	Us
Lycopersicon esculentum	Tomato	Bz, Co, Us
Malus domestica	Apple	Ci, Cn, Us
Manihot esculenta	Cassava	Bz
Medicago sativa	Lucerne	Ar, Ec, Cn, Co, Us
Musa spp.	Banana	Ec, Pu
Nicotiana tabaccum	Tobacco	Us
Olea europaea	Olive	Us
Oryza sativa	Rice	Bz, Co, Ec
Persea americana	Avocado	Us
Phaseolus vulgaris	Bean, French	Bz, Cn, Co, Us
Pinus spp.	Pines	Bz, Ci, Cn, Co
Pisum sativum	Pea	Cn, Us
Prunus amygdalus	Almond	Us

Table 3. Continued.

		Country*
P. armeniaca	Apricot	Cn, Us
P. cerasus	Cherry	Us
P. domestica	Plum	Us
P. persica	Peach	Us
Pseudotsuga menziensii	Douglas fir	Cn
Pyrus communis	Pear	Cn, Ci, Mx, Us
Raphanus sativus	Radish	Us
Rheum spp.	Rhubarb	Us
Rubus idaeus	Raspberry	Cn
Saccharum officinarum	Sugar cane	Bz, Co
Sesamum indicum	Sesame	Ve
Solanum melongena	Aubergine	Us
S. tuberosum	Potato	Bz, Co, Us
Sorghum vulgare	Sorghum	Bz, Co
Spinacia oleracea	Spinach	Us
Theobroma cacao	Cocoa	Bz, Co, Ec
Trifolium spp.	Clovers	Cn, Us
T. incarnatum	Crimson clover	Us
Triticum spp.	Wheat	Bz, Us
Vaccinium corymbosum	Blueberry	Us
Vitis vinifera	Grape	Bz, Ci, Co, Us
Vicia faba	Bean, field	Bz
Zea mays	Maize	Co, Us

* Countries: Ar, Argentina. Bz, Brazil. Cn, Canada. Ci, Chile. Co, Colombia. Cr, Costa Rica. Ec, Ecuador. Mx, Mexico. Pe, Peru. Pu, Puerto Rico. Us, United States. Ve, Venezuela.

1940/1941, 1943) and Dennis and O'Brien (1937). The first demonstration of a field response to the application of B in the USA came from experiments designed to assess crop tolerance to large amounts of borax, which were reported in the period 1920-1922, before it was known that B was essential. Although application of 2–10 kg borax ha^{-1} improved the yields of potato (*Solanum tuberosum*) and French bean (*Phaseolus vulgaris*), no attention was paid to the results as all interest was on toxicity. The toxicity studies were required because of the toxic effects of the borax-rich potash supplied to American farmers during the 1914-1918 war, when supplies of potash from the German beds were stopped. The potash, from the brine of Searles lake in California, could contain large and varying amounts (22–57%) of borax decahydrate. None of the borax-rich potash was shipped to Europe, and it is doubtful whether the American borax scare delayed the field studies on B nutrition outside the USA.

Beta vulgaris

Brandenburg (1931), in Germany, is usually credited with being the first to report on the occurrence and correction in the field of B deficiency but it is likely that experiments had been carried out earlier in Finland in 1928 even though they were reported much later (Kiviniemi, 1946). By 1935, several reports on the correction of heart rot in beet (*Beta vulgaris*) by soil application of borax had been presented in Germany and studies on beet in most European countries quickly followed. It is chastening to learn that the early workers quickly recognised that B deficiency was more likely to develop on limed land, that it was exacerbated by drought, and was worse on freely-draining sandy or gravelly soils.

Brassica rutabaga *and* B. rapa

The Brandenburg work stimulated studies on swede (*B. rutabaga*) and turnip in Canada, Finland and Wales over the period 1932–1934, which demonstrated that the brown heart symptoms that had been described

Table 4. Africa: occurrence of boron deficiency based on responses

		Country*
Acacia spp.	Wattle	Tz, Zi, Zm
Agave sisalana	Sisal	Tz
Apium graveolens var. *dulce*	Celery	Zi
Arachis hypogaea	Groundnut	Mg, Mw, Ng, Sf
Beta vulgaris	Sugar beet	Eg, Mo
Brassica rapa	Turnip	Zi
B. oleracea var. *botrytis*	Cauliflower	Mw, Zm
B. oleracea var. *capitata*	Cabbage	Zi
Camellia sinensis	Tea	Mw
Citrus spp.	Citrus	Sf, Zi
Cocos nucifera	Coconut	Iv, Mg
Coffea spp.	Coffee	Cg, Ke, Mg, Tz
Cola nitida	Kola	Ng
Cucumis melo	Melon	Sf
Elaeis guineensis	Oil palm	Ca, Cg, Iv, Mg, Ng, Za
Eucalyptus spp.	Eucalypt	Mw, Ng, Tz, Uv, Zm, Zi
Gossypium hirsutum	Cotton	An, Bb, Bt, Ca, Cd, Cr, Mg, Mw, Mz, Ng, Sg, Tz, Uv, Zi, Zm
Grevillea robusta	Grevillea	Ke
Helianthus annuus	Sunflower	Sf, Zm
Lycopersicon esculentum	Tomato	Eg
Malus domestica	Apple	Eg, Sf
Mangifera indica	Mango	Sf
Medicago sativa	Lucerne	Eg
Musa spp.	Banana	Iv, Sf
Nicotiana tabaccum	Tobacco	Mg, Mw, Sf, Tz, Zi, Zm
Parthenium argentatum	Guayule	Sf
Persea americana	Avocado	Sf
Pinus spp.	Pines	Ke, Mw, Tz, Zm
Prunus persica	Peach	Sf
Pyrus communis	Pear	Eg, Sf
Solanum tuberosum	Potato	Eg, Sf
Stylosanthes hamata	Stylosanthes	Ng
Theobroma cacao	Cocoa	Gh, Iv, Ng
Triticum spp.	Wheat	Mg, Sf, Tz, Zm
Vitis vinifera	Grape	Sf, Tu
Zea mays	Maize	Bt, Ng, Sf, Zi, Zm

*Countries: Al, Algeria. An, Angola. Bb, Benin. Bt, Botswana. Ca, Cameroon. Cd, Chad. Cg, Congo. Cr, Central African Republic. Eg, Egypt. Gh, Ghana. Iv, Ivory Coast. Ke, Kenya. Mg, Madagascar. Mw, Malawi. Mo, Morocco. Mz, Mozambique. Ng, Nigeria. Sg, Senegal. Sf, South Africa. Tu, Tunisia, Tz, Tanzania. Uv, Burkina Faso. Za, Zaire. Zm, Zambia. Zi, Zimbabwe.

some 20 years before were caused by a shortage of B. The same symptoms were also found, at about the same time in other countries, including New Zealand, Denmark and Ireland.

Crops

Boron deficiency has been reported in the field on at least 132 crops, and in 80 countries (Tables 1, 2, 3 and 4). Crop species undoubtedly plays a major part in determining whether B deficiency is a problem or not, even to the extent of obscuring soil differences in B supplying capacity. For example, beet suffers from B

Table 5. First reports of the field occurrence of boron deficiency. Based on positive responses to application of B

	Country*	
1910	Fr	Oilseed rape (*Brassica napus*), turnip (*B. rapa*), maize (*Zea mays*)
1920-1922	Us	Potato (*Solanum tuberosum*), French bean, (*Phaseolus vulgaris*)
1930	Id	Tobacco (*Nicotiana tabaccum*)
1931-1932	Gd	Fodder and sugar beet (*Beta vulgaris*)
1933-1938	Be, Cz, De, Ei, Fr, Gd, Ho, Hu, Po, Se, Si, Us, Uk	Mainly sugar beet (*B. vulgaris*)
1935-1937	Cn, De, Ei, Fi, No, Nz, Uk	Swede (*Brassica rutabaga*), turnip (*B. rapa*)
	Au, Cn, Fi, Nz, Us	Apple (*Malus domestica*)
	Us	Red beet (*Beta vulgaris*), cauliflower (*Brassica oleracea* var. *botrytis*), cabbage (*B. oleracea* v. *capitata*), carrot (*Daucus carota*), cotton (*Gossypium hirsutum*), tobacco (*Nicotiana tabaccum*)
	Zi	Orange (*Citrus* spp.)
	Ur	Flax (*Linum usitatissimum*)
	Cn, Nz	Apricot (*Prunus armeniaca*)
	Jp	Grape (*Vitis vinifera*)
	Fr, Gd, Uk	Potato (*Solanum tuberosum*)
	Cn, Us	Lucerne (*Medicago sativa*)
	Ho, Us	Celery (*Apium graveolens* var. *dulce*)
1939-1946	Us	Strawberry (*Fragaria* spp.), walnut (*Juglans regia*), olive (*Olea europaea*), red clover (*Trifolium pratense*)
	No	Barley (*Hordeum vulgare*)
	Cn	Cherry (*Prunus cerasus*)
	Au	Pear (*Pyrus communis*)
	Zi	Wheat (*Triticum* spp.)

*Countries: Au, Australia. Be, Belgium. Cn, Canada. Cz, Czechoslovakia. De, Denmark. Ei, Ireland. Fi, Finland. Fr, France. Gd, Germany. Ho, Netherlands. Hu, Hungary. Id, Indonesia. Jp, Japan. No, Norway. Nz, New Zealand. Po, Poland. Se, Sweden. Si, Switzerland. Uk, United Kingdom. Ur, USSR. Us, United States. Zi, Zimbabwe.

deficiency in nearly all countries in which it is grown, likewise lucerne (*Medicago sativa*) responds to B in over 40 states in the USA; neither crop is thus a good indicator of regional or soil differences that might be linked to B deficiency.

Crops can be rated, albeit somewhat subjectively, according to the likelihood with which they are expected to suffer from B deficiency; the crops which are most susceptible to B deficiency and which are those most responsive to B application are listed in Table 6. The validity of such a table can be questioned in view of the likely, but as yet largely unexamined, genotypic variation in the ability of cultivars of a given species to obtain and utilise B (Rerkasem and Jamjod, Chapter 11). However, the rating of crops on a sensitivity scale has been of benefit, for example in helping to set priorities for crops to study. At the practical level, a B response on a crop in the tolerant group would suggest that B responses and symptoms would be expected on crops in the sensitive group.

Plant families

Except perhaps for the Rosidae, consideration of the likely incidence of B deficiency amongst flowering plants according to taxonomic relationships reveals nothing significant. Six subclasses, 13 orders and 14 families are represented by the 16 crops in Table 6. The large subclass Rosidae is represented by seven crops in Table 6, however, no single subclass or order of plants is distinguishable as being either sensitive or tolerant to B deficiency. However, at the family level and below some differences with respect to B appear.

Among the monocotyledons, the Gramineae appear to have the lowest B requirement, due to the fact that primary cell walls of this family contain very little of the B-complexing pectin (Matoh, Chapter 5). Onion (*Allium cepa*) and asparagus (*Asparagus officinalis*) of the Liliaceae, however are pectin-rich and have high B requirements similar to, if not higher than, some of the dicotyledons. It is thus important not to think that all monocots have low B requirements as is also evidenced

Table 6. Species most sensitive to boron deficiency and most responsive to B application

Apium graveolens var. *dulce*	Celery	*Eucalyptus* spp.	Eucalypts
Arachis hypogaea	Groundnut	*Gossypium* spp.	Cotton
Beta vulgaris	Sugar beet	*Helianthus annuus*	Sunflower
Brassica spp.	Brassica	*Malus domestica*	Apple
B. rutabaga	Swede	*Medicago sativa*	Lucerne
Coffea spp.	Coffee	*Olea europaea*	Olive
Daucus carota	Carrot	*Pinus* spp.	Pines
Elaeis guineensis	Oil palm	*Vitis vinifera*	Grape

by two members of the Palmae, namely coconut (*Cocos nucifera*) and oil palm (*Elaeis guineensi*).

Amongst the dicotyledons, only two families are notable with respect to sensitivity to B deficiency, namely the Cruciferae and Chenopodiaceae. The common occurrence of B deficiency on members of the Cruciferae (*Brassica* spp. and *Raphanus* spp.) and on *Beta* spp. of the Chenopodiaceae has been recognised for over 60 years. It is likely, in the case of the Chenopodiaceae that it was the evolution of the wild sea beet (*Beta vulgaris*) under saline, B rich conditions, that resulted in the selected derivatives (sugar beet, fodder beet) having relatively high B requirements. Unfortunately, there is little opportunity to extend this hypothesis to other crop plants, as only two other members of the family are grown for human consumption, namely spinach (*Spinacia oleracea*) and quinoa (*Chenopodium quinoa*).

Soils

Soil B analysis has frequently been used to assess the B supplying capacity of soils for particular crops but there are only a few instances where sufficient data has been collected to permit soil B maps to be drawn; notable exceptions are the maps prepared by Liu Zheng et al. (1980, 1989) in China, by Bergmann et al. (1968) in the former East Germany and by Kurki (1982) in Finland.

Considerable, but locally limited data, is available to allow associations to be made between soil conditions, soil B analysis and the occurrence of B deficiency. In this section links will be attempted, using the FAO/UNESCO (1978) mapping units as being the only ones for which global maps are available.

Soil analysis and crop responses

A useful global assessment was made by Sillanpaa (1982) of soil micronutrient status by means of centralised analysis (a) of soils from 30 countries and (b) of wheat grown on these soils under identical conditions in Finland. Eight years later Sillanpaa (1990) reported on field experiments in 14 participating countries designed to assess the need for micronutrients on the main food crops [rice (*Oryza sativa*), wheat and maize].

Even though the crops used are not particularly responsive to B, responses were shown in 11 of the 14 countries (Table 7) where field trials were carried out (Sillanpaa, 1990). Acute B deficiency was rated in 10% of all trials and a latent B deficiency in a further 21%. The fact that B responses were found in trials in countries where the national mean B concentrations are undoubtedly high, is a further indication of the danger of attempting to generalise on a country or regional basis from a limited database.

Inevitably in a study of this kind, there is the problem of obtaining sufficient samples to typify any given country, but with this limitation in mind it is possible to rate countries according to the average national soil B concentrations (hot water soluble); the values extracted in Table 7 should be viewed as indications of the likely average situation on the selected cultivated soils and should be modified in the light of local field knowledge on the occurrence of B deficiency.

Sillanpaa (1982) concluded that B deficiency would be possible on crops at some locations in almost every country but the likelihood of deficiency was greater in several countries in the Far East notably, India, Korea, Nepal, Philippines and Thailand and also in Africa particularly Malawi, Nigeria, Sierra Leone and Zambia. On the other hand, B toxicity was most likely in Iraq, Mexico, Pakistan and Turkey particularly at irrigated

Table 7. National mean boron concentrations (Sillanpaa, 1982)

Country	Response to B	Sample size	Hot water soluble B (mg L^{-1})
Nepal	+	35	0.19
Zambia	–	44	0.25
Nigeria		153	0.27
Philippines	+	194	0.28
Malawi	+	97	0.29
Sri Lanka	+	18	0.32
Korea		90	0.37
Sierre Leone	–	48	0.39
Thailand	–	150	0.42
India		128	0.42
Belgium		36	0.52
Lebanon		16	0.54
Ethiopia	+	125	0.54
Finland	+	90	0.55
Ghana		93	0.57
Brazil		158	0.63
Peru		68	0.65
Pakistan	+	237	0.68
Tanzania	+	163	0.78
Argentina		208	0.74
New Zealand		35	0.76
Italy		170	0.82
Egypt		198	0.93
Hungary		201	1.02
Syria		38	1.08
Turkey	+	298	1.10
Malta		25	1.14
Mexico	+	142	1.26
Iraq	+	150	1.51

+ Positive responses to B application.
– No response to B (Sillanpaa, 1990).

sites. These are not comprehensive lists as the majority of countries did not participate in the study.

Soil groups

The limited data presented by Sillanpaa (1982) permits an initial examination of the link between soil B levels and soil groups. In Table 8 the soil groups have been listed in ascending order of soil B concentration. A few features emerge.

Soils with low B concentrations include: strongly weathered soils (Acrisols, Podzols, Ferralsols); coarse textured soils (Arenosols); shallow soils (Lithosols); thin soils over calcareous material (Rendzinas); and volcanic ash soils (Andosols).

Not unexpectedly it was the strongly weathered and coarse textured soils that contained the least available B. Whether B deficiency can be linked to any one soil unit is discussed below.

Global assessment of boron deficient areas

There are only a few regions in the world where B deficiency occurs sufficiently frequently on crops to permit relatively large areas to be mapped on a world scale. These areas are roughly identified in Figure 1 and listed in Table 9, together with the principle soil groups. Elsewhere, B deficiency exists on relatively small areas and is more dependent on local conditions and crops; this makes it impossible in the absence of

Table 8. Mean hot water soluble boron concentration (corrected for variation in CEC) for FAO/UNESCO soil groups (Sillanpaa, 1982). USDA Soil Taxonomy equivalents are shown in parentheses.

Mean hot water soluble B ($mg\ L^{-1}$)	Sample size	Soil group
0.5–0.6	48	Rendzina (Rendoll)
	73	Acrisol (Ultisol)*
	22	Arenosol (Psamment)*
	8	Andosol (Andept)
	31	Gleysol (Aquic)**
	27	Podzol (Spodosol)
	135	Vertisol (Vertisol)
	23	Lithosol (Lithic)***
	127	Ferralsol (Oxisol)
0.6–1.0	106	Nitosol (some Ultisols and Alfisols)
	217	Luvisol (Alfisol)*
	246	Cambisol (Inceptisol)
	7	Histosol (Histosol)
	307	Phaeozem (Udoll)
	64	Kastanozem (Ustoll)
	48	Chernozem (Boroll)
	11	Planosol (–)
	470	Fluvisol (Fluvent)
1.0–2.0	42	Regosol (Orthent, Psamment)
	101	Xerosol (Aridisol)*
	92	Yermosol (Aridisol)*
	60	Halosol (–)

*partial correspondence; **corresponds to aquic subgroups of various Great groups; ***corresponds to lithic subgroups of various Great groups.

Figure 1. Boron deficient areas. Areas identified on the basis of crop responses to boron and by soil B analysis. Solid lines indicate that boundaries have a firmer basis.

Table 9. Main FAO/UNESCO soil groups in principal boron deficiency areas. USDA Soil Taxonomy equivalents are shown in parentheses

		Soil group
Asia	China, south of Yangtze	Acrisols (Ultisols*)
	China, north of Yangtze	Lithosols/Cambisols (Lithic Inceptisols)
	China, north east	Lithosols/Luvisols (Lithic Fluvents)
	Korea	Lithosols/Cambisols (Lithic Inceptisols)
		Acrisols (Ultisols*)
		Cambisols (Inceptisols)
	Thailand, north west	Acrisols (Ultisols*)
	India, north east	Acrisols (Ultisols*)
		Luvisols (Alfisols*)
West Africa	Nigeria and Benin	Arenosols (Psamments*)
		Luvisols (Alfisols*)
	Ivory Coast	Acrisols (Ultisols*)
	Chad	Arenosols (Psamments*)
Central Africa	Zambia, Zimbabwe, Malawi	Luvisols (Alfisols*)
		Ferralsols (Oxisols)
Scandinavia and North Europe	Denmark, Germany, Norway, Sweden	Podzols (Spodosols)
	Finland, USSR	Luvisols (Alfisols*)
		Podzoluvisols (Alfisols*)
North America	USA, Atlantic SE coastal plain	Acrisols (Ultisols*)
	USA, Pacific Northwest	Podzols (Spodosols)
		Luvisols (Alfisols*)
		Acrisols (Ultisols*)
	USA, Great Lakes area	Podzols (Spodosols)
		Luvisols (Alfisols*)
	Canada, Atlantic	Podzols (Spodosols)
	Canada, north west	Podzols (Spodosols)
South America	Brazil	Ferralsols (Oxisols)
		Acrisols (Ultisols)
	Chile	Andosols (Andepts)

*Partial correspondence.

detailed soil studies to link B deficiency and soil B levels to soil groups. The situations discussed here are those where sufficient data is available.

Asia. China: Cultivated soils in south and east China, south of the Yangtze and many of the soils in northeast China are known, on the basis of B responses as well as soil analysis, to constitute the single largest contiguous area of B deficiency in the world (Liu Zheng et al., 1980, 1982a, 1982b, 1983). The soil B maps of China (Liu Zheng et al., 1980, 1989), show large areas of soils containing very low levels of hot water soluble B (<0.25 mg kg^{-1}). Acrisols predominate south of the Yangtze, many of which are formed on acid igneous and metamorphic rocks and on fluviatile deposits in the valleys, giving soils locally described as red earths and laterites. North of the Yangtze, B deficiency is common on soils derived from loess and calcareous alluvium of the Yellow river (Liu Zheng et al., 1980). In contrast the soils in the arid regions of western China, mostly classified as Lithic Kastanozems and Lithic Yermosols, are rich in B.

India: Soil B analyses (over 13,000 mostly from just four states) and crop responses (Table 2) show that B deficiency occurs quite commonly in several states. Links with soil groups have not been reported. Sakal and Singh (1995) highlighted northern Bihar, part of Assam, West Bengal, Meghalaya and northern Orissa, all in northeast India together with Karnataka and Gujarat as areas where B deficiency was most common. Granites and gneisses are the predominant parent material in these states, with the exception of

Gujarat. In Maharashtra, where most soils are formed from the basalts of the Deccan Trap, soil B levels seldom indicate B deficiency.

Acrisols are widespread in Assam whereas, in the other listed states Luvisols predominate. Luvisols and Acrisols are common in all but northern Karnataka. Luvisols are also prevalent in Tamil Nadu and Kerala (Acrisols also) but little soil B data is available for these states.

Sri Lanka: Boron deficiency has long been recognised on high grown tea (*Camellia sinensis*) on the Acrisols in the wetter central and south-western parts of Sri Lanka; most of these soils are derived from acid metamorphic rocks.

Africa. Cameroon, Chad, Benin, Ivory Coast, Nigeria: The occurrence of B deficiency on cotton (*Gossypium hirsutum*) has been known for many years in these countries where metamorphic and acid igneous rocks are fairly common and on which have formed Acrisols, notably in Ivory Coast, and Luvisols and Arenosols in Nigeria and Benin. Limited soil B data is available; in Nigeria Lombin (1985) reported low B levels and Odunze et al. (1992) found 0.32–0.49 mg B kg^{-1} in Luvisols from the Nigerian savannah.

Malawi, Zimbabwe, Zambia: As in west Africa recent interest has centred on cotton but tobacco (*Nicotiana tabacum*) has long been treated with boronated fertilisers in these countries. In Zimbabwe, where over 60% of the soils are derived from granitic and metamorphosed igneous rocks, Luvisols are very common in the eastern half of the country. A similar geological situation prevails in Zambia where Ferralsols have formed on the ancient acidic parent materials.

Europe. Scandinavia and Northern Europe: Podzols and Luvisols (also some Cambisols) cover much of the B responsive areas in Finland, Sweden, Denmark, Germany and Poland. In Finland and Sweden, the soil parent materials are almost exclusively acid igneous and metamorphic rocks.

Belarus, Estonia, Latvia, Lithuania and Russia: To the east in the former USSR, the soils with the lowest B levels are the Podzoluvisols and Luvisols (mainly formed on Devonian and Carboniferous sediments), whereas the Chernozems and Kastanozems to the south, formed on more recent sediments, are better supplied with B (V Ivchenko pers. comm., 1993); Pieve (1959) quoted water soluble B values (mg kg^{-1}) of 0.08–0.38 in Podzoluvisols, 0.38–1.58 in Chernozems and 0.30–0.90 in Kastanozems. About 35% of soils in the European part of the former USSR are rated as having a low B status.

Germany: As in the former USSR, many of the soils in the south of eastern Germany are Cambisols which Bergmann et al. (1968) found contained more B than the Podzols and Luvisols to the north.

Poland: Over 50% of the soils (mostly Podzols) in 11 provinces in central and eastern Poland were rated as low in B by Gembarzewski (1987).

Latvia: 40% of soils, mainly Podzoluvisols were found to be B deficient (Anspoks and Liepins, 1987).

Finland: The Podzols that cover most of southern Finland are likely to contain, on average, <0.5 mg hot water soluble B L^{-1} (Kurki, 1982); soil B levels were greater in soils with higher base exchange capacity and higher pH.

Americas. Brazil: Many of the B responsive soils in the state of Sao Paulo are known locally as various latosols (Ferralsols) and podzolic soils (Luvisols and Acrisols) (Sobrinho and Freire, 1980). Soluble B concentrations are usually <0.5 mg kg^{-1}; with the Ferralsols containing less B than the Luvisols; many of the soils are derived from basaltic parent material. Boron responsive soils also occur throughout the "Cerrado" region, which covers much of Minas Gerais, Mato Grosso and Goias (Lopes, 1984) and where 56% of soils are Ferralsols, 20% Arenosols and 10% Acrisols.

Chile: Many soils in central Chile carry B deficient crops (Tollenaar, 1970), including beet, pine (*Pinus radiata*) and grape (*Vitis vinifera*). Most are Andosols formed from basaltic-andesitic ashes, but a few are Cambisols derived from granitic and metamorphic rocks (Etchevers et al., 1983; Schlatter and Gerding, 1985).

USA: Although B deficiency has been reported in over 40 states (in most of which lucerne predominates) three areas have traditionally been cited (Anon., 1977; Berger, 1962) as being B deficient, namely the states along the Atlantic and Gulf of Mexico, the Pacific Northwest and the Great Lakes region; the occurrence of Podzols (Great Lakes and northeast), Acrisols (southeast) and Andosols and Acrisols (northwest) is noteworthy.

Canada: Boron deficiency occurs in the northwest in British Columbia on various soils, including Podzols derived from glacial till as well as some Luvisols and Kastanozems in the Okanagan valley; in the eastern Atlantic states on Podzols (Gupta, 1983, 1984).

Soil groups closely associated with boron deficiency
Boron deficiency seems to occur more frequently on certain soil groups than on others. Identification of problem areas will be facilitated if these associations are confirmed by future work. The Acrisols and Podzols are prominent as often carrying B deficient crops, less noteworthy are the Andosols and Luvisols.

Acrisols. These strongly weathered, and frequently coarse-textured soils with low base exchange capacity are, together with the Podzols, the soils on which B deficiency most often occurs; they are the soils that typically develop under humid conditions. Apart from the countries listed in Table 9, over 50% of the soils in southeast Asia are classified as Acrisols. The occurrence of B deficiency on Acrisols in Indonesia, Malaysia, Thailand, Laos and Vietnam is recognised and in Burma low soil B levels (0.1–0.7 mg kg^{-1}) have been reported (Gyul'akhmedov and Mamedov, 1984); however, in the absence of sufficient soil B data it is not possible to consider these countries as one contiguous B deficient zone.

Acrisols are found in other countries, apart from China, USA and Brazil, all of which have more than 100 million ha of Acrisols (FAO/UNESCO, 1978): 30–100 million ha (Burma, Indonesia, Tanzania and Thailand); 10–30 million ha (Australia, Bolivia, Cambodia, Colombia, India, Ivory Coast, Laos, Malaysia, Peru, Venezuela and Vietnam); and 4–10 million ha (Central African Republic, Ghana, Guinea, Japan, Mexico, Nicaragua, Nigeria, Papua New Guinea, Paraguay, Philippines and Suriname).

A small (200,000 ha) area of Acrisols in the north of South Island New Zealand is particularly noteworthy as it is the area on which Askew et al. (1936) discovered B deficiency on apple (*Malus domestica*) in 1935 and on which Rigg et al. (1937) reported wide-scale B deficiency on swede and turnip. Boron deficiency has since been found on other crops in the area including hop (*Humulus lupulus*) (Askew and Monk, 1953), raspberry (*Rubus idaeus*) (Askew et al., 1951) and pine (*Pinus radiata*) (Stone and Will, 1965). These Acrisols are formed on alluvium from metamorphic rocks and greywackes.

Sillanpaa and Vlek (1985) highlighted three countries in the tropics as B deficient zones namely Colombia, Nigeria and Zambia, the first two of which are notable for their Acrisols.

Podzols. Podzols commonly develop under coniferous forest in climates often unsuitable for agriculture but their occurrence in such regions as Scandinavia, Europe and North America appears relevant to the B situation. Elsewhere, for example in Australia (Tasmania and NSW) podzolic soils were some of the first to show B deficiency on apple, swede and turnip (Savage and Broadfoot, 1937). In northeast (Atlantic) Canada, B deficiency occurs on cabbage (*Brassica oleracea*), carrot (*Daucus carota*), lucerne and radish on podzolic soils (Gupta, 1983, 1984; Gupta and Cutliffe, 1975, 1985).

Andosols. The strong adsorption of B by allophane in soils derived from volcanic ash (Bingham et al., 1971) would appear to be the common factor in relation to the occurrence of B deficiency in such countries as Chile, Colombia, Ecuador, Japan, New Zealand and USA. The Andosols in central-southern Chile can adsorb massive amounts of B (Schalscha et al., 1973). However it may be dangerous to extrapolate to all Andosols, as demonstrated by a survey of parts of southern Chile by Rodriguez and Tomic (1984); they found wide (0.3–1.5 mg kg^{-1}) variation in hot water soluble B concentrations in Andosols in a region, where 70% were rated to be deficient for a sensitive crop such as beet. Lora (1978) obtained B responses on Andosols in Colombia on crops not normally rated as sensitive, such as potato and garlic (*Allium sativum*). Boron deficiency occurs in *Pinus radiata* (Will, 1971) on Andosols (coarse-textured pumice soils) in north island of New Zealand.

Arenosols and Lithosols. Sandy soils that have been well leached and thin rocky soils are prime candidates for B deficiency as indicated by Tables 8 and 9. However, the variable nature of Arenosols and Lithosols, particularly with regard to parent material, means that as groups they can not be classified as B deficient.

Luvisols. Luvisols, which together with the Acrisols are characterised by illuviation of clay, commonly occur throughout the world. Although in several countries they carry B deficient crops, they are not closely linked with B deficiency, probably due to varying parent materials and the fact that these soils normally have a high base exchange capacity. For example, the Luvisols in northeast India are B deficient most probably because they are formed from B poor parent materials.

Table 10. Estimated areas ('000 ha) and main crops treated annually with boron. Based on 1 kg B ha^{-1}

	'000 ha	
Europe	3000–4000	Beet (*Beta vulgaris*), oilseed rape (*Brassica napus*), grape (*Vitis vinifera*), sunflower (*Helianthus annuus*), fruit, vegetables
Australasia	3700–5000	Cotton (*Gossypium* spp.), oil palm (*Elaeis guineensis*), oilseed rape, groundnut (*Arachis hypogaea*)
Africa	600–1000	Cotton
Americas	6000–8000	Lucerne (*Medicago sativa*), coffee (*Coffea* spp.), cotton, fruit, vegetables

Extent of boron deficiency

There have been very few published statistics on the use of agricultural borates (either in terms of tonnes of products or areas treated), which makes it very difficult to assess the actual areas of different crops given B on a regular basis. Experience in the global agricultural borate market, combined with information published by the US Bureau of Mines on the use of B has permitted an estimate of the likely current use (Table 10) on a continental basis. In the USA, 5–7% of all B has historically been used in agriculture, in the rest of the world the percentage is smaller. The estimates in Table 10 are consistent with an annual global consumption for all end uses of B of 350–400,000 tonnes B, of which 6% USA and 3–4% in the rest of the world is for agricultural use. No allowance has been made, in Table 10, for B use in the former USSR.

The estimates of areas treated annually are based on a uniform application rate of 1 kg B ha^{-1} throughout the world; it is likely that in USA and Europe B rates are higher, and in the rest of the world lower, than 1 kg ha^{-1}. In the absence of survey data it is not possible to achieve greater precision.

Factors associated with the occurrence of boron deficiency

Within the few large, contiguous areas where the soils are of low B status, responses to B can be expected on many crops. However, even in such areas, local conditions of soil, climate and crop species will influence the incidence of B deficiency. In most countries in the rest of the world, it is these local conditions that play the major role and result in B deficiency occurring on particular crops and soils.

The occurrence of B deficiency at the local level can be related to several factors, including: the ability of the soil to provide sufficient available B during the growing season; the crop requirement and the removal of B; nature of the crop and of the harvested portion; climate; and management. Many of these factors, particularly those relating to soils, are discussed in detail elsewhere in this book. In this section the operation of these factors in the field will be considered, particularly their significance to the occurrence of B deficiency.

Supply of boron from soil

Boron can be considered as existing in five categories in the soil: in primary minerals such as tourmaline and B rich micas (1); in secondary minerals, particularly within the clay mineral lattice (2); adsorbed on clays, hydrous oxide surfaces and organic matter (3); in solution usually as boric acid but also as borate anions (4); and in organic matter and the microbial biomass (5).

Parent material, soil texture, pH, liming and leaching can be viewed as variously influencing B in one or more of these categories; parent material and soil texture are considered to be the dominant factors.

Parent material

Soil parent materials differ widely in their B content (Table 11), thereby having a fundamental effect on B in categories 1 and 2; they also differ in their content of clay-forming minerals which ultimately dictate how much of the B released from 1 will be held in the soil.

These differences are commonly reflected in the B contents of soils as well as in the regional occurrence of B deficiency. For example, granite-derived soils often carry B deficient crops, as in Korea where upland soils on granite were found to contain on average 0.07–0.15 mg hot water soluble B kg^{-1} (Park and Park, 1966) compared to 0.25–0.35 in soils derived from basalt and up to 0.50 mg kg^{-1} in soils from sedimentary rocks.

Table 11. Boron concentrations (mg kg^{-1}) in major rock types

		B (mg kg^{-1})[1]	B (mg kg^{-1})[2]
Igneous	Basic: gabbro, basalt	5–20	5
	Intermediate: diorite	9–25	
	Acid: granite, rhyolite	10–30	15
Metamorphic	Gneiss	10–30	
Sedimentary	Shale	120–130	100
	Sandstone	30	35
	Limestone, dolomite	20–30	20

[1] Data from Kabata-Pendias and Pendias (1992).
[2] Data from Sillanpaa and Vlek (1985).

Table 12. Total and water soluble boron contents (mg kg^{-1}) of Chinese soils[1]

Region	Soil group	Parent material	Total B (mg kg^{-1})	Water Soluble B (mg kg^{-1})
South of Yangtze	Acrisols	Granite	4-16	0.06–0.29
	Acrisols	Rhyolite	13–19	0.21–0.24
	Acrisols	Phyllite	15–40	0.02–0.19
	Acrisols	Red Sandstone	25–28	0.0–0.11
	Acrisols	Gneiss	13–22	0.0–0.13
North of Yangtze	Various	Loess	55–87	0.14–0.70
Central China just north of Yangtze	Various	Quaternary red clay	48–96	0.08–0.37
–	–	Limestone	49–145	0.14–0.34

[1] Liu Zheng et al. (1980). B deficiencies occur in all soil regions but particularly on the Acrisols.

Studies in China have linked the low B soils with parent material. For example, Liu Zheng et al. (1980, 1982a, 1982b, 1983) found that soils derived from granite and other igneous rocks, gneiss and sandstone were particularly low in both total and water soluble B (Table 12), whereas soil derived from loess contained more B. In the northeast of China, where many soils are derived from loess and calcareous alluvial deposits of the Yellow river, B deficiency is not so serious. Despite the clear links between B deficiency in China and parent material, it is not possible to assess the extent to which the current low soil B status is due to parent material, to soil age, to the intensity of weathering or to the extensive exploitation of the soil for food production.

Granitic and metamorphic rocks frequently give rise to B deficient soils and their importance as a common factor, regardless of soil group, cannot be underestimated. Boron is lost during metamorphism, some contributing to tourmaline formation, but most is released into the environment (Harder, 1961). In contrast, young soils derived from fine grained marine sediments will be B rich. The factor of greatest importance in determining the B supplying capacity of soils derived from sedimentary rocks is undoubtedly the clay content and possibly the type of clay. Marine clays and shales can accumulate much B particularly if they contain illite, which incorporates B into the clay lattice. Sediments in which glauconite forms will contain much B, as it is incorporated as the mineral forms in the presence of sea water (Harder, 1970). Leakage of B from illite and glauconite as they weather could be relevant to the soil buffering capacity for B.

Soil texture

Soil texture, which itself is very dependent on parent material and on the weathering and the illuviation of clay from the top soil, has a major influence on B in categories 3 and 4. Coarse-textured soils contain less total and water soluble B, and as a result, B deficiency commonly occurs more frequently on these soils than on fine-textured, less freely drained soils. Boron held in categories 2 and 3 comprise very important soil reserves in fine-textured soils which as a result seldom carry B deficient crops.

It seems likely that the occurrence of B deficiency in the highly weathered Acrisols and Podzols is basically a function of the loss of B reserves held with clay minerals. In the similarly weathered Luvisols, with generally higher clay contents (base exchange capacity), B deficiency is not as likely to occur.

Somewhat paradoxically, less B is required to correct B deficiency on coarse-textured soils than on fine-textured soils due to the stronger adsorption of B by the latter. An understanding of the nature of this phenomenon was provided by Wear and Patterson (1962) who showed that absorption of B by lucerne was greater from a coarse- than from a fine-textured soil containing the same concentration of hot water soluble B. Liming had a much greater impact in reducing B absorption in the coarse-textured than in the fine-textured soil. Furthermore, B losses by leaching will be greater on freely drained coarse-textured than on impervious, fine-textured soils. Coarse-textured soils will quickly dry out during droughts, making crops more reliant on deep roots growing in the lower, B poor soil horizons. In England and Wales soils which contain <0.2 mg hot water soluble B L^{-1} are almost invariably sandy in texture (Archer and Hodgson, 1987). Boron deficiency frequently occurs in England on beet and swede on soils derived from Triassic and Devonian sandstones as well as post-glacial sands (Farrer, 1976). In France the risk of B deficiency on sunflower (*Helianthus annuus*) is highest on soils containing more than 15% fine and coarse sand, on thin soils and soils with pH > 7.0 (Merrien, 1993).

Soil pH, clay minerals and hydrous oxides of iron and aluminium

It is frequently stated that soil pH is one of the most important factors affecting availability of soil B. However, except in the case of the adverse effects of liming on B availability, soil pH *per se* appears to have little practical bearing, via its effect on adsorption by clay minerals and aluminium and iron oxides. The strong adsorption of B at both high and low pH levels has not been held responsible for causing B deficiency. Despite adsorption of B by soils, clay minerals and hydrous oxides being maximal at about pH 8–10, B deficiency does not occur on soils with a naturally high pH such as the desert soils of North Africa and Saudi Arabia, or on soils with very high calcium carbonate levels. At low pH levels it will be factors associated with low fertility (e.g. leached, coarse-textured soils) that will be more relevant than any tendency for adsorption to increase.

The many adsorption studies reported in the literature have not had a major impact on the understanding of the non-toxic behaviour of either indigenous or applied B despite the large differences between adsorbents and the marked pH effects. The understanding of the toxic effects of B rich irrigation water has, however, been considerably helped by adsorption studies. It remains to be seen whether the strong adsorption (and weak desorption) by humic acid and by the "short-range" clay minerals, imogolite and ferrihydrite plays a significant part in determining B availability and deficiency in any soils. Boron can be adsorbed by calcium carbonate (Goldberg and Forster, 1991) and some workers report positive correlations between amounts of adsorbed B and calcium carbonate. However, there is little, if any evidence, to indicate that such adsorption is responsible for B deficiency anywhere in the world.

It is the increased adsorption by raising the pH within the range pH 4-7 (at which B adsorption by phylosilicate clay minerals is usually minimal) that is in practice of greatest significance in inducing B deficiency. It is important to distinguish between situations where liming has recently been carried out and where "equilibration" may not be complete, and those where the pH of a soil has been raised a considerable time before. Peterson and Newman (1976), for example, showed on the same soil that neither indigenous B uptake nor recovery of applied B by *Festuca arundinacea* was affected over a pH range of 4.7–6.3 when soils had been limed 11 years previously. However, when the pH was raised to 7.4 by liming 4 months before the experiment, B uptake and recovery were markedly reduced. Likewise Wear and Patterson (1962) found that liming just before B treatment reduced uptake of B by lucerne and that the effect was more marked on a coarse-textured soil than on a silty clay soil.

Liming

Liming of soil with marginal B content frequently induces B deficiency. The idea that freshly precipitated aluminium hydroxide is responsible for scavenging soluble B has been widely accepted for many years (Hatcher et al., 1967; Sims and Bingham, 1968a, 1968b). The fact that the ability of aluminium hydroxide to adsorb B diminishes with time, is in accord not only with the soil situation mentioned above, but it fits with the idea that it is the recent effects of liming rather than a general pH rise which markedly affects B availability.

Organic matter
Many workers (e.g. Elrashidi and O'Connor, 1982) have shown strong correlations between B and organic carbon contents of soil. Mineralisation of organic matter releases some B. The risk of B deficiency increases when organic matter contents decline. On the other hand, organic matter has the ability to complex large amounts of B (usually more than mineral soil constituents e.g. Yermiyaho et al., 1988), a feature which has sometimes (e.g. Liu Zheng et al., 1989) been held responsible for B deficiency. However there is no suggestion that Histosols or Chernozems can be considered as being basically deficient in B.

Leaching
Data is available to support the logical view that leaching losses of B will be maximal on the coarse-textured soils e.g. Kubota et al. (1948), Ouellette (1958) and Wilson et al. (1951). There have, however, been few attempts to build models that can be used to predict the vertical (and lateral) movement of B in soils (the work of Wild and Mazaheri (1979) with soil columns is an exception); the lack of information is unfortunate in view of the importance of leaching not only in the short term during agricultural use of the soil, but also during the millennia of soil formation. More is known about the behaviour of potentially toxic amounts of B supplied in irrigation water than about indigenous and corrective B applications.

Distance from sea
Significant amounts of B can be supplied in rainfall in coastal regions, for example Wikner (1983) reported that soils near the west coast of Sweden received 25–40 g B ha^{-1} annually compared to <2 g B ha^{-1} in northern Sweden far from the sea. Sea water normally contains 4.5 mg B kg^{-1}. Similar data is available for Western Australia (Hingston, 1986).

Crop requirement and crop removal of boron

Data (Shorrocks, 1991) on crop B requirements and removals as given in Table 13 should be used to indicate little more than the general order of magnitude of the amounts involved. The values must be linked to a yield level as well as to B concentrations. It is only possible to contrast these amounts with the B supplying capacity of the soil in a roughly qualitative manner. It is likely that most crops remove less than 100 g B ha^{-1} and that it is only crops such as lucerne and beet that often remove more than 300 g B ha^{-1}. Despite the small amounts of B removed by cereals in grain, B deficiencies do occur because of the particular B requirement for fertilisation and seed set.

Nature of the crop and of the harvested portion

The relative sensitivity of different crops to B deficiency was discussed in "Crops". Here the particular features of crops that are relevant to the effects of impaired B supply are considered. It is extremely rare for B deficiency to result in crop death in the field, although cases are known (e.g. *Pinus radiata* in Chile and oilseed rape in China). The economic consequences of an inadequate supply of B have to be considered in relation to whether damage is caused to a particular part of the plant, or whether overall growth is impaired. The symptoms that have the greatest impact on the performance of the crops which are most sensitive to B deficiency, are shown in Table 14. Impairment or death of the apical meristem has serious consequences for crops such as beet, *Eucalyptus* spp., *Pinus* spp. and oil palm. For the vegetable brassicas and carrot, it is essential that damage to root and stem tissues does not reduce quality.

For virtually all crops, not only cotton, sunflower, apple and grape, included in Tables 6 and 14, it is the B requirement for pollen tube growth and pollen viability, and thereby for seed and fruit development, that is of vital importance. For example, many clovers (*Trifolium* spp.) produce more seed when given B, but will seldom show increased dry matter production. Boron deficient wheat and other small grained cereals exhibit sterility without exhibiting any other symptoms (Rerkasem, 1996). Unfertilised fruit will fall and partially fertilised fruit will be misshapen [e.g. papaya (*Carica papaya*), apple and pear (*Pyrus communis*)]. The optimal concentrations of boric acid in culture media for pollen germination and growth are high for many species – for some such as chick pea (*Cicer arietinum*), coconut, eucalypts, *Hibiscus esculentis*, olive (*Olea europaea*) and almond (*Prunus amygdalus*), 100 mg B L^{-1} is required (Anon., 1996).

Table 13. Boron requirement for growth and removal in harvested crop (g B ha^{-1})

		Yield t ha^{-1}	B requirement for growth (g ha^{-1})	B removal in crop (g ha^{-1})
Beta vulgaris	Sugarbeet	50	480	300
Brassica napus	Oilseed rape	4	320	80
Gossypium hirsutum	Cotton	3	225	150
Helianthus annuus	Sunflower	3.5	400	100
Malus domestica	Apple	40	380	250
Medicago sativa	Lucerne	7		350
Triticum spp.	Wheat	3	60	25
Vitis vinifera	Grape	6	240	120
Zea mays	Maize	6	150	30

Table 14. Agronomically significant symptoms on crops most sensitive to boron deficiency

	Symptoms	Species
Impaired meristem activity	Dieback	Beet (*Beta vulgaris*)
		Eucalypt (*Eucalyptus* spp.)
		Pine (*Pinus* spp.)
	Malformed leaves	Oil palm (*Elaeis guineensis*)
		Coffee (*Coffea* spp.)
Impaired cell wall development	Cracking of stems and petioles	Celery (*Apium graveolens* var. *dulce*)
		Chinese cabbage (*Brassica chinensis*)
		Beet
Protection against phenols	Damaged tissue	Swede (*Brassica rutabaga*)
		Carrot (*Daucus carota*)
Pollen tube growth	Reduced seed set	Sunflower (*Helianthus annuus*)
		Coffee
		Oilseed rape (*Brassica napus*)
		Olive (*Olea europaea*)
		Clover (*Trifolium* spp.)
	Fruit fall	Cotton (*Gossypium hirsutum*)
		Apple (*Malus domestica*)
	Deformed fruit	Apple
		Grape (*Vitis vinifera*)
		Olive

Climate

Climate can have an impact on B deficiency, both in the short and long term. The more extreme the conditions of rainfall and temperature, and the longer the weathering of the soil, the greater the loss of soil B. Both rainfall and temperature can also play a part in the short term. Heavy rainfall, particularly on freely drained soils, can not only exacerbate B deficiency, but minimise residual effects of B application. Application of B should be timed to avoid leaching by heavy rain.

The onset of B deficiency symptoms on many crops has long been associated with hot, dry weather. Until recently, the only explanation of this association has been that as the topsoil dries out, less B can be absorbed from the relatively B rich topsoil (e.g. as demonstrated by Hobbs and Bertramson (1949) with tomatoes (*Lycopersicon esculentum*) in a split root experiment). When root growth recommences in the topsoil with the improving soil moisture conditions and more B becomes available, the symptoms disappear.

There were anecdotal cases of B deficiency developing in hot weather on irrigated crops, e.g. beet in France, but these were unfortunately ignored. Also ignored was the finding by Tanaka (1966), albeit with duck weed (*Lemna pausicostata*), that B deficiency was exacerbated and the critical B concentration increased as the light intensity was raised from 1000 to 5500 lux.

A major advance was made by Merrien (1993) on sunflower in France; he observed that the somewhat unusual necrotic symptoms on midstem/upper leaves on otherwise reasonably well grown sunflower, developed a few days after hot, dry conditions, particularly on thin and gravelly soils. Careful observation of climatic conditions allowed Merrien to specify that if the maximum temperature exceeded 35 °C and daily evapotranspiration exceeded 6 mm for several days, then the necrotic symptoms could be expected within 15 days.

Cakmak et al. (1995) reported that B deficiency symptoms develop more quickly on sunflower seedlings (at the first leaf stage) when they are grown under high (580 μmol $m^{-2}s^{-1}$) rather than low light intensity in a growth chamber. Noppakoonwong et al. (1993) showed in glasshouse trials that high light intensity increased the requirement for B in shading experiments with black gram (*Vigna mungo*). Possible involvement of B in protecting leaves against the damaging consequence of elevated phenol levels caused by high light, is discussed by Cakmak and Römheld (Chapter 6).

Further evidence of the involvement of climatic factors in B nutrition comes from the several studies on the causes of sterility in wheat (Rawson and Subedi, 1996). Rawson (1996) hypothesises that the primary cause of sterility is due to inadequate B reaching the flowers in the critical 6–10 days of pollen formation. Climatic factors that reduce transpiration such as low vapour pressure deficit and low radiation as well as inadequate B availability in the soil will be important. High temperatures, which have been reported to increase the incidence of sterility, are believed to operate by reducing the number of days available for the critical B supply to the flowers.

Boron nutrition also appears to play a part at low temperatures, insofar as frost damage can be exacerbated by B deficiency and minimised by B application. Anecdotal evidence was followed by field experiments, mostly in Yugoslavia but also in Germany and Austria, on ash (*Fraxinus*), apple, *Pinus*, poplar (*Populus*), plum (*Prunus domestic*), peach (*Prunus persica*) and grape (Anon., 1958; Beltram, 1958) which showed definite protective effects of B application against frost damage. It is not possible to assess whether the experiments, which mostly used a 0.5% w/v solution of borax decahydrate as the B treatment, were on B deficient soils on which growth/yield responses could be expected or not. In some cases rapid, almost immediate effects of spraying borax were claimed by Beltram. In the absence of proof under controlled conditions (which was attempted in Germany on vines in the 1960's) this early work has tended to be discounted.

Probably the best evidence of an involvement of B in helping to protect plants against low temperature damage was provided by work on eucalypts in field studies in Zambia (Cooling, 1967; Cooling and Jones, 1970); frost damage, which caused a dieback very similar to B deficiency, was minimised by B application in a dose-related manner.

Braekke (1979, 1983) working with birch (*Betula*), Scots pine (*Pinus sylvestris*), *P. contorta* and Norway spruce (*Picea abies*) on peat soils in Norway found that young plantations were frequently damaged not only by winter frosts, but also by radiation frosts in the growing season and that the damage could be reduced by B application. Moller (1983) in Sweden and Veijalainen (1983) in Finland both provided circumstantial evidence on conifers that the shoot dieback and "growth disturbances" that had always been attributed to frost damage, were in fact, caused by B deficiency. Hanson and Breen (1985) reported that prunes responded positively to autumn B sprays which increased fruit set only when the flowering period in the following spring was "cool" (mean 8 °C without any frost) and not when it was "warm" (mean 12 °C), when fruit set was higher. Borax application was found by Ryu et al. (1986) to increase frost tolerance of *Morus* in South Korea. Other recent examples of B benefits in relation to frost include Milovankic et al. (1990) with apple and pear in Yugoslavia, Svagzdys (1995) with apple in Lithuania and Blevins et al. (1996) with blueberry (*Vaccinium corymbosum*) in USA.

Observations indicating an involvement of B nutrition in frost resistance have continued for almost 40 years, but the basic question as to whether the protective effects of B applications only operate under deficient conditions remains unanswered. In other words, is it just a matter of B deficiency symptoms being accentuated by the low temperature stress?

Management

In a few cases, crop management may have a bearing on the occurrence of, and significance attached to, B deficiency. For example, the cutting of lucerne for consumption away from the field of production means that relatively large amounts of B will be removed (Table 13). The combination of high productivity and cutting led to lucerne being the principle crop treated with B in the USA. In contrast, in Argentina, where lucerne has been mostly grazed, there has been relatively little B deficiency; not only do the Argentinian soils contain more B than many in the USA, but B is recycled via urine and faeces.

Boron deficient leaves and petioles are often brittle due to impaired cell wall development (Table 14). Beet has traditionally been harvested in Ireland by equipment that grabs the petioles and leaves to lift the roots out of the ground. The risk of crop losses due to brittle petioles is minimised by applying B.

Correction of boron deficiency

History

Two B products, namely borax (sodium pentaborate decahydrate) and boric acid, were available to the workers in the 1930's and virtually all the early literature on B use relates to them. The easy use of boric acid in solution made it the preferred compound for spray application but there was in fact very little interest in B sprays until the 1950's, which indicates that most found soil applications to be perfectly satisfactory.

Because the first B problems in the USA were due to toxicity (see "History of responses to boron application") work continued there until the 1950's on the use of borates with low solubility, particularly colemanite and borosilicate, glassy frits (fritted trace elements) but neither have found a permanent place in the agricultural market. Fritted trace elements formed by the fusion and quenching process became popular in horticulture in several countries as a means of providing season-long and toxicity-free supply not only of B but also of other micronutrients; although similar named products are still available not all are fused glassy materials.

With the increase in use of B, after 1945, upgraded sodium borate ores (essentially anhydrous and pentahydrate borax) were offered (in 1952) as cheaper alternatives to the refined borax decahydrate and boric acid for soil application. Although partially processed, these upgraded ores contained some insoluble gangue and were accordingly not suitable for spray application. The big increase in cost of production of anhydrous sodium tetraborate following the rise in fuel prices in the 1970's effectively removed this product from the agricultural market and helped again to stimulate interest in the cheaper ores, particularly colemanite and ulexite. The upgraded pentahydrate sodium tetraborate has been replaced by a completely refined grade for economic reasons of large-scale production. The spray grade product Solubor was introduced in 1954.

The bulk blending of similar sized fertiliser granules in the USA resulted in the development, in the late 1970's, of granular formulations of borax pentahydrate; from the viewpoint of uniform application they are to be welcomed but too little seems to be known about the agronomic efficacy of such materials. A relatively small number of the 1.4–4.0 mm granules are normally applied; an application rate of 1–2 kg B ha^{-1} would involve about 80 to 160 granules m^{-2} and would seem to indicate reliance on lateral movement of B in the soil.

The expanding use of B in agriculture has resulted, in the last 20 years, in a proliferation of different formulations designed to facilitate use; a dry flowable formulation of Solubor was introduced in 1994 and there are many liquid formulations of borates, some apparently simple aqueous solutions with performance-enhancing additives and others utilising the relatively high solubility of boric acid in monoethanolamine to increase the strength of the B concentrate and so reduce handling costs.

Sources of boron

The main B products that are used as primary sources of B are listed in Table 15. Basically there are two types, firstly the refined, completely soluble materials which can be conveniently applied either in solution or as solids, and secondly the crushed ores that contain insoluble gangue and which have variable chemical and physical properties. Aspects of the solubility of different borates are discussed by Power and Woods (Chapter 1). Borax and boric acid are easily dissolved in soils and are quickly available for uptake, but at the same time the B that is not adsorbed can be leached. Pentahydrate borax dissolves more slowly than decahydrate borax and faster than anhydrous borax but there has been no indication that these differences are agronomically significant. Once in solution there are no chemical differences. In order to facilitate dissolu-

Table 15. Commonly used agricultural borates

			B (%)
Refined products	$Na_2B_4O_7 \cdot 5H_2O$	Sodium tetraborate pentahydrate	14.9
	$Na_2B_8O_{13} \cdot 4H_2O$	Solubor	20.8
	$Na_2B_4O_7 \cdot 10H_2O$	Sodium tetraborate decahydrate	11.3
	$Na_2B_4O_7$	Sodium tetraborate	21.4
	$B(OH)_3$	Boric acid	17.5
Crushed ores	$2CaO \cdot 3B_2O_3 \cdot 5H_2O$	Colemanite	variable
	$Na_2O \cdot 2CaO \cdot 5B_2O_3\ 16H_2O$	Ulexite	variable
	$2CaO \cdot B_2O_3 \cdot 2SiO_2 \cdot H_2O$	Datolite	variable
	$CaO \cdot MgO \cdot 3B_2O_3 \cdot 6H_2O$	Hydroboracite	variable
	$2MgO \cdot B_2O_3 \cdot H_2O$	Ascharite	variable

tion in cold water, the product Solubor, which in many respects can be considered a hybrid between borax and boric acid, was developed. Solubor dissolves more quickly than borax and boric acid, results in solutions with high saturated B concentrations and readily forms super-saturated solutions, which are all valuable properties for a spray-grade product.

The two crushed ores ulexite and colemanite are only used for soil application, usually after inclusion in a boronated NPK or PK fertiliser, but also sometimes on their own; there have been attempts to promote a colemanite suspension as a foliar spray product. When such ores are incorporated during the manufacture of compound boronated fertilisers they will react with any free acid, and boric acid will be formed so converting a slowly soluble borate into a quickly soluble form.

In the former USSR and eastern Europe datolite (a calcium borosilicate), ascharite (a magnesium borate) and hydroboracite (a calcium magnesium borate) have all been used for making boronated fertilisers but data on solubility of B in the finished fertiliser is not available. Pot tests with white clover (*Trifolium repens*) by M Ewens and S P McGrath (pers. comm.) showed that hydroboracite was a satisfactory and quickly available source of B when applied on its own. Muller and McSweeney (1976) found, in pot tests, that massive applications of datolite had little effect on either the B content or growth of *Brassica campestris*. In contrast, Sherrell (1983) concluded that finely-divided datolite could be an adequate source of B for lucerne and red clover (*Trifolium pratense*). Calcining has been found, in the Ukraine, to increase the solubility (in citric acid) of datolite (Polyak and Boronkova, 1958). Magnesium borate made by reacting boric acid with magnesium hydroxide has been proposed/used in the former USSR (Katalimov and Churbanov, 1959) as a slow release borate source.

Very few B rich waste products have been used as B fertilisers. The boric acid/gypsum mud, a waste from the manufacture of boric acid is an obvious candidate, but would appear to have been used only in the former east Germany (Bergmann and Buechel, 1963).

Where a long acting borate is required for application on its own, as for example in forestry, there may be a case for using a crushed ore with a large particle size. However apart from any questions about chemical and physical variability, the efficacy of the actual ore to be used needs to be assured; particle size, physical condition and whether it needs to be incorporated into the topsoil must be considered. Uncertainties about the speed of action of a borate with slow release properties are clearly of greater concern for annual than for perennial crops.

The results of comparative studies on the efficacy of different crushed ores to supply B are not apparently consistent. Currently, for example, ulexite is used as a slow acting borate in New Zealand whereas in Brazil it is viewed as being more or less identical to borax decahydrate. The overall picture that emerges is that fine particles of ulexite behave similarly to borax both when incorporated into the soil or left on the surface. On the other hand large particles of non-incorporated ulexite will be long lasting – and will also take longer to provide B initially. The situation with regard to colemanite is similar but differs insofar as colemanite is generally less reactive than ulexite, with the result that non-incorporated colemanite, especially granular, is of much less benefit as a B source – at least for annual crops. Unfortunately very few field comparative studies on borate ores have been reported; most studies on ores have been carried out in pots often with toxicity and leaching in mind rather than crop responses.

Borates can be applied alone, incorporated in or mixed with a fertiliser. For mixed fertilisers, the major concern is normally one of particle size and density in order to achieve uniform mixing and distribution. Normally the pentahydrate borax is used. Borate ores can also be used with due consideration being given to their properties. In compound fertilisers the temperature of the granulation process, and particularly of drying (which can exceed 200 °C), may also dictate the use of a borate with a higher melting temperature than pentahydrate borax, which starts to melt at 200 °C; agglomeration (balling) of granules occurs if the borate on the granule surface melts. Anhydrous borax which starts to melt at 742 °C does not give these problems, likewise colemanite which starts to disintegrate at around 400 °C. In a fertiliser such as boronated fused magnesium phosphate, popular in Japan and Brazil, the high (1300–1400 °C) fusing temperature necessitates the use of an anhydrous borate or at least one that does not disrupt production by the violent release of water.

Procedures have been developed (e.g. Mortvedt, 1991) for coating granular fertilisers with finely divided sodium borates using oil (<1.0%) as a sticker; although this system offers considerable flexibility in production, it has not, somewhat surprisingly, attracted much attention. For liquid and suspension fertiliser, the more rapidly soluble and small particle size sodium borates are preferred.

Soil application

Boron can be satisfactorily applied to the soil to provide season-long elevation of the B status of a crop. Regardless of the chemical form of the borate applied, it is likely, after dissolution in most agricultural soils, that B is absorbed as boric acid, the B species that predominates at pH <7.0.

There is very little data on the recovery of broadcast B by annual crops. M Ewens and S P McGrath (pers. comm.) measured recoveries of 20–30% of incorporated B (non-toxic rates of application) by white clover in a pot test over a 9 month period. Jin et al. (1988) reported B (applied as $B(OH)_3$) recoveries, over 5 weeks, by maize of 0.4–13.5% in 14 soils (several Acrisols). Recovery under field conditions in the year of application is likely to be in the range 5–15%. The relatively small amounts of B that are normally required to effect a significant improvement in B status of an annual crop, namely 1–2 kg B ha^{-1}, are in broad accord with such recovery rates. For many crops, absorption of 100–200 g B ha^{-1} of applied B could be expected to be sufficient.

Boron is readily absorbed by roots and is rapidly translocated to the growing points and actively transpiring tissues, and as a result becomes well distributed throughout the plant. There is very little evidence of significant transport to the roots, either from other roots or from the shoot, which emphasises the importance for all roots to obtain B from the soil they are growing in – in a way analogous to that of the pollen tube in the style. Good root growth depends on there being sufficient B in the rooting zone.

It is important, in order to minimise the risk of B toxicity hot-spots in the field, that borates are uniformly distributed. Historically this was achieved by mixing the borate with soil or sand before application. Boronated fertilisers (solid, liquid and suspension) enable good distribution as does application of a soluble (and compatible) borate with a herbicide. Application via nutrient feed and irrigation systems also affords good distribution.

Timing of soil application

Because of the over-riding importance of local seasons and climates, only the basic principles can be considered. Soluble borates should be thought of as being very similar to nitrate fertilisers, although they do not leach as readily; they should be applied early enough before growth starts in the spring to allow for the boric acid to move into the rooting zone, but they must not be applied too early in order to minimise leaching losses, especially on coarse-textured soils.

Residual effects of soil applied boron

The length of any residual effect of soil applied B will clearly depend on many soil factors, notably those that influence adsorption/desorption and leaching. It is not therefore possible to specify length of residual effects except after precise local experiments e.g. Gupta and Cutcliffe (1982, 1984) who were concerned about toxic residual affects. Mathematical modelling of the behaviour of both indigenous and applied B is required in order to improve our understanding and to facilitate predictions. The different rates of dissolution of borates and the method of application, especially whether they are incorporated or not, will also have a marked influence on the length of any residual effect, as shown for example by Wilson et al. (1951). Because B deficient soils are very often coarse-textured it is customary to think of soluble borates as being readily leached with applications of 1–2 kg B ha^{-1} being effective for only one crop with little carry-over to a second year. Where B is needed by only a single crop in the rotation this means that the each such crop should be given B. However the adsorption of B by fine-textured soils can result in significant carry-over (see "Leaching").

Foliar application

Foliar application of B provides not only a means to apply B at a particular stage in the growth cycle, but it also permits rapidly-acting remedial action to be taken soon after diagnosis of a deficiency; soil applied B would not be as quick acting. Until the recent work in the USA by Hanson (1985, 1991) and Brown and Hu (1996) there had been considerable doubt as to the mode of action of foliar applied B, and although much has been revealed, the picture is not entirely clear. The fact that root absorbed B appeared to be more or less immobile in the phloem, after it reaches the leaves, had led to the widely held conclusion that foliar applied B would be similarly immobile. Boron deficiency symptoms invariably occur in the young tissues and apical meristems.

It is still accepted that once B has been incorporated into cell walls in leaves it is effectively removed from circulation and is not available to support new growth. The rethinking concerning foliar applied B originates from the work of Hanson (1985) who showed that spray application of Solubor, in the autumn just before leaf fall, to prunes resulted in much improved fruit set in the following year indicating significant phloem mobility of B to flowering meristems from either senescing leaves and/or bark. Hanson (1991) demonstrated the rapid – within a few days – export of sprayed B from the leaves of apple, plum and pear; export was much slower in cherry, indicating that it will be difficult to generalise. Similar results were obtained by Gu et al. (1995) with *Malus prunifolia*.

An explanation of what seemed the anomalous results of Hanson with foliar applied B has been provided by the work of Brown and Hu (1996). There seems to be no impediment to the phloem movement of foliar absorbed B in the woody Rosaceae e.g. apple, *Prunus* spp. and pear, most probably because these species translocate the B-complexing sugar alcohol, sorbitol (Brown and Shelp, Chapter 7). Very few species, apart from the woody Rosaceae, use sorbitol, so it remains to be seen whether other sugar alcohols such as mannitol assist in B transport. Delgado et al. (1994) showed that B was phloem-mobile in olive and time will tell whether this can be positively linked to the fact that this species transports mannitol in the phloem.

These results have a direct bearing on the question of timing of B applications. For the woody Rosaceae the opportunity is now afforded to apply B sprays in the autumn before leaf fall as an alternative to spring application at and around the time of flowering; comparison of the efficacy of autumn and spring sprays on fruit set and B utilisation by these species is now required. Autumn sprays depend on B absorption by leaves followed by mobilisation, whereas spring sprays rely on the floral tissues receiving the B directly. Questions about the capacity of leaves to retain B sprays, which Picchioni et al. (1995) found to be the major determinant of the amount of B absorbed, and therefore exported, will need to be asked not only of species but also of genotypes. There will be little merit in a genotype having an inherent capacity to transport B out of leaves unless it exhibits good spray retention.

For annual crops and perennials, belonging to families other than the woody Rosaceae, the position is unchanged with regard to timing. For example Brown and Hu (1996) showed that there was no evidence of significant phloem transport from leaves of three species that do not use sorbitol, namely fig (*Ficus carica*), walnut (*Juglans regia*) and pistachio (*Pistacia vera*). There is a strong physiological case for applying B before, and just after, flowering as a means of ensuring that the floral tissues receive B at the time of peak demand. The poor phloem mobility of spray applied B to all but the woody Rosaceae, means that a single foliar application of B is often not sufficient; repeated sprays on cotton are particularly effective in maintaining a satisfactory B supply over the prolonged flowering period. In the case of annual crops, some of the spray applied B, which is intended for the leaves, inevitably reaches the soil surface thereby benefiting the crop via root absorption.

Solubor is applied to some crops at concentrations up to 5% w/v e.g.beet, but in the absence of tests, it is recommended that spray concentrations do not exceed 1% w/v. The main reason for using low concentrations of Solubor is to minimise the risk of applying too much B if high spray volumes are mistakenly employed, rather than to limit phytotoxic damage to the leaves; it is the amount of B applied, and absorbed, not the concentration which is important. In a Solubor spray trial in the UK (A D Scaife, pers. comm.) with several crops, red beet (*Beta vulgaris*), Chinese cabbage

(*Brassica chinensis*), cauliflower (*Brassica oleracea* var. *botrytis*), rape, swede, carrot and field bean, it was found that the extent of necrotic patches that develop where B accumulates in the leaf was related to the total amount of Solubor applied and not to the concentration of spray solution: Solubor was applied at concentrations from 0.17–8% and at rates of 1.6 to 22 kg Solubor ha^{-1} (0.3–4.8 kg B ha^{-1}). Similar results were obtained for sugar beet (*Beta vulgaris*), Chinese cabbage, rape, swede, carrot, sunflower, tomato, French bean, pea (*Pisum sativum*), radish and potato, by M Hoppe and S P McGrath (pers. comm.) and for apple (var. Cox) by M J Marks (pers. comm.).

Aerial application

There has been limited application of borates from the air. A notable exception is the application to *Pinus radiata* in New Zealand where large particle size pieces of ulexite are now preferred to the dusty colemanite and to the shorter-acting borax pentahydrate.

Suspensions of Solubor have been applied from the air in South Africa to sunflower. Sunflower in Spain is often sprayed with Solubor (2.1 kg ha^{-1} at 1.4–1.7% w/v) using ground equipment or at 3.5% w/v from aircraft.

Applications to seeds

Application of B to seeds or even application in close proximity to the seed is not advisable; root damage (albeit often ephemeral) can readily occur, as for example when boronated sugar beet fertilisers are mistakenly drilled with cereal seed. Likewise application of boronated fertilisers close to potato tubers can be extremely damaging. Soaking of seeds in dilute solutions (2 to 20 m*M*) of boric acid (e.g. 100–1000 mg L^{-1}) has been used as a research tool, almost exclusively in the former USSR, in mainly short term physiological and biochemical studies. Experiments have been carried out on a large number of crops including, beet, turnip, soybean (*Glycine max*), sunflower, barley (*Hordeum vulgare*), lucerne, and wheat. Both beneficial and harmful effects (mostly on germination) have been reported. Despite the beneficial effects there are no reports of the practical use of B as a seed treatment in the USSR – or elsewhere. Very few experiments have been carried out in which seed has been dusted, coated or pelleted with a borate, none of which practices can be recommended.

Rates of application

The current and generally recommended rates of B application given in the review "Boron Deficiency its Prevention and Cure" (Shorrocks, 1989) are based on the rates determined by trials around the world as well as those that had become official recommendations in different countries. The rates had their origin in the early trials on beet etc. (Table 5) and have been refined with time. Inevitably more data is available to substantiate soil application rates than spray-applied rates. For annual crops, similar rates of B application (1–2 kg B ha^{-1}) are recommended regardless of method of application. This is in line with the view that efficacy of foliar applications during the early growth of annual crops is likely to depend both on absorption via the roots as well as via the leaves; the poor phloem mobility of B in annual crops adds credence to this view. For perennial crops, where the target tissues are the flowers and the newly developing fruit, low concentrations, aimed at avoiding phytotoxicity, have been successfully employed; although several applications are usually needed, the total amounts of B applied are, on an area basis, normally lower than for annual crops. Local conditions dictate the severity of the deficiency and accordingly it is not possible to do more in a general review, such as Shorrocks (1989), than indicate a range of rates. On the one hand it is important to apply sufficient B to make a significant improvement in the uptake of B and on the other hand not to be wasteful or run the risk of toxicity.

References

Agulhon H 1910a Emploi du bore comme engrais catalytique. C. R. Acad. Sci. 150, 288–291.

Agulhon H 1910b Presence and utility of boron in plants. Ann. Inst. Pasteur (Paris) 24, 321–329.

Anon. 1935 Boron and plant life. The Fertilizer, Feeding Stuffs and Farm Supplies Journal, Part I, 720–752.

Anon. 1936 Boron and plant life. Part I (continued). The Fertilizer, Feeding Stuffs and Farm Supplies Journal, 4–11.

Anon. 1958 Borax for frost protection. Horticultural Advertiser, September 1958.

Anon. 1977 Boron – its unique role. Fertilizer Solutions (USA), May–June.

Anon. 1996 Boron – important for pollen, fertilization and fruit set. Boron in Agric. 16, 1, 1–2.

Anspoks P and Liepins J 1987 Mobile forms of boron, copper and molybdenum in Latvian soils. Pochvovedenie 4, 28–40.

Archer F C and Hodgson I H 1987 Total and extractable trace element contents of soils in England and Wales. J. Soil Sci. 38, 421–431.

Arnon D I and Stout P R 1939 The essentiality of certain elements in minute quantity for plant with special reference to copper. Plant Physiol. 14, 371–375.

Askew H O and Monk R J 1953 Boron deficiency and uptake of boron in the hop plant. New Zealand J. Sci. and Technol. 35, 279–300.

Askew H O, Chittenden E T and Monk R J 1951 Die-back in raspberries - a boron deficiency ailment. J. Hort. Sci. 26, 268–284.

Askew H O, Chittenden E and Thomson R H K 1936 The use of borax in the control of internal cork of apple. New Zealand J. Sci. and Technol. 18, 365–380.

Beltram V 1958 Bor als frostschutz (Boron as frost protector). Allgemeine Forstzeitschrift 10, 147–148.

Berger K C 1962 Micronutrient deficiencies in the United States. J. Agric. Food Chem. 10, 178–181.

Bergmann W and Buechel L 1963 Borax mud – a waste product of borax fertilizer production. Z. Landwirtsch. Versuhs-Untersuchungsw. 9, 297.

Bergmann W, Ebeling R, Witter B, Schnorr H, Karlheinz K, Gey S, Runge P, Wahl H and Knabe O 1968 Erster orientierender Uberblick uber die B-, Cu-, Mn- und Mo- Versorgung der Boden der DDR. Albrecht-Thaer-Archiv, 12, 1099–1116.

Bingham F T, Page A L, Coleman N T and Flach K 1971 Boron adsorption characteristics of selected amorphous soils from Mexico and Hawaii. Soil Sci. Soc. Am. Proc. 35, 546–550.

Blevins D G, Scrivner C L, Reinbott T M and Schon M K 1996 Foliar boron increases berry number and yield of two highbush blueberry cultivars in Missouri. J. Plant Nutr. 19, 99–113.

Boron in Agriculture, First Series 1948–1979, published by Borax Consolidated Ltd., London.

Boron in Agriculture, Second Series 1980–1989, published by Micronutrient Bureau, Tring, England.

Braekke F H 1979 Boron deficiency in forest plantations on peatland in Norway. Meddelesar fra Norsk Institutt for Skogforskning, 35, 3, 213–236.

Braekke F H 1983 Micronutrients – prophylactic use and cure of forest growth disturbances. Proc. Int. Workshop on Growth Disturbances of Forest Trees. Finland, October 1982. Commun. Inst. Forest. Fenniae Helsinki 116, 159–169.

Brandenburg E 1931 Die Herz- und Trockenfaule der Ruben als Bormangelerscheinung. Phytopath. 3, 499–517.

Brown P H and Hu H 1996 Phloem mobility of boron is species dependent: evidence for phloem mobility in sorbitol-rich species. Ann. Bot. 77, 497–505.

Cakmak I, Kurz H and Marschner H 1995 Short-term effects of boron, germanium and high light intensity on membrane permeability in boron deficient leaves of sunflower. Physiol. Plant. 95, 11–18.

Cooling E N 1967 Frost resistance in *Eucalyptus grandis* following the application of fertilizer borate. Rhod. Zamb. Mal. Agric. Res. 5, 97–100.

Cooling E N and Jones B E 1970 The importance of boron and NPK fertilizers to eucalypts in the southern Province, Zambia. East Afr. Agric. & For. J. – special issue, 36, 185–194.

Delgado A, Benlloch M and Fernandez-Escobar R 1994 Mobilization of boron in olive trees during flowering and fruit development. HortScience 29, 616–618.

Dennis R W G 1937 A review of developments in the applications of boron in agriculture and horticulture 1935-1937. The Fertilizer, Feeding Stuffs and Farm Supplies Journal, Sept.8&22, Oct.6&20.

Dennis R W G 1947/1948 Boron and Plant Life Part VI. Developments in agriculture and horticulture 1943-1946. The Fertilizer, Feeding Stuffs and Farm Supplies Journal, Aug.13&27, Sept.10&24, Oct.8&22, Feb.11&25, Mar.10&24.

Dennis R W G and O'Brien D G 1937 Boron in Agriculture. The West of Scotland Agric. Coll. Res. Bull. No. 5, 3–98.

Dennis A C and Dennis R W G 1939 Boron and Plant Life Part III. Developments in agriculture and horticulture 1937–1938. The Fertilizer, Feeding Stuffs and Farm Supplies Journal, Feb.8&22, Mar.8&22, Apr.5&19.

Dennis A C and Dennis R W G 1940/1941 Boron and Plant Life Part IV. Developments in agriculture and horticulture 1939–1940. The Fertilizer, Feeding Stuffs and Farm Supplies Journal, Nov.13&27, Dec.11&24, Jan.8&;22, Feb.5&19.

Dennis A C and Dennis R W G 1943 Boron and Plant Life Part V. Developments in agriculture and horticulture 1940–1942. The Fertilizer, Feeding Stuffs and Farm Supplies Journal, Mar.17&31, Apr.14&28, May 12&26.

Elrashidi M A and O'Connor G A 1982 Boron sorption and desorption in soils. Soil Sci. Soc. Am. J. 46, 27–31.

Etchevers B J, Merino H R, Vidal P I, Riqueime F E and Llanos R F 1983 Nutritional survey of vineyards in the coastal zone of the VIII region. Agric. Tecnica (Santiago) 43, 13–20.

FAO/UNESCO 1978 Soil maps of the world 1:5 000 000 Volumes 1–10. UNESCO, Paris.

Farrer K 1976 Trace deficiencies in crops. MAFF, ADAS Advisory Paper 17, 1–34.

Gembarzewski H 1987 Presence and deficiencies of secondary and trace elements in arable and grassland with high fertilizer use and varying soil conditions in Poland. *In* The Utilization of Secondary and Trace Elements in Agriculture. Ed. United Nations. pp 55–74. Martinus Nijhoff Publishers, Dordrecht, The Netherlands.

Goldberg S and Forster H S 1991 Boron sorption on calcareous soils and reference calcites. Soil Sci. 152, 304–310.

Gu M, Jiang Y, Huang H and Peng F 1995 The translocation and distribution of foliar-applied boron during blooming period of apple trees. Acta Agric. Nucleatae Sinica 9, 86–90.

Gupta U C 1983 Boron deficiency and toxicity symptoms for several crops as related to tissue boron levels. J. Plant Nutr. 6, 387–395.

Gupta U C 1984 Boron nutrition of alfalfa, red clover, and timothy grown on podzol soils of eastern Can. J. Soil Sci. 137, 16–22.

Gupta U C and Cutcliffe J A 1975 Boron deficiency in cole crops under field and greenhouse conditions. Commun. Soil Sci. Plant Anal. 6, 181–188.

Gupta U C and Cutcliffe J A 1982 Residual effect of boron applied to rutabaga on subsequent cereal crops. Soil Sci. 133, 155–159.

Gupta U C and Cutcliffe J A 1984 Effects of applied and residual boron on the nutrition of cabbage and field beans. Canada J. Soil Sci. 64, 571–576.

Gupta U C and Cutcliffe J A 1985 Boron nutrition of carrots and table beets grown in a boron deficient soil. Commun. Soil Sci. Plant Anal. 16, 509–516.

Gyul'akhmedov A N and Mamedov O K 1984 Boron content in soils of central Burma. Izv. Akad. Nauk. Az. SSR, Ser. Biol. Nauk 1, 49–51.

Hanson E J 1991 Movement of boron out of tree fruit leaves. HortScience 26, 271–273.

Hanson E J and Breen P J 1985 Effects of fall boron sprays and environmental factors on fruit set and boron accumulation in "Italian" prune flowers. J. Amer. Soc. Hort. Sci. 110, 389–392.

Harder H 1961 Contribution to the geochemistry of boron: III. Boron in metamorphic rocks and in the geochemical cycle. Nachr, Akad. Wiss, Gottingen: II. Math-Physik. Kl., 1, 1–26.

Harder H 1970 Boron content of sediments as a tool in facies analysis. Sediment. Geol. 4, 153–175.

Hatcher J T, Bower C A and Clark M 1967 Adsorption of boron by soils as influenced by hydroxy aluminium and surface area. Soil Sci. 104, 422–426.

Hingston F J 1986 Biochemical cycling of boron in native eucalypt forest of south-western Australia. Aust. For. Res. 16, 73–83.

Hobbs J A and Bertramson B R 1949 Boron uptake by plants as influenced by soil moisture. Soil Sci. Soc. Am. Proc. 14, 257–262.

Jin J Y, Martens D C and Zelazny L W 1988 Plant availability of applied and native boron in soils with diverse properties. Plant and Soil 105, 127–132.

Katalimov M V and Churbanov V M 1959 Magnesium borate as a boron fertilizer. Zhurn. Khim. 24, 25.

Kiviniemi K J 1946 Observations on the use of lime, its requirements and effects in Finland during this century. Manuscript in Finnish 73 p. Dept. of Agric. Chem. and Phys., Vantaa, Finland.

Kubota J, Berger J C and Truog E 1948 Boron movement in soils. Soil Sci. Soc. Am. Proc. 13, 130–134.

Kurki M 1982 On the fertility of Finnish tilled fields in the light of investigations of soil fertility carried out in the years 1955-1980. Suomen Peltojen Viljavuudesta III, ISBN 951-993399-0-3.

Liu Zheng, Zhu Qi-Qing and Tong Li-Hua 1980 Boron deficient soils and their distribution in China. Acta Ped. Sin. 17, 228–239.

Liu Zheng, Zhu Qi-Qing and Tong Li-Hua 1982a On the status of microelements in soils and their role in crop production of China. Int. Conf. Chem and World Food Supplies, Manila, Philippines Dec. 1982.

Liu Zheng, Zhu Qi-Qing and Tong Li-Hua 1983 Microelements in the main soils in China. Soil Sci. 135, 40–46.

Liu Zheng, Zhu Qi-Qing and Tong Li-Hua 1989 Regularities of content and distribution of boron in soils. Acta Ped. Sin. 26, 353–361.

Liu Zheng, Zhu Qi-Qing, Tong Li-Hua, Xu Jung-Xiang and Yen Chu-Liang 1982b Geographical distribution of trace elements-deficient soils in China. Acta Ped. Sin. 19, 209–223.

Lombin G 1985 Micronutrient soil tests for the semi-arid savannah of Nigeria: boron and molybdenum. Soil Sci. Plant Nutr. (Tokyo) 31, 1–11.

Lopes A S 1984 Solos sob "cerrado": Caracteristicas, propriedades e manejo. Piracicaba: Associacao Brasileira para Psquisa da Potasse e do Fosfato.

Lora R S 1978 Respuesta de los cultivos de clima frio a la aplicacion de micronutrientos. Memorias del V coloquio de suelos. Suelos Ecuatoriales. Soc. Colombiana de la Ciencia del Suelo 9, 183–191.

Maze P 1915 Determination des elements mineraux rares necessaires au developpement du mais. C.R. Acad. Sci. 160, 211–214.

Merrien A 1993 Diagnostic parcellaire sur une culture de tournesol. Perspectives Agricoles 179, 80–84.

Micronutrient News and Information 1990–1996. Publisher: Micronutrient Bureau, Tring, England.

Milovankic M, Keserovic Z and Vasiljevic Z 1990 Studies on some chemicals used for prevention of frost damage to fertile buds and flowers of fruit crops. Yugoslovensko Vocarstvo 24, 4, 21–27.

Moller G 1983 Variation of boron concentration in pine needles from trees growing on mineral soil in Sweden and response to nitrogen fertilization. *In* Proc. Int. Workshop on Growth Disturbances of Forest Trees. Finland, October 1982. Commun. Inst. Forest. Fenniae Helsinki 116, 111–115.

Mortvedt J 1991 Micronutrient fertilizer technology. *In* Micronutrients in Agriculture, 2nd ed. Eds. J Mortvedt, F R Cox, L M Shuman and R M Welch. pp 523–548. Soil Sci. Soc. of America Book Series No. 4, Madison, WI, USA.

Muller F B and McSweeney G 1976 Toxicity of borates to turnips. New Zealand J. Exp. Agric. 4, 451–455.

Noppakoonwong R N, Bell R W, Dell B and Loneragan J F 1993 An effect of shade on the boron requirement for leaf blade elongation in black gram (*Vigna mungo* L. Hepper). *In* Plant Nutrition – from Genetic Engineering to Field Practice. Ed. N J Barrow. pp 413–416. Kluwer Academic Publishers, Dordrecht, the Netherlands.

Odunze A C, Chude V O and Adeoye K B 1992 Physicochemical properties and management of alfisols in the Nigerian savanna. *In* Integrated land use for tropical agriculture. Proc. second international symposium, Queensland, Brisbane, Australia. Module 2, 12.1–12.8.

Ouellette G J 1958 Persistence of borax in sandy soils. Can. J. Soil Sci. 38, 77–84.

Park C S and Park N J 1966 Studies on the available boron content of soils in the upland crop area in Korea. Ann. Res. Report, Office of Rural Development 9, 1, 163–174.

Peterson L A and Newman R C 1976 Influence of soil pH on the availability of added boron. Soil Sci. Soc. Am. Proc. 40, 280–282.

Picchioni G A, Weinbaum S A and Brown P H 1995 Retention and the kinetics of uptake and export of foliage- applied, labelled boron by apple, pear, prune and sweet cherry leaves. J. Amer. Soc. Hort. Sci. 120, 28–35.

Pieve J V 1959 Content of forms of trace elements that are available to plants in the soils of the USSR. World Congress of Agric. Res. FAO Rome, 1270–1272.

Polyak A M and Boronkova M E 1958 Production of boron fertilizers from datolite from the Far East. Zhurn. Khim. 22, 25.

Rawson H M 1996 Hypothesis for why sterility occurs in wheat in Asia. *In* Sterility in wheat in subtropical Asia: extent, causes and solutions. Eds. H M Rawson and K D Subedi. Proc. workshop 18–21 Sept, Lumle Agric. Res. Cent., Pokhara Nepal. pp 132–134. ACIAR Proc. No. 72.

Rawson H M and Subedi K D (Eds) 1996 Sterility in wheat in subtropical Asia: extent, causes and solutions. Proc. workshop 18–21 Sept, Lumle Agric. Res. Cent., Pokhara Nepal. ACIAR Proc. No. 72.

Rerkasem B 1996 A general survey of the incidence of wheat sterility. *In* Sterility in wheat in subtropical Asia: extent, causes and solutions. Eds. H M Rawson and K D Subedi. 1996 Proc. workshop 18–21 Sept, Lumle Agric. Res. Cent., Pokhara Nepal. pp 8–12. ACIAR Proc. No. 72.

Rigg T, Askew H O and Chittenden E 1937 Brown-heart of swedes and turnips in Nelson district: a boron deficiency ailment. New Zealand J. Sci. Tech. 18, 750–755.

Rodriguez J S and Tomic M T Y 1984 Micronutrient availability in andosols and ultisols of Los Lagos region. Cienc. Invest. Agrar. 11, 3, 169–178.

Ryu K S, Kim K R and Kim R S 1986 Studies on the cause and control of non- sprouting in the mulberry fields of Kyiungpook province. Han'guk Chamsa Hakhoechi 28, 2, 1–8.

Sakal R and Singh A P 1995 Boron research and agricultural prodcution. *In* Micronutrient Research and Agricultural Production. Ed. H L S Tandon. pp 1–31. Fetilizer Development and Consultation Organization, New Delhi, India.

Savage C G and Broadfoot H 1937 Internal cork of apples, soil dressings of borax give excellent results. Agric. Gaz. N.S.W. 48, 387–390 and 447–452.

Schalscha E B, Bingham F T, Galindo G G and Galvan H P 1973 Boron adsorption by volcanic ash soils in southern Chile. Soil Sci. 116, 2, 70–76.

Sherrell C G 1983 Comparison of materials of different solubility as sources of boron for plants. New Zealand J. Exp. Agric. 11, 325–329.

Schlatter J E and Gerding V 1985 Boron deficiency in *Pinus radiata* plantations in Chile 1. Distribution and origin. Bosque 6, 24–31.

Shorrocks V M 1989 Boron Deficiency – Its Prevention and Cure. Published by Borax Consolidated Ltd., London.

Shorrocks V M 1991 Boron – a global appraisal of the occurrence, diagnosis and correction of boron deficiency. *In* Proc. Int. Symp. on the Role of Sulphur, Magnesium and Micronutrients in Balanced Plant Nutrition. Ed. S Portch. pp 39–53. Potash and Phosphate Institute, Hong Kong.

Sillanpaa M 1982 Micronutrient and the nutrient status of soils - a global study. FAO Soils Bulletin No. 48, FAO, Rome, Italy.

Sillanpaa M 1990 Micronutrient assessment at the country level: an international study. FAO Soils Bulletin No. 63, FAO, Rome, Italy.

Sillanpaa M and Vlek P L G 1985 Micronutrients and the agroecology of tropical and Mediterranean regions. *In* Micronutrients in Tropical Food Crop Production, Dev. Plant Soil Sci. 14, 151–167.

Sims J R and Bingham F T 1968a Retention of boron by layer silicates, sesquioxides and soil materials: II Sesquioxides. Soil Sci. Soc. Am. Proc. 32, 364–369.

Sims J R and Bingham F T 1968b Retention of boron by layer silicates, sesquioxides and soil materials: III Iron- and aluminium-coated layer silicates and soil materials. Soil Sci. Soc. Am. Proc. 32, 369–373.

Sobrinho M O C and Freire O 1980 Boron in some soils of the state of Sao Paulo. Availability by chemical methods. Rev. Agric. (Piracicaba, Brazil) 55, 83–105.

Stone E L and Will G M 1965 Boron deficiency on *Pinus radiata* and *Pinus pinaster.* For. Sci. 11, 425–433.

Svagzdys S 1995 Use of boron fertilizers in apple orchards of Lithuania. Acta Hort. 383, 487–490.

Tanaka H 1966 Response of *Lemna paucicostata* to boron as affected by light intensity. Plant and Soil 25, 425–434.

Tollenaar H 1970 Boron deficiency in pine plantations in central Chile. Agric. Tecnica (Santiago) 29, 85–88.

Veijalainen H 1983 Preliminary results of micronutrient fertilization experiments in disordered Scots Pine stands. Proc. int. workshop on growth disturbances of forest trees. Finland, October 1982. Commun. Inst. Forest. Fenniae Helsinki 116, 153–159.

Warington K 1923 The effect of boric acid and borax on the broad bean and certain other plants. Ann. Bot. 37, 457–466.

Wear J I and Patterson R M 1962 Effect of soil pH and texture on the availability of water-soluble boron in the soil. Soil Sci. Soc. Am. Proc. 26, 344–346

Wikner B 1983 Distribution and mobility of boron in forest ecosystems. Proc. Int. workshop on growth disturbances of forest trees. Finland, October 1982. Commun. Inst. Forest. Fenniae Helsinki 116, 131–141.

Wild A and Mazaheri A 1979 Prediction of the leaching rate of boric acid under field conditions. Geoderma 22, 127–36.

Will G M 1971 The occurrence and treatment of boron deficiency in New Zealand pine forests. Research Leaflet 32, NZ Forest Service.

Wilson C M, Lovvorn R L and Woodhouse W W 1951 Movement and accumulation of water soluble boron within the soil profile. Agron. J. 43, 363–367.

Yermiyaho U, Keren R and Chen Y 1988 Boron sorption on composted organic matter. Soil Sci. Soc. Am. Proc. 53, 1309–1313.

Plant and Soil **193:** 149–168, 1997.

Chapter 10

Diagnosis and prediction of boron deficiency for plant production

Richard W. Bell
*School of Biological and Environmental Sciences, Murdoch University, Perth 6150, Australia**

Abstract

Boron deficiency has been diagnosed or predicted in a variety of plants using a combination of approaches including plant response to B fertiliser, plant symptoms, and soil and plant analysis. Local knowledge of soils, crops, and factors known to accentuate B deficiency provide important background for its diagnosis and prognosis. However, accurate diagnosis and prognosis of B deficiency depends on calibrated standards for plant and soil testing. For diagnosis of B deficiency the plant part sampled needs to reflect current B supply, and will therefore usually be an actively expanding plant part such as a young leaf. Soil and plant analysis are probably more useful when used to predict the likelihood of B deficiency. Increased reliability of prediction can be achieved when soil and plant analysis values are calibrated for particular soil type and crop species combinations. In addition to calibrating soil and plant analysis against yield responses, standards established for pollen viability and vigour, seed viability, seed or fruit quality, root, stem or tuber quality or lignification of wood could be used more often for B.

Introduction

Low B soils are prevalent globally (Shorrocks, Chapter 9) so that in a great variety of soil-crop-climate combinations reliable means of diagnosis and prognosis of the incidence of B deficiency are necessary for good crop management. A combination of approaches including crop response to B fertiliser, plant symptoms, and soil and plant analysis is most reliable especially when supplemented by local knowledge of soils, crops, and factors known to accentuate B deficiency. Accurate diagnosis and prognosis of B deficiency increasingly depends on calibrated procedures for soil and plant testing. The present review will concentrate on the use of soil and plant analysis for diagnosis and prediction, emphasising the principles underlying the development of tests and their use. However, the methods and procedures, and instruments required for B analysis of soil and plants are well covered in other places (Bingham, 1982; Gupta, 1993; Sah and Brown, 1996; Sah and Brown, Chapter 2) and will not be reviewed here.

Plant and soil analysis are used in a number of different ways that are essentially complementary in managing crop nutrition. The key distinction to recognise for the effective use of soil and plant analysis is the difference between their use for diagnosis compared to prognosis. Failure to do so may lead to disappointing outcomes from the tests or worse, erroneous information.

Diagnosis

When there is a perception that growth of a plant or crop is poor or in some way impaired, there is a need to diagnose the existing cause. Diagnosis is the determination of nutrient status in the plant at the time of sampling (Smith, 1986) with the object of positively identifying a disorder. Analysis of the B concentration in the plant or a specified part of the plant can provide such a positive diagnosis if the B concentration is below a defined standard B concentration. For diagnosis, the standard B concentration should be one related to the physiological requirement for B in the plant part. That is, the standard needs to be established by relating the internal B concentration in the plant part to a physiological or growth function which is clearly regulated by B concentration in that plant part. Such standards were defined by Loneragan (1968) as functional nutrient requirements.

* E-mail: rbell@essun1.murdoch.edu.au

Functional nutrient requirements

Kirk and Loneragan (1988) applied the concept of functional nutrient requirements to the setting of a standard for the diagnosis of B deficiency in soybean (*Glycine max*). They based their test on the known requirement for B in cell elongation (Marschner, 1995) and reasoned that the rate of leaf expansion ought to be closely related to its B concentration. In soybean, the functional B requirement for leaf blade elongation was shown to be 10–12 mg B kg^{-1} dry wt. At leaf B concentrations above that range, leaf elongation was unimpaired: at leaf B concentrations below that range, leaf elongation was depressed. Application of Kirk and Loneragan's standard to soybean in the field suggests that it does indeed diagnose existing B deficiency in soybean crops (Rerkasem et al., 1993; Woodruff, 1979). By contrast, the critical concentration developed for soybean in the conventional way (Kirk and Loneragan, 1988) failed to do so.

Critical concentrations

The conventional approach to setting standards for diagnosis is based on establishing a critical concentration or critical range from a relationship between shoot dry matter and B concentration in a plant part (Ulrich and Hills, 1967). In culture systems where there is a significant decline in external B supply over time this approach generates critical concentrations that do not reflect the functional B requirement. In soybean, the critical B concentration derived from the relationship between shoot dry matter and B concentration in the youngest open leaf (YOL) was 2 mg B kg^{-1} dry wt. By contrast, leaf elongation was impaired at <10 mg B kg^{-1}. The discrepancy between the two values can be attributed to the rapid decline in B concentration in the YOL, and the slow response of shoot dry matter to declining B supply. By contrast, when Huang et al. (1996a) used a programmed nutrient addition to maintain deficient and adequate B supply levels, they obtained critical B concentrations in the YOL of oil seed rape (*Brassica napus*) which were similar to the functional B requirement established for leaf blade elongation. Thus the establishment of a critical concentration or critical range from the relationship between shoot dry matter and leaf B concentration can still be a valid approach for establishing a critical value for B deficiency diagnosis provided precautions are adopted to prevent a rapid decline in B supply during plant growth.

Paired comparisons of plants with good and poor growth

Plant analysis can also provide evidence that B is the cause of a plant growth disorder even if there are no defined functional or critical B concentrations for diagnosis. Boron deficiency may be inferred where there is a lower B concentration in tissues of plants exhibiting depressed growth. However, this is in itself not proof of a B deficiency, but may provide an indication of this disorder.

None of the above diagnostic approaches necessarily identifies the underlying cause of the depressed B concentration in the plants. Whilst low B in the soil is usually the cause of low B in the plant, other environmental factors that restrict B uptake can also induce B deficiency in the plant (Rawson, 1996). Low soil water, for example, has been reported to induce B deficiency in plants in dry periods when in years of higher, uniform rainfall, the same levels of soil B would result in adequate B uptake into the plant (Kluge, 1971; Noppakoonwong et al., 1997).

Prognosis

Prognosis is the prediction of the possibility of a B deficiency impairing plant growth at some later stage in the growth cycle after the sample was taken (Smith, 1986). Prognosis has features in common with diagnosis leading some authors to confuse the two by using the term diagnosis loosely to refer to both applications (e.g. Bates, 1971; Moraghan, 1985). Both are based on use of critical B concentrations for the interpretation of plant or soil B status. However, failure to recognise the distinctions between diagnosis and prognosis may lead to unrealistic expectations about the use of plant or soil analysis for prognosis (Smith and Loneragan, 1997), or the use of inappropriate standards for diagnosis or prediction.

Prognosis of B response can be made from either soil or plant analysis. By contrast, in the strict sense, usually only plant analysis can diagnose B deficiency. Since the soil does not require B *per se*, the diagnosis of B deficiency relates to the plant's internal need for B. Hence soil analysis for B cannot diagnose B deficiency except in the circumstance where there is an established minimum external B requirement in the soil solution for root growth and function. However, since there are very few reported measurements of soil solution B and none in the deficient range (Aitken et al., 1987; Loneragan, 1975), this particular use of soil analysis for the diagnosis of B deficiency can largely be overlooked at present.

Primarily, prognosis differs from diagnosis because of the time that elapses between the collection of a sample, and the measurement of the effect on later plant growth. In the case of soil analysis, the time interval is the entire crop growing season. In the case of plant analysis, the time interval is often shorter but still a significant period of time from at least vegetative or early flowering stages until harvest. Uncertainty is inherent in prognosis because of the time lapse between sampling and measurement of effects on plants (Smith and Loneragan, 1997). Unfortunately, this characteristic of prognosis, and its distinction from diagnosis is often not recognised. The growth of the plant is affected by agronomic, biotic and environmental factors that determine actual yield in a given season and environment. The total demand for B in a crop is a function of the yield realised. That is, a crop which yields 1 tonne of grain ha^{-1} will have a lower total B requirement than one which yields 3 tonnes ha^{-1}. The probability of a B response will differ accordingly even when the soil B level at sowing was the same in both cases. Moreover, the pattern of B supply from the soil to the plant over time is a function of the soil properties, particularly those that affect B leaching, mineralisation, and sorption. Since the above range of factors affects the probability of a B response, it suggests that the accuracy of predictions about B deficiency can be improved when the plant or soil analysis is supplemented by information on soil type (e.g. Sillanpaa, 1982; Snowdon, 1982) and probable weather between sampling and harvest (Smith and Loneragan, 1997).

Monitoring

Plant and soil analysis is increasingly used for monitoring B status in areas where research has diagnosed the possibility of a deficiency and has established means to correct it. Monitoring is used to verify that longstanding B management practices are still satisfactory for modern cultivars (e.g. Heitholt, 1994) or that practices should be modified. It uses both diagnostic and prognostic standards and is likely to become a routine part of agricultural practice as community concerns rise over the environmental and economic consequences of over fertilising .

Methods of diagnosis and prognosis

Alleviation of a B deficiency depends firstly on being able to correctly identify it. The most convincing diagnosis of a B deficiency is to apply a corrective treatment in the field and demonstrate alleviation of the constraint by the crop response (Craswell et al., 1985). However, this approach is rather costly when many sites are involved. In addition, inappropriate selection of treatments or unseasonal conditions may cause total failure of the experiment. Hence this approach to diagnosis, although indispensable, must be used sparingly. When completed, field response experiments serve as the benchmark against which other forms of diagnosis or prognosis are calibrated (Lindsay and Cox, 1985; Peck and Soltanpour, 1990).

Field response to boron fertiliser

Several precautions are necessary for field experiments to accurately diagnose B deficiency. The basis principle is that growth will be limited by the most limiting factor. Unless the most limiting factor has been identified or added to the experiment as a basal nutrient or treatment then the experiment can be expected to achieve little (Craswell et al., 1985).

Field experiments may also fail to diagnose a B deficiency if adequate rates for nutrients added as basal fertiliser are not used. This is especially important on sites which suffer from multiple deficiencies because the alleviation of only one or some of them will fail to give a true indication of the responsiveness of plant growth on the site to B supply (e.g. Bell et al., 1990b). In such cases, a preliminary subtractive design experiment comprising a complete nutrient treatment and the omission of single nutrients can be an effective approach (Bell et al., 1990b; Craswell et al., 1985).

In cases of multiple deficiency, care must be taken with the source of nutrients added. Sulfate forms of N or K often give responses because of their supply of S on sites that are not only deficient in N or K but also in S. Less well recognised is the variable content of S in compound fertilisers. In Thailand, for example, compound fertilisers marketed with the same NPK content varied in S from 2–20% (Bell et al., 1990b). Other examples exist for micronutrients which occur in some macronutrient fertilisers in significant amounts (e.g. B in triple superphosphate, Bell et al., 1990b). For the correction of deficiencies, the additional B in some macronutrient fertilisers is a bonus, but for the diagnosis of B deficiency it may cause erroneous conclusions about the most limiting factor, or variable responses from site to site and year-to-year that are difficult to explain.

Since B field experiments will often use local seed collected from low B soils, seed B levels may need to be considered when setting up field experiments to test B response. Low soil B levels which do not depress seed yield may nevertheless depress seed B concentrations sufficiently to impair seed viability and the vigour of seedlings which germinate and emerge (Bell et al., 1989; Rerkasem et al., 1990). When seed B is low enough to depress germination, the sowing of low B seed will depress plant density in plots (Rerkasem et al., 1997). In black gram (*Vigna mungo*), seed with 6–14 mg B kg^{-1} dry wt germinated without impairment but the seedlings developed B deficiency in shoot apices when sown into low B soil (Rerkasem et al., 1990). Thus part of the response to B when using low B seed may be associated with depressed early plant growth at low soil B only. Rerkasem et al. (1997) found that the use of low B soybean seed (7–14 mg B kg^{-1} dry wt) depressed yield potential compared to high B seed (20–23 mg B kg^{-1}) in two field experiments.

Plant symptoms

Symptoms of B deficiency have been described for a wide range of plant species (e.g. Bergmann, 1992; Gupta, 1993; Shorrocks, 1989). Mostly the B deficiency symptoms are found in the young growing parts of the plant including the vegetative and reproductive apices and in flowers, fruits and seeds. Underground storage organs of rutabaga (*Brassica napobrassica*) and sugar beet (*Beta vulgaris*) also appear to be especially susceptible to B deficiency symptoms (Gupta and Cutcliffe, 1972; Vlamis and Ulrich, 1971). The form of symptoms, and their distribution on the plant are attributed in most species to the phloem immobility of B (Gupta, 1993). However, symptoms of B deficiency are also observed in growing plant parts of species in the *Prunus*, *Pyrus* and *Malus* genera in which B is phloem mobile (Brown and Hu, 1996), probably because the primary function of B is in cell wall development in growing plant parts, regardless of the phloem mobility of B (Brown and Shelp, Chapter 7).

Reproductive growth of some plants is impaired by low soil B even when the vegetative growth of the same plant is apparently unaffected by B deficiency (Mozafar, 1993). Thus, in the field, it is commonly observed that plants remain free of B deficiency symptoms until reproductive growth commences when a variety of B deficiency symptoms may appear including: abortion of flower buds or flowers (Bell et al., 1990a); male sterile flowers (Rerkasem and Loneragan, 1994); aborted fruit or pods (Rerkasem et al., 1988); pods with reduced numbers or no seeds (Bergmann, 1992; Dear and Lipsett, 1987); reduced seed size (Keerati-Kasikorn et al., 1993); or fruit with impaired quality from B induced lesions [e.g. hollow heart in peanut (*Arachis hypogaea*), Harris and Brolman, 1966; internal fruit necrosis in mango (*Mangifera indica*) Ram et al., 1989].

In peanut, so distinctive is the hollow heart symptom, and specific to B deficiency, that it was used for a extensive survey of farmers' peanut crops across several regions of Thailand with the object of identifying the incidence and severity of low B in soils (Bell et al., 1990b). In other cases, plants may suffer from B deficiency and exhibit no symptoms that can be unambiguously recognised as B deficiency. Plant and soil analysis which are described below are necessary in these cases.

The varied expression of B deficiency symptoms in plants which do not retranslocate B can be attributed to the fact that a continuous supply of B from the soil is necessary for new growth (Bell et al., 1990a; Kirk and Loneragan, 1988; Mozafar, 1987; Oertli, 1993). For example, Blamey et al. (1987) reported sunflower (*Helianthus annuus*) grown on low B soil exhibited B deficiency symptoms in leaves that developed during a period of drought but not in leaves produced subsequently during periods of adequate rainfall. Bergmann (1992) suggested that symptoms described in the literature from B deficiency induced in non-renewed solution culture or in sand cultures may be limited in their diversity because they typically represent very severe symptoms at a single time in plant growth. By contrast, in the field the phase of plant growth affected by B deficiency varies spatially and temporally and hence the diversity of symptoms is greater, including a greater range of mild to moderate symptoms.

Some symptoms attributed to B deficiency are particularly difficult to ascribe with certainty to B [e.g. nubbins in maize (*Zea mays*), Mozafar, 1993; sterility in wheat (*Triticum aestivum*), Rawson, 1996]. The symptoms observed are consistent with B deficiency since they are restricted to the growing reproductive organs and result in depressed seed set. However, when B is determined in the affected ears, concentrations may not differ measurably from those in plants with no symptoms (Rerkasem and Lordkaew, 1996). In order to diagnose a deficient current supply, smaller plant parts such as pollen, anthers, or carpels may be necessary. Huang (pers. comm.) has shown that withdrawing B supply for only 4–7 days during the period after double ridge stage for the wheat ear (Bennett et al., 1973) was sufficient to induce severe male sterility and strongly depress grain set. Thus very short periods of inadequate B supply are sufficient to impair reproductive growth but by the time symptoms are observed the temporary deficiency has disappeared and B concentration in the affected organs restored to adequate levels. In such cases of temporary B deficiency, traditional plant analysis of large plant parts which accumulate B over a long period spanning the B deficient period, and periods before and after it, will fail to confirm the cause of the symptom observed.

Examples of the B concentrations associated with B deficiency symptoms in various plant parts are shown below (Table 1). In a number of dicotyledons, leaf B concentrations < 10 mg B kg^{-1} dry wt were generally associated with deficiency symptoms in the sampled leaves. By contrast, wheat leaves only exhibited symptoms when they contained < 1 mg B kg^{-1} dry wt. In the carpel and pollen of wheat, symptoms of B deficiency were evident at much higher B concentrations than those which impaired leaves.

Plant analysis for diagnosis

The use of elemental analysis of plants to quantitatively assess plant nutrient requirements was first proposed by Macy (1936). Its subsequent development as a methodology, and the refinement of its theory can be traced through key papers including Bates (1971), Smith (1986), and Smith and Loneragan (1997), Ulrich (1952), Ulrich and Hill (1967).

Ulrich (1952) described plant analysis as largely an empirical exercise because of a rudimentary understanding of the physiology of elements in plants and it largely remains so today. He also noted that advances in plant analysis would follow new insights into the biochemical and physiological behaviour of nutrients in plants. Several significant recent advances in understanding the role and behaviour of B in plants have opened the way for improvement in plant analysis for B. These are described in detail in preceding chapters.

Physiological and functional basis for plant analysis

The following account briefly outlines those aspects of B behaviour fundamental to the development of plant analysis for B deficiency diagnosis. The aim of the account below is to identify fruitful areas for future research and development of plant analysis as a tool for managing B nutrition of plants.

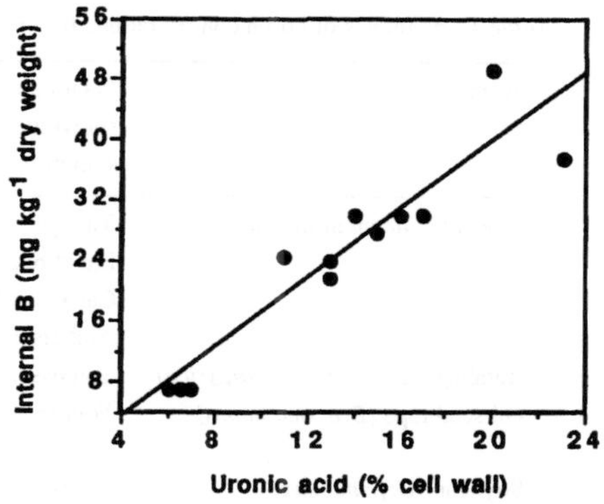

Figure 1. Relationship between internal boron requirements of 13 species and uronic acid concentration of cell wall (% cell walls). Plotted from Hu et al. (1996). $Y = 2.26\,X - 5.27$: $r^2 = 0.85$.

Most of the tissue B is found in the cell wall (Hu and Brown, 1994; Matoh et al., 1992). Since the fraction of tissue B not bound with cell walls increases only after the requirements for cell walls are met (Hu et al., 1996), it has been suggested that the level of water soluble B extracted from plants parts could be a sensitive indicator of adequate B status (Skok and McIlraith, 1958). Moreover, since there is a close correlation ($r^2 = 0.85$) between the pectin content of the cell wall and the internal B requirement of the species, across a range of species (Hu et al., 1996; Figure 1), the possibility of using pectin determination as a means of establishing functional B requirements for diagnosis warrants further exploration.

Whilst most attempts to use plant analysis have relied on the B concentration alone in plants, some authors, starting with Reeve and Shive (1944), have suggested the use of the Ca:B ratio instead. According to Marschner (1995), there is probably a physiological basis for Ca–B interaction based on the fact that both have a primary role in cell walls and at the cell wall–plasma membrane interface. However, results for diagnosis based on Ca:B ratios have so far proved inconclusive (Gupta, 1979). Hence without a better understanding of the physiological basis for establishing a critical Ca:B ratio, it appears that this approach would merely lead to empirical standards for plant analysis which have only very local or experiment-specific validity.

Table 1. Examples of boron concentrations (mg B kg^{-1}dry wt) in plant parts exhibiting B deficiency symptoms

Species	Plant part showing symptom	B in affected plant part	Source
Wheat *(Triticum aestivum)*	Youngest emerged leaves	< 1	Huang et al. (1996b)
	Ear at booting	3–7	Rerkasem and Loneragan (1994)
	Carpels at booting	<6	Rerkasem and Lordkaew (1996)
	Anthers at anthesis	<9	Rerkasem and Lordkaew (1996)
Rutabaga *(Brassica napobrassica)*	Youngest mature leaves	2–7	Gupta and Cutcliffe (1972)
Oil seed rape *(Brassica napus)*	Youngest open leaves	<10	Huang et al. (1996a)
	Flowers	11	cited in Holmes (1980)
Soybean *(Glycine max)*	Youngest open leaves	<10	Kirk and Loneragan (1988)
	Seeds	<9–13	Rerkasem et al. (1993)
Sunflower *(Helianthus annuus)*	Upper fully expanded leaves	<10–13	Bergmann (1992)
Mango *(Mangifera indica)*	Fruit	<20	Ram et al. (1989)

Table 2. Boron distribution (mg B kg^{-1}) in shoots of field grown apple and walnut. From Brown and Hu (1996)

Leaf age	Apple (*Malus domestica*)	Walnut (*Juglans regia*)
Old	50	304
Matured	57	225
Young expanded	56	127
Expanding	73	62
Meristematic	70	48

Whilst the flux of water through the plant clearly has a significant bearing on the uptake and distribution of B in the plant (Raven, 1980), water movement alone is clearly insufficient to explain the patterns of B distribution observed in plants (Brown and Shelp, Chapter 7). In plants that retranslocate B, its distribution will clearly not correlate with water flux patterns (Table 2). Developing diagnostic criteria for plant analysis may be assisted by a better understanding of the nexus between water uptake, B uptake and B deficiency in growing plant parts.

Rawson (1996) has hypothesised that a temporary interruption in xylem water flux to the developing ear of wheat may induce sterility in the ear when a transient B deficiency coincides with a period of pollen development which has a critical B requirement. He outlined several environmental factors, including heavy fog and low vapour pressure deficit, which may trigger wheat sterility. Such cases of B deficiency will prove intractable to diagnosis by traditional forms of plant analysis which involve sampling of whole leaves, usually before flowering or at the commencement of flowering.

There is strong evidence that membrane functions are impaired by B deficiency (Loomis and Durst, 1992; Parr and Loughman, 1983). Within minutes of withdrawing B from solutions, increases in leakage of K^+ from leaves has been measured (Cakmak et al., 1995). The possibility of using this property as a means of setting functional B requirements has some promise. Recalculating the data of Cakmak et al. (1995) who measured the effect of external B on K^+ leakage from sunflower leaves, an approximate functional B requirement for membrane function of 16–17 mg B kg^{-1} dry wt is established corresponding with a K leakage rate of 50 μg K^+ g^{-1} fresh wt 2 h^{-1}. This estimate of a functional B requirement for sunflower is consistent with reports that leaves exhibiting B deficiency symptoms contained <10–13 mg B kg^{-1} dry wt at the time of sampling (Bergmann, 1992; Table 1).

Standards for diagnosis

Plant analysis can be used to diagnose B deficiency when standards have been established to distinguish adequate from deficient B concentrations. As discussed above, such standards are in the form of a critical concentration or critical range which has been established from a correlation between B concentration in a plant part and plant growth, or preferably with the response of a physiological function to B.

For diagnosis, an ideal plant part for sampling would have a constant critical concentration with plant age. In addition, B concentrations in the plant part should respond to increasing B supply, but have a narrow transition zone between deficient and adequate concentrations, and an absence of Piper–Steenbjerg curvature (Bates, 1971; Ulrich and Hills, 1967). There have been no studies of the effect of plant age on functional B concentrations in defined plant parts. However, in two experiments reported by Huang et al. (1996a) the functional B requirement for leaf elongation in the YOL of oil seed rape at an early vegetative stage was closely similar to the critical B concentration in the YOL of older plants immediately before stem elongation.

Cases of Piper–Steenbjerg curvature for the response of leaf B to declining B supply are unusual (Bates, 1971), with one notable exception reported by Vlamis and Ulrich (1971) for sugar beet. In this experiment sugar beet plants were grown in non-renewed solutions for 7 weeks. Boron deficiency appeared over time at progressively higher levels of initial B supply. In the young leaves, which were a composite of the shoot tip, rapidly expanding leaves, and mature leaves yet to reach full size, the deficient plants apparently continued to accumulate B after deficiency depressed apical growth. Avoidance of Piper–Steenbjerg curvature may have been achieved in this case by sampling defined young growing plant parts as recommended for phloem immobile elements (Smith and Loneragan, 1997), and for B in particular (Huang et al., 1996a). The above case of Piper–Steenbjerg curvature may also suggest that leaf sampling for diagnosis should occur soon after deficiency symptoms appear, as was recommended by Ulrich (1952).

Most plant analysis standards have been derived from the relationship between B concentration and shoot dry matter. However, for many crops, quality of the harvested product is also depressed by low B and should be considered in the setting of standards for diagnosis. Crop quality may well be a more sensitive indicator of low soil B than shoot dry matter *per se*. In peanut, for example, Rerkasem et al. (1988) reported that B deficiency induced the hollow heart symptom in kernels but had no effect on seed yield. By contrast, a similar hollow heart symptom also appeared in seed of B deficient soybean but the seed deformity was no more sensitive to low B that the yield of seed (Rerkasem et al., 1993). Bell et al. (1989) found that low B in soils depressed B concentration in black gram seed but had no effect on seed yield. When the seed containing <6 mg B kg^{-1} dry wt were germinated *in vitro* in a B treated medium, a high proportion of the seeds were non-viable or produced seedlings with aborted or deformed shoots.

Quality issues are also of particular importance for vegetables such as cauliflower (*Brassica oleracea*) (Kotur, 1991), fruit trees such as mango (Ram et al., 1989), and for timber trees (Stone, 1990). When dry matter is no longer the criterion of plant response, different forms of relationship between B concentration in a plant part and plant response may be evident. For example, in the case of internal necrosis in mango fruit, an inverse linear relationship was found between the rating of symptoms and the B concentration in the fruit (Ram et al., 1989).

Factors affecting interpretation

Plant part. In species which cannot retranslocate B, young growing plant parts appear to be preferable to recently matured parts for B deficiency diagnosis. The growing plant parts are the only ones which reflect current B supply and therefore in the strict sense are the only plant parts in which B deficiency can be diagnosed. Recently matured plant parts can be expected to contain B concentrations which correlate with those in the growing tissues under conditions where B supply is relatively constant over time, but not when B supply changes rapidly (e.g. Huang et al., 1996a). However, whilst Huang et al. (1996a) recommended the use of the YOL for diagnosis of B deficiency in oil seed rape, they nevertheless recognised that on occasions it is necessary to make a diagnosis based on B in a recently matured leaf. They proposed a critical range for diagnosis of B deficiency in the youngest mature leaf (YML) which was about half that in the YOL (6–8 vs 10–14 mg B kg^{-1}) with the caveat that values in the YML can be expected to be less reliable for diagnosis. Furthermore, the large difference in critical ranges for B in the YML and YOL suggest that great care is needed in sampling leaves in the field.

For elements which are phloem immobile or variably mobile, mature or old leaves are unsuitable for diagnosis. This is illustrated by studies with B where the transfer of tomato (*Lycopersicum esculentum*) plants from solutions with excess B to those with no B results in the curious situation where young leaves develop B deficiency symptoms and contain very low B concentrations whereas in the same plant the old leaves still contain very high B concentration and B toxicity symptoms (Oertli, 1993). Similarly, Rerkasem and Lordkaew (1996) drew attention to the fact that sampling whole ears of wheat plants failed to diagnose a known B deficiency which depressed seed set because B concentrations in the ears did not reflected current B supply. However, because the affected organs, anthers and the carpel, comprised only 10% of the dry matter of the ear, change in their B concentration which diagnosed B deficiency was masked by the B content of the remainder of the ear, which reflected previous B supply.

In species which are able to retranslocate B, it is expected that recently matured leaves will reflect the current B status of the plant although this remains to be tested.

Boron concentrations associated with impaired function in reproductive plant parts exceed those required in vegetative plant parts. In maize, for example, Agrawala et al. (1981) found that maximum pollen number in the mother plant, and pollen germination *in vitro* were associated with a pollen concentration of 11 mg B kg^{-1} dry wt. By contrast at 8 mg B kg^{-1} dry wt both pollen number and pollen germination were depressed. However, in maize leaves, <2 mg B kg^{-1}dry wt was associated with deficiency symptoms (Jones, 1967). Similarly, in wheat, Rerkasem and Lordkaew (1996) reported that maximum levels of grain set were associated with >8 mg B kg^{-1} dry wt in pollen, and >6 mg B kg^{-1} in the carpels. By contrast, leaf B concentrations of the same plants were about 3 mg B kg^{-1}dry wt and unchanged by B supply, and their vegetative growth was not measurably impaired by low B. In other experiments with wheat, no impairment of leaf expansion or membrane function, as measured by K^+ leakage, was evident even when leaf B concentrations were as low as 1 mg B kg^{-1} (Huang et al., 1996b). Gartel (1974) reported that the pistil of grape (*Vitis vinifera*) flowers required 50–60 mg B kg^{-1} dry wt for optimum germination of pollen on the stigma: by contrast, at <20 mg B kg^{-1} in the pistil, fertilisation was impaired leading to floral abortion.

Thus whilst there is evidence that reproductive plant parts in fact require higher internal B concentrations than vegetative plant parts of the same plants, the great diversity of B-deficiency symptoms during reproductive growth among species has not been adequately explained. It is possible that different stages of reproductive growth are sensitive to B deficiency in different species but this does not appear to have been demonstrated.

Higher pectin levels in cell walls in reproductive plant parts could also account for their higher tissue B requirements but this does not appear to have been examined. Similarly differences in other B complexing compounds such as phenols and sugar alcohols could account for the higher B requirements of floral organs compared to vegetative plant parts in the same plants. Further investigation in this area could be fruitful in predicting species and genotype differences in sensitivity to B deficiency. For plant analysis, the existence of higher B requirements for growth of reproductive organs means that leaf analysis may be less effective for diagnosis of B deficiency in those species than analysis of the reproductive organs.

Plant age. The effect of plant age on functional B requirements has not been established. When diagnostic standards have been developed by estimating the functional B requirement for a plant part they can be expected to vary little, if at all, with age provided the plant part is consistently sampled. However, Wilson (1976) reported that the cell wall composition of leaves changes with their position on the stem axis. Given that most plant B is associated with the cell wall (Matoh, Chapter 5), the possibility that functional B requirements may change with plant age needs to be assessed. Obviously in determinate plants, diagnosis can no longer be carried out using standards for growing leaves once the terminal leaf on the stem has fully expanded. In the case of oil seed rape, after stem elongation there is a distinct change in leaf form (Sylvester-Bradley, 1985) suggesting that a different critical range may apply to the sampling of a YOL after stem elongation compared to before.

Species. For diagnosis of B deficiency, internal B requirements of species have been shown to vary (Table 3). Since very few studies have established functional B requirements, published data on species differences in internal B requirements are difficult to judge because of differences in plant parts, and stage of growth sampled, and the B supply regime prior to sampling. Studies that have used whole shoots to establish critical concentration are not considered reliable because of: the different distribution of B in parts of the plant (Table 2) which will result in critical values changing in shoots of different age as the proportions of plant parts change over time, and; the effect that variation in prior B supply regime can have on the B distribution. Unfortunately, many reports of critical concentrations reported for field crops, pastures and vegetables are of this type (e.g. Reuter and Robinson, 1986). Studies where no procedure has been followed to maintain B supply throughout the duration of the experiment should also be used with caution (e.g. Vlamis and Ulrich, 1971). Again, many reported

Table 3. Critical boron concentration (mg B kg^{-1} dry wt) or concentration ranges for diagnosis of B deficiency in plants

Species	Plant part	Critical B concentration range	Source
Soybean *(Glycine max)*	Youngest open leaf	10–12	Kirk and Loneragan (1988)
Oil seed rape *(Brassica napus)*	Youngest open leaf	10–14	Huang et al. (1996a)
Black gram *(Vigna mungo)*	Youngest open leaf	15–17	Noppakoonwong et al. (1993)
Green gram *(Vigna radiata)*	Youngest open leaf	16-18	Bell et al. (1990a)
Cassava *(Manihot esculenta)*	Youngest mature blade	35	Howeler et al. (1982)

B critical concentrations from solution culture studies, sand culture and some soil culture studies will suffer from these drawbacks (e.g. Reuter and Robinson, 1986). The remaining studies in which greater confidence can be held are relatively few in number (Table 3). Apart from those which determined a functional B requirement, other studies which maintained B supply throughout the experimental period by programmed nutrient addition, or by frequent solution replacement in solution and sand culture can be considered reliable estimates of the functional B requirements.

The similar functional B requirements for green (*Vigna radiata*) and black gram suggest the possibility that functional B requirements might be similar for other groups of closely related species. Smith and Loneragan (1997) drew attention to this possibility for other elements as well. If it is possible to establish groups of plant species with similar functional B requirements then the enormous task of establishing functional B requirements for all economically important plants may be short circuited. Examination of cell wall pectin content and composition may well establish whether similarities between closely related species and cultivars are sufficient to anticipate similar functional B requirements.

Cultivars. As yet there is no evidence that functional B requirements vary among cultivars of the same species. With other micronutrients, cultivars with vastly different nutrient efficiency had the same critical internal concentration (e.g. Cu – Reuter et al., 1983: Mn – Marcar and Graham, 1987). Moreover, for B toxicity there appears to be no differences in internal tolerance to B (Nable et al., 1989).

Environmental conditions. For diagnosis there is little evidence yet that environmental conditions apart from light affect functional B requirements. As early as 1933, Warington observed that B deficiency symptoms in plants were enhanced in the summer when light intensity was high and photoperiods long, leading to continued speculation that exposure to high light intensity increases B requirements of plants. Experimental results from Noppakoonwong et al. (1993) indicated that decreasing sunlight intensity from 70 to 35% of full sunlight by shading, decreased B concentrations required for normal leaf elongation of black gram plants from 15 to 10 mg B kg^{-1} dry wt. Noppakoonwong et al. (1993) did not examine the possibility of differences in B distribution in mature and old leaves of the shaded and unshaded plants. Neither did their work distinguish between the possibility that the difference in the B concentration required for leaf elongation between shaded and unshaded plants was the consequence of differences in external B supply in the nutrient solutions or changes in the internal B requirements for leaf growth. Cakmak et al. (1995) also demonstrated that increasing light intensity or photoperiod enhanced the effects of B deficiency on cell membrane permeability in leaf discs of sunflower plants which had been subject to B deficiency treatment for 10 days. Physiological and biochemical mechanisms by which light intensity and B deficiency interact in plants remain unknown although Marschner (1995), and Cakmak and Römheld (Chapter 6) favoured the hypothesis that high light increases phenol levels in tissues and hence increases the B required to complex with phenols.

Several reports have linked low temperature damage in plants and B deficiency (Shorrocks, Chapter 9). The substance of these reports is limited to observations of increased frost or low temperature damage in shoots when plant B status is low, and in some cases, alleviation of the injury with foliar B sprays. Experimental evidence that low temperatures increase sensitivity to B deficiency is limited to one brief report by Parr and Loughman (1983). Parr and Loughman (1983) examined the response of P uptake by maize to decreasing temperature in the presence or absence of B in solution. In solutions supplied with B, the uptake of P declined with decreasing temperature but below 20 °C there was a distinct inflection in the curve implying a temperature dependent change in membrane conformation from a fluid to a gel state. That low temperature causes a change in membrane properties has been reported previously. What had not been previously reported was the findings of Parr and Loughman (1983) that the critical temperature at which membrane properties changed, depressing P uptake, was 2 °C higher in low B solutions than in B adequate solutions. These results, limited though they are, imply that the low B tissues might be more sensitive to cold temperature damage to membranes than B adequate tissues.

Plant analysis for prognosis

A comprehensive listing of standards for the prediction of B deficiency by plant analysis was prepared by Bradford (1966) and subsequently updated by Bergmann (1992), Jones et al. (1991), Martin-Prevel et al. (1987), Reuter and Robinson (1986) and Shear and Faust (1980). For critical values appropriate for particular crop-soil-environment combinations the above references are still useful sources of information.

Plant parts

Plant parts that best serve the function of predicting B deficiency should meet the same criteria as discussed above for diagnosis. They will not necessarily be the same plant parts as those which diagnose deficiency (Smith and Loneragan, 1997). For prognosis, plant parts that reflect nutrient supply from the soil as well as nutrient reserves within the plant are most likely to predict subsequent deficiencies. Few studies have systematically examined a range of plant parts to assess their relative efficacy in predicting yield. In almond (*Prunus amygdalus*) which retranslocates B, leaf B concentration predicts only severe B deficiency whereas hulls are more predictive of B deficiency at all levels of severity (Brown and Shelp, Chapter 7). For prognosis, the store of B in the water-soluble fraction may be a good predictor of the likelihood of a subsequent B deficiency in species that retranslocate B. However, in most species that do not retranslocate B, the existence of a pool of water-soluble B in mature leaves will provide no insurance against B deficiency occurring later (e.g. Oertli, 1993). Most studies have simply used the same plant part identified from earlier diagnostic standards, or used a standard leaf such as the youngest mature leaf or its equivalent. A possible weakness of this approach is suggested in a study of potato (*Solanum tuberosum*) by Pregno and Armour (1992) who sampled the youngest fully expanded leaf (YFEL) and found its B concentration at 7 weeks correlated with tuber yield at maturity. However, the sufficiency range between deficiency and toxicity was 24–26 mg B kg^{-1} dry wt raising questions as to how useful plant analysis might be for potato in that situation apart from diagnosing severe B deficiency or toxicity. A different plant part may well have allowed a clearer distinction to be draw between deficient and toxic B concentrations.

Plant age

In peanut grown on two sandy soils in northeast Thailand, Keerati-Kasikorn (pers. comm.) compared the relationship between B concentrations in the youngest folded leaf (YFL) and the YFEL and hollow heart in kernels at maturity for two leaf sampling times, early flowering (R1-2: after Boote, 1982) and early seed filling (R5-6). The level of hollow heart was well correlated with B concentration in both leaf types, although the critical range was higher in the YFL (22–23 mg B kg^{-1} dry wt) than in the YFEL (17–19 mg B kg^{-1}). Between the two sampling times which were about 30 days apart there was no change in the critical range of B concentrations in the leaves which predicted 5% incidence of hollow heart in kernels.

For fruit trees, the relationship between leaf age and the prognosis of B deficiency has received little attention. Shear and Faust (1980) in their review of nutritional ranges of deciduous fruit and nut trees state that the relationship between leaf age and leaf B concentrations is variable.

Species and cultivars

Apart from field studies with sunflower (Blamey et al., 1979) and bean (*Phaseolus vulgaris*) (Howeler et al., 1978), the question of whether critical B concentrations for prognosis change among cultivars does not appear to have been examined. In two sunflower cultivars grown on two soil types in two seasons, the critical range in the topmost mature leaf was the same at 31–38 mg B kg^{-1} dry wt. The cultivars, one of which was open pollinated whilst the other was a hybrid, produced similar maximum yields even though among the three site x season combinations, seed yield varied two-fold from about 1.5 to 3 tonnes ha^{-1}. That the critical range remained unchanged despite the fact that low rainfall in the second year depressed yield by 30–50%, and that the two soils varied markedly in clay content (37 vs 13%) and in reaction to B fertiliser, suggests that the test for B deficiency by plant analysis was fairly robust.

In a study with two bean cultivars (Howeler et al., 1978), like that with sunflowers there was no major differences in B efficiency between the cultivars although they differed markedly in responsiveness to B fertiliser and maximum yield. Despite these differences in response to B fertiliser, the apparent critical B concentration in uppermost mature leaves were similar for the two cultivars. Critical ranges of 20–24 at 37 days and 16–18 mg B kg^{-1} at 75 days after sowing were proposed for prognosis of B deficiency. Thus in two species, cultivars with similar B efficiency had the same critical B concentration in leaves for prognosis of B deficiency. It remains to be seen whether cultivars which differ greatly in B efficiency, unlike those used by Howeler et al. (1978) and Blamey et al. (1979), will also have similar critical B values for prognosis of B deficiency.

Among other groups of plants substantially higher and lower critical B concentrations have been reported for the prediction of B deficiency or B response. In wheat, critical B ranges in leaves, and in ears for the prognosis of B deficiency were much lower than values derived for legumes and sunflower (Table 4). Wheat, along with other cereals appears to have a much lower internal B requirement on account of lower uronic acid levels in the cell walls (Figure 1).

By contrast in fruit trees, critical leaf B concentrations are substantially higher than those of legumes and oil seed crops. Stephenson and Gallagher (1987) obtained yield increases in macadamia (*Macadamia ternifolia*) from foliar B sprays which increased leaf B concentrations from 30 to 65 mg kg^{-1}. In Hawaii, 75 mg B kg^{-1} was reported to be needed in leaves for maximum yield (Warner and Fox, 1972). In almond and pistachio (*Pistacia vera*), responses to foliar B sprays can be obtained even when leaf B concentrations are within the range regarded as adequate for maximum yield (Brown, 1995). In pistachio, leaf B concentrations in mid-summer (July) in the range 100-150 mg kg^{-1} are critical for seed yield (Brown et al., 1994): similar B concentrations are required in almond also but hulls rather than leaves are a more reliable plant part for sampling.

Indeed with trees of the *Prunus*, *Pyrus* and *Malus* genera an entirely new approach to plant analysis may emerge from current research on their response to foliar B application. Boron foliar sprays in autumn after harvest are effective in increasing fruit set and yield in the following year by increasing B stored in floral buds (Hanson, 1991; Nyomora et al., 1997). Thus leaf analysis in one year can be used to determine the need for foliar B sprays in that year in order to increase fruit yield the following year.

Environmental factors

The same environmental factors that affect critical values for diagnosis may also affect those for prognosis. In addition, those environmental factors that depress the uptake of B disproportionately to their effects on plant growth will alter the probability of B deficiency. However, because environmental factors such as low soil water, low temperature, and low vapour pressure deficit act intermittently, or temporarily during the growing season, their consequences for B nutrition are difficult to predict. Since the pattern of change in leaf B concentrations induced by unusual environmental conditions does not follow the normal pattern on which the critical values are predicated, the normal critical values may not apply in those unusual conditions.

Low soil water depresses B uptake (Hobbs and Bertramson, 1949) but its effects on critical B values are complicated by when the dry period occurs, and whether or not rainfall alleviates the period of drought. Noppakoonwong et al. (1997) found that a 30-day period of drought strongly depressed B uptake in black gram plants and induced severe B deficiency symptoms, particularly pod abortion. However, in that particular field experiment, rainfall occurred during the mid-pod setting period so that as B uptake increased with increasing soil water content, pod setting resumed and largely compensated for the loss of pods that occurred during the dry period. Critical B concentrations for prognosis in that season declined from about 30 mg B kg^{-1} before flowering to 9 mg B kg^{-1} during the period of drought and increased again for leaf samples collected after the drought period. By contrast, Bell et al. (1990b) reported that in irrigated black gram grown on similar soils in the same environment that the critical B concentration in YFEL for the prognosis of B deficiency declined from about 45 mg B kg^{-1} before flowering to 20 during late pod setting. The differences between the two sets of results only serves to illustrate that a critical value for prognosis is

Table 4. Critical boron concentration (mg B kg^{-1}) or concentration ranges in leaves of plants for the prognosis of B deficiency

Species	Leaf and plant age or growth stage[1]	Critical B concentration or range	Country and Source
Bean (*Phaseolus vulgaris*)	YFEL – 37 days after sowing	20–24	Columbia; Howeler et al. (1978)
Bean	YFEL – 75 days after sowing	16–18	Columbia; Howeler et al. (1978)
Broccoli (*Brassica oleacea*)[2]	YFEL blade when 5% heads formed	9–13	Canada; Gupta and Cutcliffe (1973, 1975)
Brussel sprouts (*B. oleacea*)[2]	YFEL blade when 5% heads formed	7–10	Canada; Gupta and Cutcliffe (1973, 1975)
Cauliflower (*B. oleacea*)[2]	YFEL blade when 5% heads formed	8–9	Canada; Gupta and Cutcliffe (1973, 1975)
Peanut (*Arachis hypogaea*)[3]	Youngest open leaf at flowering	26–30	USA; Morrill et al. (1977)
Peanut[4]	YFEL at flowering	22-23	Thailand; Bell et al. (1990b)
Potato (*Solanum tuberosum*)	YFEL – 7 weeks after sowing	24	Australia; Pregno and Armour (1992)
Rutabaga (*B. napobrassica*)[5]	Youngest mature leaf blade at 5–6 leaves	37–44	Canada; Gupta and Cutcliffe (1972)
Sugar beet (*Beta vulgaris*)	Recently matured leaf blades	35–40	USA; Vlamis and Ulrich (1971)
Sunflower (*Helianthus annuus*)	YFEL – flowering	31–38	South Africa; Blamey et al. (1979)
Wheat (*Triticum aestivum*)	YEB – booting	3–7	Thailand; Rerkasem and Loneragan (1994)

[1] YFEL – youngest fully expanded leaf blade; YEB – youngest emerged blade.
[2] Critical range obtained by combining the lowest B concentrations from Gupta and Cutcliffe (1973) where leaf B was adequate for yield and symptoms were absent, with the highest values of Gupta and Cutcliffe (1975) where yield was depressed or symptoms were present.
[3] Critical values for the elimination of hollow heart disorder in kernels.
[4] Critical values for <5% hollow heart incidence in kernels.
[5] Critical values for predicting the absence of brown heart disorder in roots at maturity.

a prediction of the possibility of a deficiency occurring for yield.

Soil analysis for prognosis

Soil B analysis has been in use for several decades, beginning with the development by Berger and colleagues of a soil B extraction method based on boiling soils in water for 5 min (Berger and Truog, 1939). Soil analysis has proved reasonably successful in predicting B response in plants (Moraghan and Mascagni, 1991), although not universally so (Sims and Johnson, 1991). Most widely used is the hot water extraction or its variants such as the hot 0.01 *M* $CaCl_2$ method now favoured in most laboratories (Shorrocks, 1992). However, it is increasingly obvious that no single critical B concentration can be used to predict B deficiency: critical values will vary with the environmental factors affecting B uptake such as soil water, soil texture and soil pH (Shorrocks, 1992; Sims and Johnson, 1991) and with species and cultivars (Table 5).

Boron behaviour in soils

Boron absorption rates by roots are determined by the B concentration in the soil solution (Aitken and McCallum, 1988; Bingham et al., 1970; Keren et al., 1984). Recently, Asad et al. (1997) showed that the external B requirement for 90–95% of maximum growth of oil

Table 5. Critical boron concentrations (mg B kg^{-1} soil) or concentration ranges or concentration ranges in soils for the prognosis of B deficiency

Species	Method of extraction: soil sampling depth (cm)[1]	Soil pH	Critical B concentration or concentration range	Country and source
Bean (*Phaseolus vulgaris*)	HWS [2]: not reported	7.6	0.4–0.5 [3]	Columbia; Howeler et al. (1978)
Black gram (*Vigna mungo*)	HWS: pot expt	7.7–10	0.53	India; Sakal et al. (1985)
Black gram	HWS: 0–25 cm	5.5–6.5	0.08–0.13	Thailand; Bell et al. (1990b)
Broccoli, Brussel sprouts, Cauliflower (*Brassica oleacea*)[4]	HWS: not reported	5.8–6	0.28–0.34	Canada; Gupta and Cutcliffe (1973, 1975)
Peanut (*Arachis hypogaea*)[5]	HWS: not reported	5.0–7.7	0.15	USA; Morrill et al. (1977)
Peanut[6]	HWS: 0–10 cm	–[7]	0.14–0.16	Thailand; Bell et al. (1990b)
Rutabaga(*Brassica napobrassica*)[8]	HWS: not reported	5.2–5.4	0.65–0.9	Canada; Gupta and Cutcliffe (1972)
Sunflower (*Helianthus annuus*)	Hot 0.01 *M* $CaCl_2$; pot expt	4.7–7.4	0.14	Australia; Aitken et al. (1987)
Wheat (*Triticum aestivum*)	HWS; not reported	6	0.12–0.15	Thailand; Rerkasem et al. (1988,1989)
Wheat	HWS; not reported	Alkaline	0.32–0.38	PR China; Li et al. (1978)

[1]Unless otherwise stated results are all from field studies.
[2]HWS – hot water soluble.
[3]Howeler et al. (1978) suggested a critical value of 0.65 mg B kg^{-1} soil but examination of their data suggests a lower range is critical.
[4]Critical range obtained by combining the lowest B concentrations from Gupta and Cutcliffe (1973) where leaf B was adequate for yield and symptoms were absent, with the highest values of Gupta and Cutcliffe (1975) where yield was depressed or symptoms were present.
[5]Critical values for the elimination of hollow heart disorder in kernels.
[6]Critical values for <5% hollow heart incidence in kernels.
[7]pH of soils in the study not reported.
[8]Critical values for brown heart disorder in roots at maturity.

seed rape was 0.5–0.6 μM B. Simple calculations show that for a C3 plant which accumulates 1 g dry matter for each L of water transpired, a minimum of 11 mg B kg^{-1} dry wt requires 1 μM in the transpiration stream (Raven, 1980). At present, there are no reliable measurements of soil solution B concentrations <1 μM to confirm the estimates of Asad et al. (1997) in the field. Soils from Hawaii have been reported to contain solution B concentrations ranging from 1 to 10 μM (Hue et al., 1988). Similarly, Aitken et al. (1987) extracted soil solutions from 20 Queensland soils and found that soil solution B ranged from <1 to 8 μM. Sunflower growth was depressed after 14 days growth on 3 of the 20 soils with soil solution B values of <1–3 μM but no growth response was observed in 17 soils when soil solution values at harvest ranged from <1 to 8 μM B.

In low B soils, the amount of B in the soil solution at any one time is a small fraction of the total B uptake by most crops during the growing season. At B concentrations < 3 μM, B in the soil solution must be replenished several times during a growing season. Thus a soil test for B deficiency must assess the potential of the soil to continually replenish soil solution B during the season (Aitken and McCallum, 1988).

The capacity of a soil to buffer soil solution B over time against losses by plant uptake or leaching is determined by soil texture, soil mineralogy, pH and organic matter. Clay soils react more strongly with B than sandy soils (Keren et al., 1984). Whilst they buffer B in the soil solution better than sandy soils they require higher fertiliser B additions to correct deficiency: boron is also less likely to leach through clay soils or clay rich horizons than sandy soils or horizons (e.g. Pinyerd et al., 1984). Organic matter also increases the B sorption capacity of soils (Yermiyahu et al., 1995).

At pH >8, when the proportion of borate [$B(OH)_4^-$] increases significantly relative to boric acid [$B(OH)_3$], higher soil solution B concentrations are necessary to achieve the same rates of B uptake. In addition, B sorption increases with pH and peaks at 7.5–9.5. Hence the critical range of soil B levels will generally be higher on an alkaline soil than an acid soil as suggested by the comparison between the data for wheat from alkaline clays soils in northern China compared to that on loams in northern Thailand (Table 5).

Whereas Gupta et al. (1985) suggested that hot water extracted B from soluble inorganic, organic and adsorbed inorganic pools, recent attempts to fractionate soil B and to relate plant uptake to supply from specific fractions (Jin et al. 1987; Tsadilas et al. 1994) provide more compelling evidence about the plant available B pools. Such studies establish a sound basis for soil B testing which has been previously lacking. Boron fractionation procedures used by Hou et al. (1996), Jin et al. (1987) and Tsadilas et al. (1994) have extracted the following fractions of soil B: soil solution B; non specifically adsorbed B; specifically adsorbed B; and B occluded in Mn oxyhydroxides; B in amorphous Al, Fe oxyhydroxides, and crystalline Al, Fe oxyhydroxides, and; B in silicate minerals. In a range of 20 soils from Greece, hot water B was correlated with all B fractions extracted apart from the Mn oxyhydroxide fraction (Tsadilas et al., 1994). A multivariate regression model suggested that plant B uptake was best explained by variation in the cold water soluble B, mannitol extractable B and $NH_2OH.HCl$ extractable B. These fractions correspond approximately to the soil solution, specifically adsorbed B and B occluded in Mn oxyhydroxide fractions, respectively. That hot water seemed to extract B from the first two fractions but not the third suggests a possible reason why it may fail on some soils to predict plant response to B. Moreover, the fact that B uptake by olive (*Olea europea*) trees was correlated with B in the Mn oxyhydroxide fraction, but that by barley (*Hordeum vulgare*) was not, suggests a possible reason why hot water soluble B may be better correlated with B response in some species than others.

Various methods are available for boiling soil samples to extract B (Bingham, 1982). Whilst boiling in glass flasks fitted with a reflux apparatus was the original method, more recent variations include boiling in long tubes in an aluminium block (Spouncer et al., 1992), in flasks on a hot plate (Gupta, 1979), and by microwave heating in sealed plastic pouches (Mahler et al., 1984). There is increasing evidence that the 5 min boiling used in the original hot water extraction is a significant source of variability with that method. Not only is it difficult for an operator to be certain that all solutions in a batch boil for exactly 5 min, but there is substantial variation in the heating and cooling rates of samples which changes the effective extraction time for different samples. Several variations of the hot water and hot 0.01 *M* $CaCl_2$ methods of soil B extraction have proposed increases in boiling time to 10 (Jeffrey and McCallum, 1988; Odum, 1980) or 30 min (Spouncer et al., 1992). Whereas Jeffrey and McCallum (1988) noted no difference in the amount of B extracted as boiling increased from 5 to 15 min they nevertheless preferred the 10 min boiling because it minimised the errors due to mis-timing the boiling period. By contrast, on soils that were generally alkaline and higher in clay than those used by Jeffrey and McCallum (1988), Spouncer et al. (1992) found that 30 min boiling was necessary for the amounts of B extracted to reach a plateau.

Aitken et al. (1987) showed that the 10 min boiling in 0.01 *M* $CaCl_2$ extracted amounts of B from Queensland soils that were well correlated with the growth response, B uptake and leaf B concentrations of sunflower plants in pots. They also found a close correlation between hot water extraction and sunflower growth. Boron extracted by hot water was closely correlated with that extracted by hot 0.01 *M* $CaCl_2$. This approach to the modification of an established method of soil B testing is to be commended because it not only establishes that the new test has practical advantages for analysis, but that it is at least as good at predicting plant response as the old method and that there is a mathematical relationship between the two sets of soil B values. This allows users to adopt the new method with confidence and to relate soil test results for the new methods with published results from the previous method.

Other extractants including 0.05 *M* HCl (Renan and Gupta, 1991), Mehlich-3 (Shuman et al., 1992), NH_4

acetate (Jin et al., 1988), mannitol and sorbitol (Vaughan and Howe, 1994) have been suggested as extractants for soil B. Whilst all may have a place in testing for plant available forms of B on some soils, such verification needs to be completed for particular crop–soil combinations. Most importantly, the usefulness of the extractant can only be determined from how well the soil test values correlate with plant response. Unfortunately many soil tests are developed and recommended without demonstration that they indeed predict plant response. Even to correlate the soil test values with B uptake by the plant is not necessarily a valid basis for selection of the soil extractant. Moreover, the fact that a modification to the boiling time increased the amount of B extracted from a soil, may not necessarily improve the ability of the test to predict B deficiency in a plant. It is the correlation between the amount of B extracted and the growth response around the range of the critical value that is central to the argument of whether an extractant is suitable for distinguishing potentially B deficient from adequate soils. When soils with a wide range of extractable B levels are examined, the overall correlation between leaf B and extractable B may be reasonably good, but significant deviations from the regression line around the range of the critical value will render the test of limited value for predicting B deficiency. There appears to be an inherent flaw in the approach often used for correlating soil B tests with B uptake. Such studies often include at least some soils with luxury levels of B with the expressed intention of demonstrating that the test is valid on a wide range of soils. When the relationship between soil test values is plotted against B uptake, the existence of a few data points with very much higher B uptake and soil B will usually ensure that a reasonable correlation exists. However, such correlations may mask the fact that the correlation at low soil B in the range where deficiency is likely is rather poor or non-existent (e.g. Tsalidas et al., 1994). In practice, only soils which are close to or below the critical range are ever analysed for the purpose of predicting B deficiency. Correlation of soil test values should be established for soils in this range (Sims and Johnson, 1991). Moreover they should be correlated against plant growth or yield or product quality rather than with B uptake.

Factors affecting interpretation

Since soil tests aim to extract soil B pools accessed by the plant, they should also sample to depths where a predominance of B is absorbed by plant roots. Shallow soil samples to 7.5 cm depth were poorly correlated with symptoms of B deficiency assessed on pine (*Pinus radiata*) in the field, and only marginally better at predicting B deficiency symptoms and dry matter in plants in pots in the glasshouse (Snowdon, 1982). With alfalfa (*Medicago sativa*), Baker and Mortensen (1966) obtained a better prediction of B response when soil was sampled from the 25–41 cm layer that from the surface layers. Bell et al. (1990b) reported that correlation between hollow heart incidence in peanut kernels and hot water soluble B was closer and the critical range narrower when soil samples were taken from 0–25 cm than from 0–10 cm.

Several studies have examined correlation between soil test values for different extractants and extraction methods for soil B (e.g. Aitken et al., 1987; Jin et al., 1988; Shuman et al., 1992; Vaughan and Howe, 1994). Clearly, amounts of B extracted vary greatly among extractants. Whilst this should ensure that critical values also vary for different extractants no evidence of this has been published. In the soils tested by Jin et al. (1988), the amounts of B extracted followed the order: Mehlich-3 > hot water = NH_4 acetate > mannitol. In a comparison involving mostly acid, sandy Queensland soils, hot 0.01 *M* $CaCl_2$ (10 min boiling) extracted very similar amounts of B to hot water (5 min boiling; Jeffrey and McCallum, 1988). By contrast, on a different range of soils that were predominantly alkaline sodic soils with high clay and extractable B content, Spouncer et al. (1992) found that hot 0.01 *M* $CaCl_2$ (30 min boiling) extracted twice as much B as hot water (5 min boiling). It is unclear whether the different relationship between hot $CaCl_2$ extractable B and hot water extractable B is due to soil type differences or to the longer boiling time used by Spouncer et al. (1992). What it does emphasise is the empirical nature of such relationships and the need for correlations to be established for particular soil types and extraction methodologies.

In published data where comparisons can be made, critical ranges for the same species appear to increase greatly on alkaline compared to acid soils (Table 5). Relationships between extractable soil B levels and wheat yield appear to differ between soils of contrasting properties for B sorption. Rerkasem et al. (1988) showed that wheat yields were not depressed on a sandy loam soil at Chiang Mai, Thailand when soil hot water soluble B levels were 0.15 mg kg^{-1}, but on a similar soil with a hot water soluble B level of 0.12 mg kg^{-1}, grain yield was depressed by 45% (Rerkasem et al., 1989). By contrast, in northeast Chi-

Table 6. Variation among species in boron concentrations (mg B kg^{-1} dry wt) of plants grown at the same site. (Sorauer et al. cited by Bergmann, 1992)

Species	B concentration	Species	B concentration
Barley (*Hordeum vulgare*)	2.3	White mustard (*Sinapis alba*)	22
Rye (*Secale cereale*)	3.1	Carrot (*Daucus carota*)	25
Leek (*Allium porrum*)	3.1	Tobacco (*Nicotiana tabacum*)	25
Wheat (*Triticum aestivum*)	3.3	Lucerne (*Medicago sativa*)	25
Onion (*Allium cepa*)	4.3	White cabbage (*Brassica oleracea*)	37
Maize (*Zea mays*)	5.0	Soybean (*Glycine max*)	37
Flax (*Linum usitatissimum*)	7.1	Lentil (*Lens culinaris*)	41
Spinach (*Spinacia oleracea*)	11	Turnip (*Brassica rapa*)	49
Celery (*Apium graveolens*)	12	Black mustard (*Brassica nigra*)	53
Potato (*Solanum tuberosum*)	14	Radish (*Raphanus sativus*)	65
Tomato (*Lycopersicum esculentum*)	15	Beet (*Beta vulgaris*)	76
Broad bean (*Vicia faba*)	15	Dandelion (*Taraxacum* spp.)	80
Pea (*Pisum sativum*)	22	Poppy (*Papaver somniferum*)	95

na, complete sterility of wheat was reported when soils contained 0.18–0.22 mg B kg^{-1} and partial sterility at 0.32–0.38 mg kg^{-1} (Li et al., 1978).

Clearly plant species differ markedly in their ability to absorb from the same soil (Gupta, 1979, Table 6). Whilst there was a 30-fold difference in B concentrations among species, the cereals were lowest in the range. Amongst the dicotyledons, a 14-fold range in B concentrations was observed. Most of the variation in B concentration among species can be attributed to variation in cell wall composition with species containing high pectin levels in their cell wall having the higher internal B concentrations (Hu et al., 1996).

Further information on soil B levels found to be deficient and adequate for plant growth are listed in Bradford (1966), Gupta et al. (1985) and Sims and Johnson (1991).

Future developments and trends in soil and plant analysis relevant to boron

Recent advances in understanding the function and physiology of B in plants have yet to be incorporated into the practise of plant analysis. A few studies have established functional B requirements for leaf elongation and shown that they are valid for the diagnosis of B deficiency. However, several other physiological roles of B have promise for establishing functional B requirements including uronic acid levels in cell walls, and solute leakage. The phloem mobility of B in some species suggests the need for a different approach to plant analysis than for species in which it is not phloem mobile. Brown and Shelp (Chapter 7) suggest that leaf analysis was less sensitive for predicting effects of low B on kernel yield of almond than hulls. Moreover, the response of fruit trees to foliar B even when there is no evidence of B deficiency in leaves suggests that new calibrations of leaf B concentrations for response to B need to be established.

The increased use of instruments that conduct simultaneous multi-element analysis, and the need to reduce the cost and turnaround time for analyses are the main driving factors behind a trend towards universal extractants for soil analysis. Such extractants have the potential for predicting deficiencies of several elements rather than a single element. Shuman et al. (1992) reported on the calibration of Mehlich 3 compared to hot water for the extraction of soil B in several eastern states of the USA. Unfortunately, this study, like many others (e.g. Mahler et al., 1984) which propose new methods of soil B analysis failed to recalibrate the new test with plant response. Their recommendation is based on a correlation with the amounts of B extracted by the current standard test which may not be the same thing as a correlation with plant response. Further improvement in soil B tests appears possible through two approaches. Firstly, fractionation of soil B, and then establishing the correlation between plant response and B fractions may provide an improved basis for soil B testing. Secondly, the incorporation of other sets of data into multi-variate regresssion models may improve the prediction of B deficiency. Sillanpaa (1982) for example reported that response to B was better predicted when cation exchange capacity was taken in account that when hot water soluble B alone was used.

One of the major difficulties with B deficiency diagnosis and prognosis is that environmental factors such as low soil water, low temperature, high light and low vapour pressure deficit can induce a temporary B deficiency. Better definition of the effect of each of these environmental stresses on B response is needed in the first instance. However, prediction of the risks of B deficiency in these cases may also be improved by measuring environmental data. For example, Lambert et al. (1997) have reported that prediction of B deficiency in pine trees can be improved when rainfall, and age of trees are considered together with leaf B concentration. Rainfall is used as an indicator of seasonal droughts which depress B uptake, the consequence of which varies with tree age. Similarly, Rawson (1996) suggests that at pan evaporation rates >2.5 mm per day, B uptake by wheat in sub-tropical Asian environments was generally sufficient to prevent sterility except when soil B was very low; by contrast at < 1.5 mm daily pan evaporation, he predicted that sterility risk would be high especially at high temperatures during critical stages of pollen development.

Acknowledgements

The author is grateful to Associate Professor P Blamey and Dr VM Shorrocks for helpful comments which have improved the present paper.

References

Agrawala S C, Sharma P N, Chatterjee C and Sharma C P 1981 Development and enzymatic changes during pollen development in boron deficient maize plants. J. Plant Nutr. 3, 329–336.

Aitken R L, Jeffrey A J and Compton B L 1987 Evaluation of selected extractants for boron in some Queensland soils. Aust. J. Soil Res. 25, 263–273.

Aitken R L and McCallum L E 1988 Boron toxicity in soil solution. Aust. J. Soil Res. 26, 605–610.

Asad A, Bell R W and Dell B 1997 External boron requirements for canola (*Brassica napus* L.) in boron buffered solution culture. Ann. Bot. (*In press*).

Baker A S and Mortensen W P 1966 Residual effect of single borate application on western Washington soils. Soil Sci. 102, 173–179.

Bates T E 1971 Factors affecting critical nutrient concentrations in plants and their evaluation: a review. Soil Sci. 112, 116–130.

Bell R W, McLay L, Plaskett D, Dell B and Loneragan J F 1989 Germination and vigour of black gram (*Vigna mungo* (L.) Hepper) seed from plants grown with and without boron. Aust. J. Agric. Res. 40, 273–279.

Bell R W, McLay L, Plaskett D, Dell B and Loneragan J F 1990a Internal boron requirements of green gram (*Vigna radiata*). *In* Plant Nutrition – Physiology and Application. Ed. M L van Beusichem. pp 275–280. Kluwer Academic Publishers, Dordrecht, The Netherlands.

Bell R W, Rerkasem B, Keerati-Kasikorn P, Phetchawee S, Hiranburana N, Ratanarat S, Pongsakul P and Loneragan J F 1990b Mineral Nutrition of Food Legumes in Thailand with particular reference to Micronutrients. ACIAR Technical Report 19, 52 p.

Bennet M D, Finch R A, Smith J B and Rao M K 1973 The time and duration of female meiosis in wheat, rye and barley. Proc. Royal Soc. London Ser. B 183, 301–319.

Berger K C and Truog E 1939 Boron determination in soils and plants using quinalizarin reaction. Ind. Eng. Chem. Anal. Ed. 11, 540–545.

Bergmann W (Ed.) 1992 Nutritional Disorders of Plants Colour Atlas. Gustav Fischer Verlag Jena, Stuttgart. 386 p.

Bingham F T 1982 Boron. *In* Methods of Soil Analysis, Part 2. Chemical and Microbiological Methods, 2nd ed. Ed. A L Page. Agronomy 9, 431–447.

Bingham F T, Elseewi A and Oertli J J 1970 Characteristics of boron absorption by excised barley roots. Soil Sci. Soc. Amer. Proc. 34, 613–617.

Blamey F P C, Edwards D G and Asher C J 1987 Nutritional Disorders of Sunflower. Department of Agriculture, University of Queensland, St Lucia, Queensland. 72 p.

Blamey F P C, Mould D and Chapman J 1979 Critical boron concentrations in plant tissues of two sunflower cultivars. Agron. J. 71, 243–247.

Boote K J 1982 Growth stages of peanut (*Arachis hypogaea* L.). Peanut Sci. 9, 35–40.

Bradford G R 1966 Boron. *In* Diagnostic Criteria for Plants and Soils. Ed. H D Chapman, pp 33–61. University of California, Riverside California.

Brown P H 1995 Boron nutrition of pistachio. Acta Hort. 419, 77.

Brown P H and Hu H 1996 Phloem mobility of boron is species dependent: evidence for phloem mobility in sorbitol-rich species. Ann. Bot. 77, 497–505.

Brown P H, Ferguson L and Picchioni G 1994 Boron nutrition of pistachio: Final report. *In* California Pistachio Industry. Ann. Rep. Crop Year 1993–94. pp 57–59. California Pistachio Commission, Fresno.

Cakmak I, Kurz H and Marschner H 1995 Short-term effects of boron, germanium and high light intensity on membrane permeability in boron deficient leaves of sunflower. Physiol. Plant. 95, 11–18.

Craswell E T, Loneragan J F and Keerati-Kasikorn P 1985 Mineral constraints to food legume crop production in Asia. *In* Food Legume Improvement in Asian Farming Systems. Eds. E S Wallis and D E Byth. pp 99–111. ACIAR Proceedings Series No. 18. Ramsay Ware Printing, Melbourne.

Dear B S and Lipsett J 1987 The effect of boron supply on the growth and seed production of subterranean clover (*Trifolium subterraneum* L.). Aust. J. Agric. Res. 38, 537–546.

Gartel W 1974 Die mikronahrstroffe- ihre Bedeutung fur die Rebenernahrung unter besonderer Berucksichtigung der Managel- und Ueberschusserscheinungen. Weinberg Keller 10–11, 435–449.

Gupta U C 1979 Boron nutrition of crops. Adv. Agron. 31, 273–307.

Gupta U C (Ed.) 1993 Boron and its Role in Crop Production. CRC Press, Boca Raton, Florida.

Gupta U C and Cutcliffe J A 1972 Effects of lime and boron on brown heart, leaf tissue calcium/boron ratios, and boron concentrations of rutabaga. Soil Sci. Soc. Amer. Proc. 36, 936–939.

Gupta U C and Cutcliffe J A 1973 Boron nutrition of broccoli, brussel sprouts, and cauliflower grown on Prince Edward Island soils. Can. J. Soil Sci. 53, 275–279.

Gupta U C and Cutcliffe J A 1975 Boron deficiency in cole crops under field and glasshouse conditions. Commun. Soil Sci. Plant Anal. 6, 181–188.

Gupta U C, James Y W, Campbell C A, Leyshon A J and Nicholaichuk W 1985 Boron toxicity and deficiency: a review. Can. J. Soil Sci. 65, 381–409.

Hanson E J 1991 Movement of boron out of tree leaves. Hort. Sci. 26, 271–273.

Harris H C and Brolman J B 1966 Comparison of calcium and boron deficiencies of the peanut. I. Physiological and yield differences. Agron. J. 58, 575–578.

Heitholt J J 1994 Supplemental boron, boll retention percentage, ovary carbohydrates, and lint yield in modern cotton genotypes. Agron. J. 86, 492–497.

Hobbs J A and Bertramson B R 1949 Boron uptake by plants as influenced by soil moisture. Soil Sci. Soc. Amer. Proc. 13, 257–261.

Holmes M R J 1980 Nutrition of the Oilseed Rape Crop. Applied Science Publishers, London. 158 p.

Hou J, Evans L T and Speiers G A 1996 Chemical fractionation of soil boron: I. Method development. Can. J. Soil Sci. 76, 485–491.

Howeler R H, Edwards D G and Asher C J 1982 Micronutrient deficiencies and toxicities of cassava plants grown in nutrient solutions. I. Critical tissue concentrations. J. Plant Nutr. 5, 1059–1076.

Howeler R H, Flor C A and Gonzalez C A 1978 Diagnosis and correction of B deficiency in beans and mungbeans in a Mollisol from Cauca valley of Columbia. Agron. J. 70, 493–497.

Hu H and Brown P H 1994 Localisation of boron in cell walls of squash and tobacco and its association with pectin. Evidence for a structural role of boron in the cell wall. Plant Physiol. 105, 681–689.

Hu H, Brown P H and Labavitch J M 1996 Species variability in boron requirement is correlated with cell wall pectin. J. Exp. Bot. 47, 227–232.

Huang L, Ye Z Q and Bell R W 1996a The importance of sampling young leaves for the diagnosis of boron deficiency in canola (*Brassica napus* L.). Plant and Soil 183, 187–198.

Huang L, Pant J, Bell R W, Dell B and Deane K 1996b Effects of boron deficiency and low temperature on wheat sterility. *In* Sterility in Wheat in Sub-tropical Asia: Extent, Causes and Solutions. Ed. H M Rawson and K D Subedi, pp. 90–102. ACIAR Proceedings Series No. 72.

Hue N V, Hirunburana N and Fox R L 1988 Boron status of Hawaiian soils as measured by B sorption and plant uptake. Commun. Soil Sci. Plant Anal. 19, 517–528.

Jeffrey A J and McCallum L E 1988 Investigation of a hot 0.01M $CaCl_2$ soil boron extraction procedure followed by ICP-AES analysis. Commun. Soil Sci. Plant Anal. 19, 663–674.

Jin J Y, Martens D C and Zelazny L W 1987 Distribution and plant availability of soil boron fractions. Soil Sci. Soc. Amer. J. 51, 1228–1231.

Jin J Y, Martens D C and Zelazny L W 1988 Plant availability of applied and native boron in soils with diverse properties. Plant and Soil 105, 127–132.

Jones J B Jr 1967 Interpretation of plant analysis for several agronomic crops. *In* Soil Testing and Plant Analysis, Part 2. Ed. G W Hardy, pp 49–85. Soil Science Society America Special Publication No 2.

Jones J B Jr, Wolf B and Mills H A 1991 Plant Analysis Handbook. Micro-Macro Publishing, Athens Georgia. 213 p.

Keerati-Kasikorn P, Bell R W, Panya P, Gilmour R, and Loneragan J F 1993 Comparison of seed yield and quality of peanut (*Arachis hypogaea* L.) cultivars in low fertility soils and their response to boron and complete fertiliser. *In* Plant Nutrition – from Genetic Engineering to Field Practice. Ed. N J Barrow. pp 409–412. Developments in Plant and Soil Sciences 54. Kluwer Academic Publishers, Dordrecht, The Netherlands.

Keren R, Bingham F T and Rhoades J D 1984 Plant uptake of boron as affected by boron distribution between liquid and solid phases. Soil Sci. Soc. Amer. Proc. 48, 297–302.

Kirk G J and Loneragan J F 1988 Functional boron requirement for leaf expansion and its use as a critical value for diagnosis of boron deficiency in soybean. Agron. J. 80, 758–762.

Kluge R 1971 Contribution to the problem of drought-induced boron deficiency of agricultural crops. Arch. Acker. Pflanzenbau Bodenkd. 15, 749–754.

Kotur S C 1991 Effect of boron, lime and their reside on yield of cauliflower, leaf composition and soil properties. *In* Plant-Soil Interactions at Low pH. Ed. R J Wright. pp 349–354. Kluwer Academic Publishers, Dordrecht, The Netherlands.

Lambert M, Turner J and Knott J 1997 Boron nutrition of radiata pine plantations in Australia *In* Boron in Soils and Plants. Proceedings. Eds. R W Bell and B Rerkasem. Kluwer Academic Publishers, Dordrecht, The Netherlands. (*In press*)

Li B H, Li W H, Kui M C, Chao WS, Jern H P, Li C R, Chu W J and Wang C L 1978 Studies on the cause of sterility of wheat. J. Northeast Agric. Coll. 3, 1–19 (*In Chinese with English abstract*).

Lindsay W and Cox F R 1985 Micronutrient soil testing for the tropics. Fert. Res. 7, 169–199.

Loneragan J F 1968 Nutrient requirement of plants. Nature (London) 220, 1307–1308.

Loneragan J F 1975 The availability and absorption of trace elements in soil- plant systems and their relation to movement and concentrations of trace elements in plants. *In* Trace Elements in Soil-Plant-Animal Systems. Eds. D J D Nicholas and A R Egan, pp 109–134. Academic Press, New York.

Loomis W D and Durst R W 1992 Chemistry and biology of boron. Biofactors 3, 229–239.

Macy P 1936 The quantitative mineral requirements of plants. Plant Physiol. 11, 749–764.

Mahler R L, Naylor D V and Fredrickson G A 1984 Hot water extraction of boron from soils using sealed plastic pouches. Commun. Soil Sci. Plant Anal. 15, 479–492.

Marcar N E and Graham R D 1987 Genotypic variation for manganese efficiency in wheat. J. Plant Nutr. 10, 2049–2055.

Marschner H 1995 Mineral Nutrition of Higher Plants, 2nd ed. Academic Press, New York. 889 p.

Matoh T, Ishigaki K, Mizutani M, Matsunaga W and Takabe K 1992 Boron nutrition of cultured tobacco BY-2 cells. I. Requirement for and intracellular localisation of boron and selection of cells that tolerate low levels of boron. Plant Cell Physiol. 33, 1135–1141.

Martin-Prevel P, Gagnard J and Gautier P (Eds.) 1987 Plant Analysis As a Guide to the Nutrient Requirements of Temperate and Tropical Crops. Lavoisier Publishing, New York. 722 p.

Moraghan J T 1985 Plant tissue testing for micronutrient deficiencies and toxicities. Fert. Res. 7, 201–219.

Moraghan J T and Mascagni H J 1991 Environmental and soil factors affecting micronutrient deficiencies and toxicities. *In* Micronutrients in Agriculture, 2nd ed. Eds. J J Mortvedt, F R Cox, L M

Shuman and R M Welch. pp 371–426. Soil Science Society America Book Series No.4.

Morrill L G, Hill W E, Chrudilmsky W W, Ashlock L O, Tripp L D, Tucker B B and Weatherly L 1977 Boron requirements of Spanish peanuts in Okalahoma: effects on yield and quality and interaction with other nutrients. Agric. Exp. Stn Okalahoma State University Report No. MP–99.

Mozafar A 1987 Effect of boron on ear formation and yield components of two maize (*Zea mays* L.) hybrids. J. Plant Nutr. 10, 319–332.

Mozafar A 1993 Role of boron in seed production. *In* Boron and its Role in Crop Production. Ed. U C Gupta. pp 186–206. CRC Press, Boca Raton, FL, USA.

Nable R O, Cartwright B and Lance R C M 1989 Genotypic differences in boron accumulation in barley: relative susceptibilities to boron deficiency and toxicity. *In* Genetic Aspects of Plant Mineral Nutrition, Eds N Bassam, M Dambroth and B C Loughman. pp 243–251. Kluwer Academic Publishers, Dordrecht, The Netherlands.

Noppakoonwong R, Bell R W, Dell B, and Loneragan J F 1993 An effect of light on the B requirement for leaf blade elongation in black gram (*Vigna mungo*). Plant and Soil 155/156, 413–416.

Noppakoonwong R N, Rerkasem B, Bell R W, Dell B and Loneragan J F 1997 Prognosis and diagnosis of boron deficiency in black gram (*Vigna mungo* L. Hepper) in the field by using plant analysis. *In* Boron in Soils and Plants. Proceedings. Eds R.W. Bell and B. Rerkasem. Kluwer Academic Publishers, Dordrecht, The Netherlands. (*In press*)

Nyomora A M S, Brown P H and Freeman M 1997 Fall foliar applied boron increases tissue boron concentration and nut set of almond. J. Amer. Soc. Hort. Sci. 122, 405–410.

Odum J W 1980 Kinetics of the hot water soluble boron test. Commun. Soil Sci. Plant Anal. 11, 759–765.

Oertli J J 1993 The mobility of boron. *In* Plant Nutrition – from Genetic Engineering to Field Practice. Ed. N J Barrow. pp 393–396. Developments in Plant and Soil Sciences 54. Kluwer Academic Publishers, Dordrecht, The Netherlands.

Parr A J and Loughman B C 1983 Boron and membrane function in plants. *In* Metals and Micronutrients Uptake and Utilisation by Plants. Eds. D A Robb and W S Pierpoint. pp 87–107. Academic Press, New York.

Peck T R and Soltanpour P N 1990 The principles of soil testing. *In* Soil Testing and Plant Analysis, 3rd ed. Ed. R L Westermann. pp 1–9. Soil Science Society America Book Series No. 3.

Pinyerd C A, Odum J W, Long F L and Dane J H 1984 Boron movement in a Norfolk loamy sand. Soil Sci. 137, 428–433.

Pregno L M and Armour J D 1992 Boron deficiency and toxicity in potato cv. Sebago on an oxisol of the Atherton Tablelands, North Queensland. Aust. J. Exp. Agric. 32, 251–253.

Ram S, Bist L D and Sirohi S C 1989 Internal fruit necrosis of mango and its control. Acta Hort. 231, 805–813.

Raven J A 1980 Short- and long-distance transport of boric acid in plants. New Phytol. 84, 231–249.

Rawson H M 1996 Hypothesis for why sterility occurs in wheat in Asia. *In* Sterility in Wheat in Sub-tropical Asia: Extent, Causes and Solutions. Eds. H M Rawson and K D Subedi, pp 132–134. ACIAR Proc. No. 72.

Reeve W and Shive J W 1944 K-B and Ca-B relationship in plant nutrition. Soil Sci. 57, 1–14.

Renan L and Gupta U C 1991 Extraction of soil boron for predicting its availability to plants. Commun. Soil Sci. Plant Anal. 22, 1003–1012.

Rerkasem B and Loneragan J F 1994 Boron deficiency in two wheat genotypes in a warm, subtropical region. Agron. J. 86, 887–890.

Rerkasem B and Lordkaew S 1996 Tissue boron. *In* Sterility in Wheat in Sub-Tropical Asia: Extent, Causes and Solutions. Eds H M Rawson and K D Subedi. pp 36–38. ACIAR Proc. No. 72.

Rerkasem B, Bell R W and Longeragan J F 1990 Effect of seed and soil boron on early seedling growth of black and green gram. *In* Plant Nutrition – Physiology and Application. Ed. M L Van Beusichem. pp 281–285. Kluwer Academic Publishers, Dordrecht, The Netherlands.

Rerkasem B, Saunders D A and Dell B 1989 Grain set failure and boron deficiency in wheat in Thailand. J. Agric. (Chiang Mai University) 5, 1–10.

Rerkasem B, Bell R W, Lordkaew S and Loneragan J F 1993 Boron deficiency in soybean (*Glycine max* L. Merr.), peanut (*Arachis hypogaea* L.), and black gram (*Vigna mungo* L. Hepper): symptoms in seeds and differences among soybean cultivars in susceptibility to boron deficiency. Plant and Soil 150, 289–294.

Rerkasem B, Bell R W, Lordkaew S, and Loneragan J F 1997 Relationship of seed boron concentration to germination and growth of soybean (*Glycine max* L. Merr.). Nutr. Cycling in Agroecosystems. (*In press*).

Rerkasem B, Netsangtip R, Bell R W, Loneragan J F 1988 Comparative species responses to boron on a Typic Tropaqualf in Northern Thailand. Plant and Soil 106, 15–21.

Reuter D J and Robinson J B (Eds) 1986 Plant Analysis. An Interpretation Manual. Inkata Press, Melbourne.

Reuter D J, Loneragan J F, Robson A D and Tranthin-Fryer D J 1983 Intraspecific variation in the external and internal copper requirements of subterranean clover. Agron. J. 75, 45–49.

Sah R N and Brown P H 1996 Boron determination – a review of analytical methods. Microchemical J. (*In press*).

Sakal R, Singh B P, Singh A P and Sinha R B 1985 Critical limit of boron in soils and plant for the response of black gram to applied boron in calcareous soil. J. Indian Soc. Soil Sci. 33, 725–727.

Shorrocks V M 1989 Boron Deficiency – its Prevention and Cure. Borax Consolidated, London.

Shorrocks VM 1992 Boron – A global appraisal of the occurrence, diagnosis and correction of boron deficiency. *In* Proc Int Symposium on the Role of Sulphur Magnesium and Micronutrients in Balanced Plant Nutrition. Ed. S Portch. pp 39–53. Potash and Phosphate Institute, Hong Kong.

Shear C B and Faust M 1980 Nutritional ranges in deciduous tree fruits and nuts. Hort. Rev. 2, 142–163.

Shuman L M, Bandel V A, Donohue S J, Isaac R A, Lippert R M, Sims J T and Tucker M R 1992 Comparison of Mehlich-1 and Mehlich-3 extractable soil boron with hot water-water extractable boron. Commun. Soil Sci. Plant Anal. 23, 1–14.

Sillanpaa M 1982 Micronutrients and the Nutrient Status of Soils. FAO (Rome) Soils Bull. 48.

Sims S R and Johnson G V 1991 Micronutrient soil tests. *In* Micronutrients in Agriculture 2nd ed. Eds. J J Mortvedt, F R Cox, L M Shuman and R M Welch. pp 427–476. Soil Science Society America Book Series No. 4.

Skok J and McIlraith W J 1958 Distribution of boron in cells of dicotyledonous plants in relation to growth. Plant Physiol. 33, 428–431.

Smith F W 1986 Interpretation of plant analysis: concepts and principles. *In* Plant Analysis An Interpretation Manual. Eds. D J Reuter and J B Robinson. pp 1–12. Inkata Press, Melbourne.

Smith F W and Loneragan J F 1997 Interpretation of plant analysis: concepts and principles. *In* Plant Analysis An Interpretation Manual, 2nd ed. CSIRO, Melbourne. (*In press*).

Snowdon P 1982 Diagnosis of boron deficiency in soils by pot experiments with *Pinus radiata*. Aust. For. Res. 12, 217–229.

Spouncer L R, Nable R O and Cartwright B 1992 A procedure for the determination of soluble boron in soils ranging widely in boron concentrations, sodicity, and pH. Commun. Soil Sci. Plant Anal. 23, 441–453.

Stephenson R A and Gallagher E C 1987 Effects of foliar boron sprays on yield and quality of macadamia. Sci. Hort. 32, 97–103.

Stone E L 1990 Boron deficiency and excess in forest trees: a review. For. Ecol. Management 37, 49–75.

Sylvester-Bradley R 1985 Revision of a code for stages of development in oilseed rape (*Brassica napus* L.). Aspects Appl. Biology 10, 395–400.

Tsadilas C D, Yassoglou N, Kosmas C S and Kallianou C 1994 The availability of soil boron fractions to olive trees and barley and their relationships to soil properties. Plant and Soil 162, 211–217.

Ulrich A 1952 Physiological bases for assessing the nutritional requirements of plants. Ann. Rev. Plant Physiol. 3, 207–228.

Ulrich A and Hills F J 1967 Principles and practises of plant analysis. *In* Soil Testing and Plant Analysis, Part 2. Ed. G W Hardy. pp 11–24. Soil Science Society America Special Publication No 2.

Vaughan B and Howe J 1994 Evaluation of boron chelates in extracting soil boron. Commun. Soil Sci. Plant Anal. 25, 1071–1084.

Vlamis J and Ulrich A 1971 Boron nutrition in the growth and sugar content of sugar beet. J. Amer. Soc. Sugar Beet Technol. 16, 428–439.

Warner R M and Fox R L 1972 Concentration and distribution of S, Mg, and five micronutrients in macadamia in relation to yields. *In* Proc. Hawaiian Macadamia Producers Assoc. 12th Ann. Meet. pp 26–37. Hawaii Macadami Producers Association, Hilo, Hawaii.

Warington K 1933 The influence of length of day on the response of plants to boron. Ann. Bot. 47, 429–457.

Westermann R L (Ed.) 1990 Soil Testing and Plant Analysis, 3rd ed. Soil Science Society of America Book Series No. 3.

Wilson J R 1976 Variation of leaf characteristics with level of insertion on a grass tiller. I. Development rate, chemical composition and dry matter digestibility. Aust. J. Agric. Res. 27, 343–354.

Woodruff J R 1979 Soil boron and soybean leaf boron in relation to soybean yield. Commun. Soil Sci. Plant Anal. 10, 941–952.

Yermiyahu U, Keren R and Chen Y 1995 Boron sorption by soil in the presence of composted organic matter. Soil Sci. Soc. Amer. J. 59, 405–409.

Plant and Soil **193:** 169–180, 1997.

Chapter 11

Genotypic variation in plant response to low boron and implications for plant breeding

Benjavan Rerkasem and Sansanee Jamjod
*Agronomy Department, Faculty of Agriculture, Chiang Mai University, Thailand 50200**

Abstract

Plant response to low B in the soil varies widely among species, and among genotypes within a species. Boron efficient genotypes are those that are able to grow well in soils in which other genotypes are adversely affected by B deficiency. This review considers the extent of variation in B efficiency in plant species and genotypes, the physiological nature of the efficiency mechanisms, what is known of the genetic basis for inheritance, screening techniques and the practical implications of the genotypic variations.

Frequently, B efficiency is the sole reason for a difference between an average yield and complete crop failure. Severe yield losses can be effectively prevented by the inclusion of B efficiency as a selection criterion in crop breeding and improvement programmes for regions with low B soils. In addition, the expression of B deficiency primarily through male sterility, which is common in many species, creates opportunities for outcrossing in normally self-fertilised species. This, in turn, leads to two possibilities. Firstly, self fertilisation, and therefore maintenance of pure lines, cannot always be assumed in self pollinated species where B efficient and inefficient genotypes are grown side by side on low B soils. Secondly, B deficiency, in soil or artificial media, may be used as a fertility selective medium in which the male sterile B inefficient genotypes and the male fertile B efficient genotypes could hybridise naturally. This would be useful as a simple and economical method for creating heterozygous populations in breeding programmes as well as for producing hybrid seeds. Now that the roles of B in plant growth and development are beginning to be clarified, the efficiency mechanisms as well as the governing genetics can be explained. Practical benefits from the genetic diversity of B efficiency will be enhanced by a better understanding of B efficiency mechanisms and the molecular bases for their genetic control.

Introduction

According to some authors (e.g. Gupta, 1979; Sillanpaa, 1982), B deficiency is one of the most widespread of all micronutrient deficiencies. The use of B fertilisers can often be the simplest and most cost effective solution to the problem in crop production on soil with low available B. However, the continuing occurrence of B deficiency in crops throughout the world indicates that for various reasons this solution is not always available to farmers. On the other hand, there are numerous reports of differences among plant genotypes in their response to low external B supply. Those genotypes that are able to grow well in soils in which standard genotypes are adversely affected by B deficiency are here termed B efficient. Graham (1984) defined Zn efficiency as the ability of a cultivar to grow and yield well in soils too deficient in Zn for a standard cultivar. Since many papers on genotypic responses to B have not followed this definition, we will use efficiency in a broader sense in this paper. It would seem that selection for B efficiency can offer a real and feasible alternative solution to the problem of B deficiency in the field. Furthermore, the expression of B deficiency in certain plant species and the associated genotypic variation in the B response have led to a possible method that utilises B deficiency in plant breeding procedures.

* E-mail: benjavan@cmu.chiangmai.ac.th

Evidence of genetic variation

The difference in responses to low soil B among plant species has long been known. Requirements and sensitivity to low levels of B in soils are commonly referred to by groups of plants. Dicotyledons are said to require more B than monocotyledons, and vegetables of the Cruciferous and Umbelliferous families to have high B requirement (Martens and Westermann, 1991). Plants normally considered to be sensitive to low soil B include lucerne *(Medicago sativa)*, *Brassica* spp., beet (*Beta vulgaris)*, celery (*Apium graveolens*), grape (*Vitis vinifera*), apple (*Malus sylvestris*), pear (*Pyrus communis*), cotton (*Gossypium hirsutum*), and sunflower (*Helianthus annuus*) (Jones, 1991). Among forage legumes, red clover (*Trifolium pratense*) is most sensitive, followed by lucerne, with Alsike clover (*Trifolium hybridum*) and white clover (*Trifolium repens*) the most tolerant (Sherrell, 1983).

A wide range of genotypic variation in B efficiency has been found in numerous species of crops and other domesticated plants (Table 1). The list, while not exhaustive, includes monocotyledons and dicotyledons, herbaceous plants and trees, field crops, vegetables, fruits and timber species.

The range of genotypic variation within species can sometimes be so wide that ranking of species may change with the genotypes being compared (Table 2). For example, the order of B efficiency among soybean (*Glycine max*), green gram (*Vigna radiata*) and black gram (*Vigna mungo*), was changed by selecting different genotypes for comparison (Rerkasem, 1990; Rerkasem et al., 1989, 1993a). Similarly, the generally held view that wheat has a high tolerance to low soil B (Martens and Westermann, 1991) has to be revised following identification of a wide range of B efficiency among wheat genotypes (Rerkasem, 1990; Rerkasem and Loneragan, 1994; Rerkasem et al., 1993a), some of which are more inefficient than dicotyledons.

The range of B efficiency in individual species has been established with extensive screening of large numbers of genotypes on low B soil (Table 3). Field evaluation in Thailand found several fold differences in yield between B-inefficient and efficient genotypes of legumes, including soybean (Rerkasem et al., 1993a), peanut (*Arachis hypogaea*) (Keerati-Kasikorn et al., 1993), green gram and black gram (Rerkasem, 1990).

For wheat (*Triticum aestivum*), a wide range of B efficiency has been reported from Brazil (da Silva and de Andrade, 1980), China (Li et al., 1978), India (Tandon and Naqvi, 1992), Nepal (Subedi et al., 1993) and Thailand (Rerkasem, 1993; Rerkasem and Jamjod, 1997). The full array of responses to low external B supply among wheat genotypes is illustrated by results from the screening of 68 genotypes in sand culture and in the field in Thailand (Figure 1) and 44 genotypes in the field in Nepal (Figure 2). In Thailand, low external B supply was obtained in sand culture without the addition of B, and this was compared to moderately low (0.2 μM) and sufficient (1.0 μM B in the nutrient solution) treatments (Rerkasem and Jamjod, 1997). In the field, B deficiency was accentuated by an application of lime to a soil with hot water soluble (HWS) B of about 0.1 mg kg^{-1}, (Rerkasem and Jamjod, 1997). Relative responses of B inefficient SW41 and the more efficient Sonora 64 (Rerkasem and Loneragan, 1994) confirmed the condition of low soil B in the field screening in Nepal (Subedi et al., 1993). At these low levels of external B, responses ranged from complete male sterility and no grain set in the very inefficient genotypes, to those partially affected (inefficient to moderately inefficient), through to the efficient genotypes which exhibited no deficiency symptom.

Mechanisms for efficiency

Genotypic differences in B efficiency may be assumed to operate in the same way as described for general "nutrient efficiency" (Marschner, 1995). That is, a genotype may be B efficient or inefficient due to one or more of these mechanisms: the ability to acquire B from the soil, the way that B is distributed, and its utilisation within the plant. For a given species, the mechanism for B efficiency may differ with the genotypes being compared and with the intensity of deficiency. Efficiency mechanisms may also vary among different species. Further complicating the explanation of the B efficiency mechanism is the fact that some crucial aspects of B physiology have begun to be clarified only recently. Most notable are the role of B in the cell wall (Matoh, Chapter 5) and B mobility in the phloem and hence potential of remobilisation (Brown and Shelp, Chapter 7).

Levels of B in plant tissues associated with sufficiency, e.g. 5–10 mg B kg^{-1} dry wt for wheat, 6–12 for ryegrass (*Lolium* sp.) compared with 20–80 for cotton, 35–80 for alfalfa, and 40–100 for sugar beet (*Beta vulgaris*) (Bergmann, 1992) are the basis on which a lower B "requirement" of grasses has been determined (Marschner, 1995). Relative B requirements among the species appears now to be related to the B in cell wall (Hu and Brown, 1994; Matoh et al., 1993, 1996), and B concentration in the cell wall of grasses was found to be much lower than that in dicotyledons (Hu

Table 1. Crop and other domesticated plants reported to have genotypic differences in their response to low boron

Plant	Reference
Apple (*Malus sylvestris*)	Tagliavini et al. (1992)
Barley (*Hordeum vulgare*)	Rerkasem and Jamjod (1989)
Beet (*Beta vulgaris*)	Kelly and Gabelman (1960)
Black gram (*Vigna mungo*)	Rerkasem (1990)
Oil seed rape (*Brassica napus*)	Yang et al. (1993)
Broccoli (*Brassica oleracea*)	Shelp et al. (1992)
Cotton (*Gossypium hirsutum*)	Smithson and Heathcote (1974)
Celery (*Apium graveolens*)	Pope and Munger (1953)
Cacao (*Threobroma cacao*)	Tollenaar (1966)
Citrus (*Citrus* spp.)	Smith (1966)
Green gram (*Vigna radiata*)	Rerkasem (1990)
Grape (*Vitis vinifera*)	Scott (1944)
Maize (*Zea mays*)	Mozafar (1993)
Peanut (*Arachis hypogaea*)	Keerati-Kasikorn et al. (1993)
Pinus caribaea	Martinez et al. (1989)
Pinus ponderosa	Wright et al. (1969)
Pinus radiata	Lambert and Turner (1977)
Plum (*Prunus domestica*)	Benson et al. (1966)
Pomegranate (*Punica granatum*)	Singh et al. (1990)
Strawberry (*Fragaria vesca*)	Willis (1945)
Sunflower (*Helianthus annuus*)	Blamey et al. (1979)
Soybean (*Glycine max*)	Rerkasem et al. (1993a)
Tomato (*Lycopersicon esculentum*)	Wall and Andrus (1962)
Wheat (*Triticum aestivum*)	Singh et al. (1976)

Table 2. Relative seed yield of boron efficient and inefficient cultivars of several crops grown on a low boron soil (0.1 mg hot water soluble B kg^{-1}) at Chiang Mai, Thailand

Crop	Genotype/cultivar	Yield, % yield in sufficient B
Soybean[a] (*Glycine max*)	NW1	39
	SJ5	70
Barley[b] (*Hordeum vulgare*)	BRB 2	53
	BRB1	83
Wheat[c] (*Triticum aestivum*)	SW41	11[e]
	Fang 60	97[e]
Green gram[d] (*Vigna radiata*)	CMU 55	34
	Kamphangsan 1	87
Black gram[d] (*Vigna mungo*)	Regur	9
	Bangrakum	71

Source: [a]Rerkasem et al. (1993a); [b]Rerkasem and Jamjod (1989); [c]Rerkasem et al. (1993b); [d]Rerkasem (1990); [e]Grain Set Index.

et al., 1996). A suggestion made by Hu et al. (1996) that B requirement might be predicted by pectin contents provides a possible tool with which to determine if requirement is a basis for differential B efficiency among genotypes. Determination of requirement for other functions will have to wait for clearer definition of the particular role of B.

Table 3. Differential sensitivity to low boron among genotypes of different crop species. Only studies involving a large number of genotypes are reported

Crop species	Number of genotypes	Range of sensitivity[a]	Attribute measured	References
Black gram (*Vigna mungo*)	10	9–71	Seed yield	Rerkasem (1990)
Green gram (*Vigna radiata*)	14	34–100	Seed yield	Rerkasem (1990)
Peanut (*Arachis hypogaea*)	14	11–71	Seed yield	Keerati-Kasikorn et al. (1993)
Soybean (*Glycine max*)	19	0–75	% hollow heart	Rerkasem et al. (1993a)
Wheat (*Triticum aestivum*)	5	37–94 (6-63)[b]	male fertility[b]	da Silva and de Andrade (1980)
Wheat (*T. aestivum*)	45	5–98	% GSI[c]	Subedi et al. (1993)
Wheat (*T. aestivum*)	15	64–100	Seed yield	Tandon and Naqvi (1992)
Wheat (*T. aestivum*)	253	0–100	% GSI	Rerkasem and Jamjod (1997)

[a] Relative to performance in B sufficiency (%).
[b] Converted from male sterility rating, in brackets.
[c] Grain Set Index. Rerkasem and Loneragan (1994).

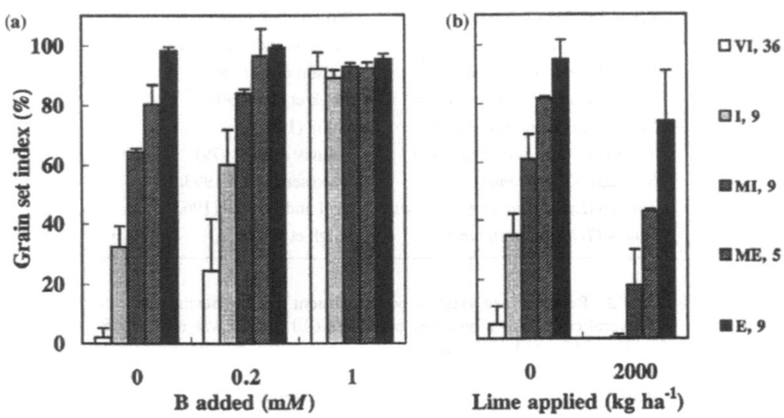

Figure 1. Boron efficiency in 68 wheat genotypes (% Grain Set Index with standard error bars) in five groups: very inefficient (VI), inefficient (I), moderately inefficient (MI), moderately efficient (ME) and efficient (E), at various levels of external B supply: (a) in sand culture with three levels of added B: 0, 0.2 and 1 μM B and (b) in a low B soil (0.1 mg hot water soluble B kg^{-1}) with and without lime. (Number of genotypes after legend in each class.) Source: Adapted from Rerkasem and Jamjod (1997).

Flowering and fruit set have often been observed to be sensitive to B deficiency. The possibility that plants may require more B for reproductive development than for vegetative growth adds to the complication in assessing B requirement as a basis for efficiency. Reports of B deficiency affecting fertilisation include those that affect the development of the female and male gametes and the fertilisation process. Crucial to the analysis of B requirement for reproduction in relation to B efficiency is the recognition of the specific role of B in reproduction.

The effect of B deficiency on grain set failure is well documented for cereals, i.e. barley (*Hordeum vulgare*) (Ambak and Tadano, 1991; Simojoki, 1972), rice (*Oryza sativa*) (Garg et al., 1979), maize (*Zea mays*) (Agrawala et al., 1981; Vaughan, 1977) and wheat (da Silva and de Andrade, 1983; Li et al., 1978; Rerkasem et al., 1989). In many of these studies, the effect of B deficiency on grain set failure was associated with male sterility, i.e. poorly developed anthers and pollen. In maize, however, B deficiency did not depress grain set through impaired development of the male gametes, but through pollen germination, which was influenced

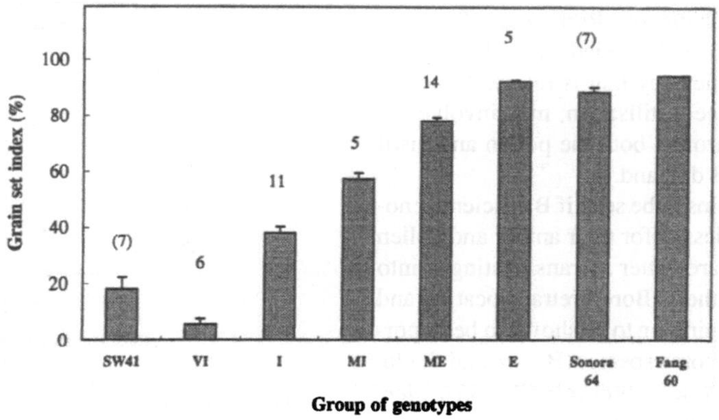

Figure 2. Grain Set Index (%, with standard error bars) in 44 wheat genotypes on a low boron soil in Nepal, in five groups of B efficiency: very inefficient (VI), inefficient (I), moderately inefficient (MI), moderately efficient (ME) and efficient (E) with 7 replicates each of SW41, Fang 60 and Sonora 64 as checks. (Number of genotypes in each group above the standard error bar.) Source: Adapted from Subedi et al. (1993).

by B concentration in the silk (i.e. the stigma and the style) (Agrawala et al., 1981; Vaughan, 1977). Only in wheat has a specific effect of B deficiency on development of the pollen been shown by cross pollination experiments (Rerkasem and Jamjod, 1997; Rerkasem et al., 1993b). No difference was observed in the *in vitro* germination response of pollen to external B from two wheat genotypes, B inefficient SW41 and the more efficient Sonora 64, in relation to the B status of the plant (Cheng and Rerkasem, 1993). That is, differential B requirement for pollen tube growth is unlikely to be the basis for B efficiency, at least between these two genotypes. The range of genotypic variation in B efficiency reported in wheat as shown in Figures 1 and 2, and possibly in all other cases (see above for references), is largely associated with the effect of B deficiency on male gamete development.

Boron inefficient and efficient wheat genotypes are not distinguishable by the B concentration in their flag leaf (Rerkasem and Loneragan, 1994) or whole ear (Rerkasem and Lordkaew, 1992; Rerkasem et al., 1997). The ranking of wheat genotypes for their vegetative responses to low external B has not shown any correlation with their ranking in reproductive responses (Rerkasem et al., 1993b). It appears that in relating B requirement with efficiency, it may sometimes be necessary to consider vegetative and reproductive efficiency separately.

In wheat, higher functional B requirement for reproductive development than that for vegetative growth is indicated in tissue B concentrations in B deficient plants. Male sterility and grain set failure have been associated with 7–8 mg B kg^{-1} dry wt in the anthers and 5–6 mg B kg^{-1} in the carpel, while no evidence of B deficiency has been seen in wheat leaves containing 4 mg B kg^{-1} dry wt, or in the external appearance of the ear with 2 mg B kg^{-1} in its vegetative parts, i.e. glumes, rachis (Rerkasem et al., 1997). Rye (*Secale cereale*) is another species that has been shown to concentrate B in its anthers (Löhnis, 1940).

Among dicotyledons, B application is often found to increase fruit set, especially in fruit trees, e.g. in prune (*Prunus domestica*) (Hanson and Breen, 1985), peach (*Prunus persica*) (Kamali and Childers, 1970), hazelnut (*Corylus avellana*) (Shrestha et al., 1987), almond (*Prunus dulcis*) (Nyomora et al., 1997) and avocado (*Persea* sp.) (Robbertse et al., 1990; Smith et al., 1997a). Notable is a common observation that the response was often seen without associated vegetative symptoms. This could also indicate a higher B requirement for reproduction, but there is also a possibility of other mechanisms limiting supply to the flowers. In many of these studies, fertilisation was shown to be affected through B supply in the pistil which affected pollen germination. It is well established that B is essential for pollen tube growth (Dickinson, 1978), but its exact role in still unclear. Data on B concentrations relating to the fertilisation process are scarce, and none can be found comparing genotypes. In grape, maximum fertilisation was reported to require 50–60 mg B kg^{-1} dry wt in the stigma (Marschner, 1995). *In vitro* pollen germination in mango (*Mangifera indica*) (de Wet and Robbertse, 1989), avocado (Smith et al.,

1997a), and wheat (Cheng and Rerkasem, 1993) has been shown to respond to B supplied internally through the plant. Thus, B efficiency that is related to pollen germination, and hence fertilisation, may involve the ability to supply B through both the pollen and pistil as well as the actual B demand.

For wheat, it remains to be seen if B efficient genotypes simply require less B for their anther and pollen development or they are better at translocating B into the nontranspiring anthers. Boron retranslocation and phloem mobility is beginning to be shown to be important to B efficiency in some species (Brown and Shelp, Chapter 7). One of these is broccoli (*Brassica oleracea*), in which B efficient genotypes were shown to be those that were better able to redistribute B into young leaves and florets (Shelp et al., 1992). The effectiveness of foliar B application on fruit set in almond has also been suggested to be related to translocation and remobilisation of a B–sorbitol complex (Nyomora et al., 1997). However, although foliar B application was found to increase yield differently in almond genotypes, there is yet no evidence that phloem mobility is the basis for their relative B efficiency.

Boron efficient genotypes have been identified with higher B concentration in certain tissues. In sunflower, efficient genotypes were those with higher B concentration in the uppermost mature leaves (Blamey et al., 1979). Similarly, in black gram and green gram, genotypes which were less severely affected in low B soil also had higher B concentration in their youngest fully expanded leaf (Rerkasem, 1990). Also with oil seed rape (*Brassica napus*) growing in a low B soil, the tolerant traditional variety 92-13 contained twice as much B in its leaves as the improved variety 51 which was sensitive to B deficiency (Yang et al., 1993). Greater uptake ability (discussed below) may explain greater B efficiency in some of these species, but B retranslocation and phloem mobility that have recently been demonstrated (Brown and Hu, 1996; Shelp, 1988; Shelp et al., 1995) may also play a role.

Tissue concentrations >20 mg B kg^{-1} dry wt in dicotyledons compared with 5–10 mg B kg^{-1} in gramineous monocotyledons growing on the same soil are often interpreted to indicate greater uptake ability in dicotyledons (Gupta, 1979). A recent flowing solution culture experiment (Chapman et al., 1997) showed that wheat indeed had a lower B absorption rate per unit root wt than field pea (*Pisum sativum*) and white lupin (*Lupinus albus*), but only slightly lower than lentil (*Lens culinaris*), another dicotyledon. The higher B concentration in the lentil leaf compared to

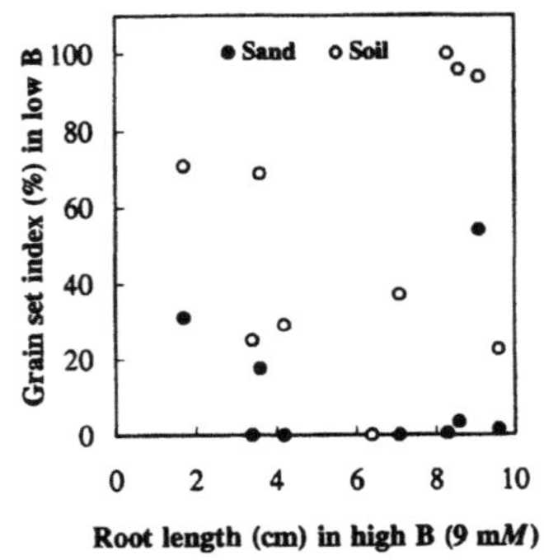

Figure 3. The relationship between the response to high boron, expressed as root length in 9 m*M* B, and to low B, in soil (0.1 mg hot water soluble B kg^{-1}, open symbol) and sand culture (with 0.2 μM B, filled symbol), expressed as Grain Set Index, in 10 wheat genotypes. Sources: Rerkasem (1993) and Chantachume (1995).

the wheat leaf may have resulted from a greater rate of assimilation or a higher rate of B uptake.

For individual cells (Huang and Graham, 1990; Seresinhe and Oertli, 1991) as well as whole plants (Nable et al., 1990), the rate of B uptake is linearly related to the external B concentration. In spite of this, distinct genotypical differences have been observed. At low B, the barley genotype Sahara 3771 grew more poorly and took up less B than the genotype Schooner (Nable et al., 1990). At toxic B levels, accumulation has been established as the primary mechanism determining differential responses among genotypes (Huang and Graham, 1990; Nable et al., 1990; Nable et al., Chapter 12). Since B efficiency in wheat genotypes has been related to their reproductive development, as discussed earlier, whereas B toxicity affects early vegetative growth, it is not surprising that no simple relationship emerges when responses of genotypes to low and high external B supply are correlated (Figure 3).

Plant responses to B have been reported to be influenced by environmental factors. Genotypic variations in B efficiency are likely to vary even further by interactions with these various factors. For certain sets of genotypes, B efficiency ranking has been consistent over a wide range of environments. For example, B efficiency of the wheat genotypes Fang 60 and Sonora 64 compared with inefficiency of SW41 and BL1022 has been reproduced over many sites, years and countries (Rerkasem et al., 1993b; Subedi et al., 1993). In such cases the same set of efficiency mechanism(s) seems to be indicated. The genotype×boron×environment effect, however, has been suggested with other sets of genotypes in other sets of environments. For example, soil water deficit (Pant, 1993) as well as excess (waterlogging) (Pintasen, 1997), have been identified as two of the environmental factors which affect genotype efficiency rankings. Reports of the interaction between aluminium toxicity and boron deficiency (LeNoble et al., 1996a, 1996b; Lukaszewski and Blevins, 1996) suggest that aluminium toxicity could be identified as another component of the environment. Root geometry has been suggested as a mechanism for zinc efficiency in the wheat cultivar Excalibur (Graham and Rengel, 1993). There is a strong possibility that the soil–root interaction, including rhizosphere effects, could also play a role in B efficiency, although direct evidence is still not available.

The chronological order of developmental events, their relative sensitivity to low B and after effects add another complication to the assessment of B efficiency. In sunflower, the effect of B deficiency on an early development of corky stem and peduncle may result in the whole flower head being severed (Fernandez et al., 1985; Rerkasem, 1986), thus overriding any effect on all subsequent stages. Similarly, flower buds in B deficient black gram may begin to shed as soon as they are formed (Rerkasem et al., 1988), so making irrelevant any question on B requirement of reproductive tissues.

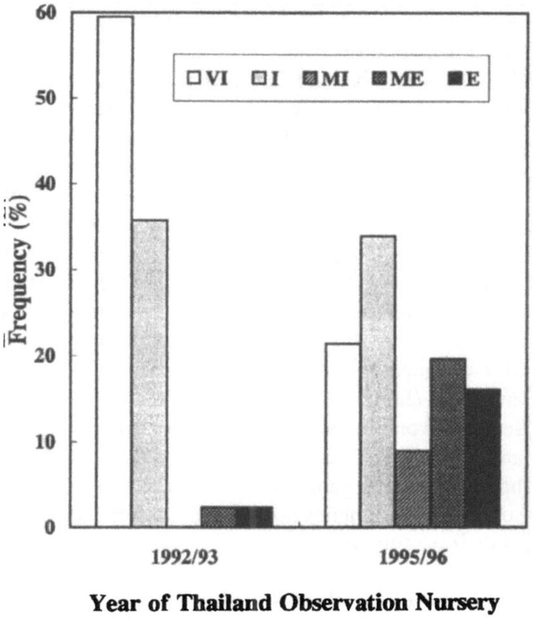

Figure 4. Frequency distribution of wheat genotypes in different classes of boron efficiency: very inefficient (VI), inefficient (I), moderately inefficient (MI), moderately efficient (ME) and efficient (E), from Wheat Thailand Observation Nurseries 1992/93 and 1995/96. Sources: Adapted from Rerkasem (1992) and Anantawiroon (1997).

Genetics of the variation in low boron responses

Genetic control of B efficiency has been studied in a number of species, with major and minor controlling genes reported. In celery (Pope and Munger, 1953), tomato (*Lycopersicon esculentum*) (Wall and Andrus, 1962) and red beet (*Beta vulgaris*) (Tehrani et al., 1971) responses were reported to be under monogenic control. Kelly and Gabelman (1960) found that the genetic control of B efficiency in table beet was complex. Blamey et al. (1984) studied the genetics of B efficiency in sunflower with half diallel analysis. They found that the most efficient varieties and their offspring had the highest B concentrations in the uppermost mature leaf. General combining ability effects were significant for leaf B concentration, deficiency symptoms and grain yield. These three characters were found to be highly heritable, with additive or additive epistatic gene action predominant. It was suggested that responses to low B of the offspring could be predicted from the performance of the parents (Blamey et al., 1984). A similar response to B has been observed in bread wheat (Jamjod et al., 1992a). Seven genotypes which differed in response to low B were intercrossed and analysed for F1 performance compared with their parents. General and specific combining ability effects for grain set in low B were significant, and it was suggested that both additive and dominant gene action were involved in controlling efficiency.

In the case of wheat, the complicated genetics of B efficiency is also suggested by the array of responses that have been demonstrated (Figures 1 and 2). The frequency of genotypes from each class found in the wheat germplasm from each wheat growing area and country probably reflects the prevailing local soil B status as well as efforts to breed and select for efficiency. For example, from Australia where no incidence of B deficiency in wheat has been reported, 44 of the 45 wheat genotypes (released cultivars, advanced lines and accessions) screened were found to be very inefficient, and one moderately inefficient (Rerkasem and Jamjod, 1997). Boron efficient genotypes have been more frequently found in the wheat germplasm from Thailand (Rerkasem, 1993; Rerkasem and Jamjod, 1997), Nepal (Subedi et al., 1993) and India (Tandon and Naqvi, 1992) where B deficiency in wheat has been identified and active selection for B efficiency has been conducted since the 1970s in India and the end of the 1980s in Thailand and Nepal. A large increase in the frequency of B efficient genotypes in Thailand Observation Nursery (annual screenings of 80–100 entries nominated by various stations involved in wheat research in Thailand) from 1992/93 to 1995/96 (Figure 4) invites a closer examination. The marginal B status of wheat plants growing at most of the sites where the genotypes have been selected from and then screened (Rerkasem et al., 1989) may have a significant bearing on this shift in B efficiency. Two of the B efficient genotypes identified in Thailand are Fang 60 and #144 (Jamjod et al., 1992b). Fang 60, a cultivar selected from an earlier introduction from CIMMYT and released in Thailand, and #144, were identified as having the same pedigree as PI/FD//PI/MZ3/MXP. A source for B efficiency may be provided in one or more of their parental lines.

Implications for plant breeding and screening methods

Evaluating germplasm for boron efficiency

Despite the relatively simple and cost effective solution of applying B as a fertiliser, incidences of deficiency continue to be observed in farmers' fields. Frequently, deficiency occurs in newly introduced germplasm where it had not previously been seen in older established cultivars. The almost concurrent observation of B deficiency in wheat in the 1960s in India (Singh et al., 1976), Brazil (da Silva and de Andrade, 1983) and Nepal (Misra et al., 1992) coincided with the spread of the new semi-dwarf wheat (*Triticum aestivum*) into new areas. Similarly, first time observations of sensitivity to B deficiency in newly introduced cultivars and genotypes were also made with cacao (*Threobroma cacao*), potato (*Solanum tuberosum*) and lucerne in Ecuador (Tollenaar, 1966), soybean and green gram in Thailand (Rerkasem et al., 1993a) and oil seed rape in China (Yang et al., 1993).

It would seem that screening for B efficiency is desirable where low B soils are common even when B application would be the eventual solution. On the other hand, when B application is for some reason not feasible, yield loss may be effectively prevented by the use of efficient genotypes. In Bangladesh it has been estimated that yield losses due to B deficiency was equivalent to approximately 60% of the yield gains from plant breeding efforts in the last 20 years (D A Saunders, pers. comm.).

After the first field observations in the 1960s, incidences of B deficiency-induced sterility in wheat, however, continue to occur in farmers' fields in China (Yang, 1992), India (Tandon and Naqvi, 1992) and Nepal (Pant, 1993), as well as Bangladesh. This is not surprising considering that as of 1992, virtually all of the released cultivars and advanced lines from Asia (ca. early 1990s) ranged from B inefficient to very inefficient (Rerkasem and Jamjod, 1997). The level of B efficiency found in wheat germplasm already in the region strongly suggests the feasibility of overcoming B deficiency-induced male sterility in wheat in Asia by genetic means. Another argument supporting the use of B efficient genotypes as a means to overcome B deficiency is the low residual value of applied B in the humid tropics where B deficiency is frequently found. On the light soils of Bangladesh, for example, B applied to wheat was not detectable in the crop in the year immediately following the application (S Lordkaew and M Saifuzzaman, pers. comm.). On the other hand, there is a real danger in the identification of B deficiency that leads to the application of B fertiliser only in experimental stations where germplasm is screened and breeding programmes are carried out but not on farmers' fields.

Screening procedures

In those traits for which screening can be done early, i.e. at the seedling stage, such as screening for B toxicity (Paull et al., 1992) and aluminium toxicity (e.g. Aniol, 1990), large numbers of genotypes can be screened quickly and economically. Low B has been shown to depress early vegetative growth such as leaf blade elongation (Kirk and Loneragan, 1988; Noppakoonwong, 1993) and root growth (Bohnsack and Albert, 1977). There is a possibility for a quick screening if these traits are the ones that are desired or if they correlate with the yield response as in the case of sunflower (Blamey et al., 1984), green gram, black gram (Rerkasem et al., 1993a) and oil seed rape (Yang et al., 1993). The possibility that cell wall pectin content could be used to indicate actual B requirement of tissues has been suggested from the close correlation found between cell pectin content and cell wall B (Hu et al., 1996). This could become a useful screening tool if efficiency is related to the function of B in the cell wall.

The criteria that can be used for quick and simple screening, however, do not always correlate with whole plant efficiency and eventual economic yield, which must integrate the influence of various factors that can affect related processes such as B uptake from the soil, its translocation, distribution, remobilisation, and utilisation. An example of wheat illustrates this point. SW41 can be said to be a B efficient genotype since its leaf blade elongation was least restricted by low external B supply among many other wheat genotypes; but when evaluated for reproductive success (male fertility and grain set), it was among the least efficient (Rerkasem et al., 1993b). In contrast, the reproductively B efficient Fang 60 was most inefficient according to its vegetative response, i.e. vegetatively inefficient.

For plant breeding, where the goal is in maximising economic yield, a quick screening method has to either indirectly correlate with the economic yield or be involved in the first (in the chronological order of development) physiological process most sensitive to B deficiency that is limiting the yield. It is clear, from the above review, that male fertility is the B sensitive yield limiting step in wheat, and possibly other cereals too. In other species, fruit set is associated with B nutrition of the pistil instead of the anthers as reported for avocado (Coetzer and Robbertse, 1987); B efficiency with respect to supply of B in the pistil, i.e. affecting pollen germination and pollen tube growth, might be relevant in these cases. However, it is sometimes important to identify the yield limiting steps from the many different B sensitive processes. For example, although B deficiency may depress fertilisation and fruit set, the real economic significance of B nutrition in some fruit crops may be in its effects on fruit size and quality, as shown in apple (Dong et al., 1997) and avocado (Smith et al., 1997a, 1997b).

The understanding of the mechanism of B efficiency and the identification of the related yield limiting process, ultimately with molecular basis for its genetic control and the identification of DNA markers, would undoubtedly help to improve and simplify the procedure for screening germplasm. In the meanwhile, simple measures of the outcome of B efficiency that are directly related to the economic yield such as the Grain Set Index in wheat (Anantawiroon et al., 1997; Rerkasem and Loneragan 1994) may be used in breeding and selection programmes. The intensity of the selection pressure as well as the relative responses of the germplasm being screened would be indicated by the inclusion of standard genotypes covering a range of B efficiency (Anantawiroon et al., 1997).

In areas with low soil B where deficiency has been established, sufficient 'natural' selection pressure may have already been provided in the unamended soil so that there is a built-in screening for the plant introduction and breeding programme with no additional cost. Unfortunately, this ideal situation can be jeopardised by the presence of high levels of B 'contamination', in the order of 200–600 mg B kg^{-1}, found in many fertilisers in Thailand (Bell et al., 1990) and other countries in Asia (Lordkaew, 1995).

Boron deficiency induced male sterility

In addition to the case of wheat described earlier, B deficiency-induced male sterility has also been found in rice (Garg et al., 1979), maize (Agrawala et al., 1981) and barley (Ambak and Tadano, 1991; Rerkasem and Jamjod, 1988; Simojoki, 1972). In situations where the female fertility is not depressed to the same extent as male fertility, the condition is set for outcrossing if a pollen donor is flowering nearby. It follows that self fertilisation, and therefore maintenance of pure lines, cannot always be assumed in self pollinated species where genotypes are grown side by side on soils with sufficiently low B to cause male sterility in some and not in others. In wheat the number of grains set in B deficiency-induced male sterile plants has been shown to be significantly increased by outcrossing: by more than 100% when B inefficient to very inefficient, male sterile, lines were sown in alternate rows (25 cm apart) with B efficient Fang 60 as pollen donor (Rerkasem and Jamjod, 1997); and by 20–100% with SW41, a B inefficient line in 1.5 m×10 m blocks that were paired lengthwise with the same size blocks of Fang 60 (Rerkasem et al., 1997).

Another significant implication of this is the possibility to use low B soils or artificial media (e.g. sand culture) in which B level can be precisely controlled, as 'fertility selective media' (Rerkasem and Jamjod, 1997). In such media very inefficient to inefficient genotypes may be selected for female parents and B efficient genotypes as male parents, and the parents grown together so that hybridisation may take place naturally. The yield of hybrid seeds can be enhanced by a B spray on the ears of the female parent within 1–2 days prior to anthesis (Figure 5, Rerkasem and Jamjod, 1997). This would provide a simple and novel method for hybridisation which can be used either in normal breeding programmes or for hybrid seed pro-

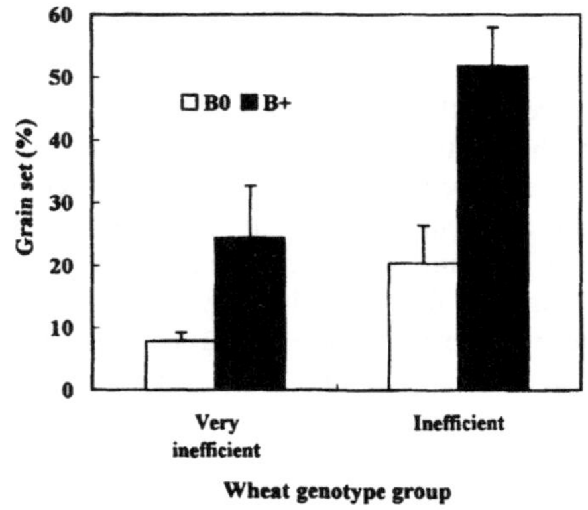

Figure 5. Effect of B application to the ear 1–2 days before pollination on grain set (% cross fertilised florets with grain) of two groups of wheat genotypes (very inefficient and inefficient) grown in the field at Chiang Mai University, Thailand. Plots were established in low B soil with B efficient and inefficient cultivars in rows side by side allowing cross pollination. Note: ears were clipped and emasculated before manually pollinated, but in most cases anthers were small, shrivelled or absent. Sources: Adapted from Rerkasem and Jamjod (1997).

duction. The chance for successful development of such a method might be strengthened by better understanding of the physiological, genetic as well as molecular basis for the genotypic variation in B efficiency.

In conclusion, the range of genotypic variation in plant response to low B in soil is clearly very large. This variation should be included in breeding programmes for regions with low B soil. Also, as B deficiency promotes male sterility in some self-fertile species, this has practical consequences. Firstly, more care will have to be exercised in the maintenance of homozygous populations in nurseries and breeding programmes in which B efficient and inefficient genotypes are grown side by side on low B soils. Secondly, B deficiency may be used as a fertility selective medium for hybrids between genotypes. A simple and economical method for creating heterozygous populations in breeding programmes as well as for producing hybrid seeds may thus be provided. Given the widespread occurrence of low B soils, especially in Asia, there is a strong case for investigating the mechanisms of B efficiency and their genetic control.

Acknowledgments

The authors wish to thank the Australian Centre for International Agricultural Research (ACIAR) and Australian International Development Assistance Bureau (now AusAID) for their long support of the boron research at Chiang Mai University. Collaborations with CIMMYT, Murdoch University, the Waite Agricultural Research Institute and national wheat programmes in Bangladesh, China, India, Nepal and Thailand which provided germplasm, comments and suggestions, and sharing of results and ideas are acknowledged. We are grateful to Robin Graham, Bernie Dell and Richard Bell for their valuable comments in the preparation of this manuscript.

References

Agrawala S C, Sharma P N, Chatterjee C and Sharma C P 1981 Development and enzymatic change during pollen development in boron deficiency maize plants. J. Plant Nutr. 3, 329–336.

Ambak K and Tadano T 1991 Effect of micronutrient application on the growth and occurrence of sterility in barley and rice in a Malaysian deep peat soil. Soil Sci. Plant Nutr. 37, 715–724.

Anantawiroon P 1997 Sensitivity to low boron of wheat genotypes from the Thailand Observation Nursery 1995/96. Agronomy Special Report. Agronomy Department, Chiang Mai University, Thailand.

Anantawiroon P, Subedi K and Rerkasem B 1997 Screening wheat for boron efficiency. *In* Boron in Soil and Plants. Proceedings. Eds. R W Bell and B Rerkasem. pp 101–104. Kluwer Academic Publishers, Dordrecht, the Netherlands.

Aniol A 1990 Genetics of tolerance to aluminium in wheat (*Triticum aestivum* L. Thell). Plant and Soil 123, 223–227.

Bell R W, Rerkasem B, Keerati-Kasikorn P, Phetchawee S, Hiranburana N, Ratanarat S, Pongsakul p and Loneragan J F 1990 Mineral Nutrition of Food Legumes in Thailand, with particular reference to micronutrients. ACIAR Tech. Rep. No. 16, 52 p. ACIAR, Canberra, Australia.

Benson N R, Lindner R C and Bullock R m 1966 Plum, prunes and apricot. *In* Nutrition of Fruit Crops. Ed. N F Childers. pp 504–517. Horticultural Publications. Rutgers University, New Brunswick, New Jersey.

Bergmann W 1992 Nutritional Disorders of Plants – Development, Visual and Analytical Diagnosis. Fisher Verlag, Jena.

Blamey F P C, Mould D and Chapman J 1979 Critical boron concentration in plant tissues of two sunflower cultivars. Agron. J. 71, 243–247.

Blamey F P C, Vermuelen W J and Chapman J 1984 Inheritance of boron status in sunflower. Crop Sci. 24, 43–46.

Bohnsack C W and Albert L S 1977 Early effects of boron deficiency on indolacetic acid oxidase levels of squash root tips. Plant Physiol. 59, 1047–1050.

Brown P H and Hu H 1996 Phloem mobility of boron is species dependent: evidence for phloem mobility in sorbitol-rich species. Ann. Bot. 77, 497–505.

Chantachume Y 1995 Genetic studies on the tolerance of wheat to high concentrations of boron. Ph D Thesis, University of Adelaide, Australia.

Chapman V J, Edwards D G, Blamey F P C and Asher C J 1997 Challenging the dogma of a narrrow supply range between deficiency and toxicity of boron. *In* Boron in Soils and Plants. Proceedings. Eds. R W Bell and B Rerkasem. pp 151–155. Kluwer Academic Publishers, Dordrecht, the Netherlands.

Cheng C and Rerkasem B 1993 Effects of boron on pollen viability in wheat. Plant and Soil 155/156, 313–315.

Coetzer L A and Robbertse P J 1987 Pollination biology of *Persea americana* Fuerte. Sth. Africa Avocado Grow. Assoc. Yearbook 10, 43–45.

da Silva A R and de Andrade J M V 1980 A Cultura do Trigo nas Varzeas de Minas Gerais – possibilidades e dificuldades. Embrapa, Centro de Pesquisa Agropecuária dos Cerrados - CPAC. Circular TÈcnica no. 2. 69 p. *(In Portuguese)*

da Silva A R and de Andrade J M V 1983 Efeito micronutrientes no chochamento do trigo de seqqueiro e nas culturas de soja e arroz, em latossolo vermelho-amarelo. [Influence of micronutrients on the male sterility, on upland wheat and on rice and soybean yields, in red-yellow latosol] Pesq. agropec. bras., Brasilia 18, 593–601. *(In Portuguese with English Abstract).*

de Wet E and Robbertse P J 1989 The influence of temperature and boron on pollen germination in *Mangifera indica* L. Sth. Africa J. Plant Soil 6, 228–234.

Dickinson D B 1978 Influence of borate and pentaerythriol concentrations on germination and tube growth of *Lilium longiflorum* pollen. J. Amer. Soc. Hort. Sci. 103, 413–416.

Dong R, Noppakoonwong R, Song M and Rerkasem B 1997. Boron and fruit quality of apple. *In* Proc. in Soils and Plants. Eds. R W Bell and B Rerkasem. pp 125–129. Kluwer Academic Publishers, The Netherlands.

Fernandez P G Baudin C G, Esquinas T M and Vara J M M 1985 Boron deficiency in sunflower in Spain. Proc XI Int. Sunflower Conf. pp 1–8. Borax Holdings Ltd., London.

Garg O K, Sharma A N and Kona G R S S 1979 Effect of boron on the pollen vitality and yield of rice plants (*Oryza sativa* L. var. Jaya). Plant and Soil 52, 591–594.

Graham R D 1984 Breeding for nutritional characteristics in cereals. Adv. Plant Nutr. 1, 57–102.

Graham R D and Rengel Z 1993 Genotypic variation in zinc uptake and utilisation in plants. *In* Zinc in Soils and Plants. Ed. A D Robson. Dev. Plant Soil Sci. 55, 107–118.

Gupta U C 1979 Boron nutrition of crops. Adv. Agron. 31, 273–307.

Hanson E J and Breen P J 1985 Effects of fall boron sprays and environmental factors on fruit set and boron accumulation in "Italian" prune flowers. J. Amer. Soc. Hort. Sci. 110, 389–392.

Hu H and Brown P H 1994 Localisation of boron in cell walls of squash and tobacco and its association with pectin. Plant Physiol. 105, 681–689.

Hu H, Brown P H and Labavitch J M 1996 Species variability in boron requirement is correlated with cell wall pectin. J. Exp. Bot. 47, 227–232.

Huang C and Graham R D 1990 Resistance of wheat genotypes to boron toxicity is expressed at the cellular level. Plant and Soil 126, 295–300.

Jamjod S, Mann C E and Rerkasem B 1992a Combining ability of the response to boron deficiency in wheat. *In* Genetic Aspects of Plant Mineral Nutrition. Eds. P J Randall, E Delhaize, R A Richards and R Munns. Dev. Plant Soil Sci. 50, 359–362.

Jamjod S, Mann C E and Rerkasem B 1992b Screening for boron deficiency in wheat. *In* Boron Deficiency in Wheat. Eds. C E Mann and B Rerkasem. pp 79–82. CIMMYT Wheat Spec Rep 11. CIMMYT, Mexico.

Jones J B Jr 1991 Plant tissue analysis in micronutrients. *In* Micronutrients in Agriculture. 2nd ed. Eds. J J Mordtvedt, F R Cox, L M Shuman and R M Welch. pp 523–548. SSSA Book Series no. 4. SSSA, Madison, WI.

Kamali A and Childers N F 1970 Growth and fruiting of peach in sand culture as affected by boron and a fritted form of trace elements. J. Amer. Soc. Hort. Sci. 95, 652–656.

Keerati-Kasikorn P, Bell R W, Panya P, Gilmour R F and Loneragan J F 1993 Comparison of seed yield and quality of peanut (*Arachis hypogaea* L.) cultivars in low fertility soils and their response to boron and complete fertiliser. *In* Plant Nutrition from Genetic Engineering to Field Practice. Ed. N J Barrow. Dev. Plant Soil Sci. 54, 409–412.

Kelly J F and Gabelman W H 1960 Variability in the tolerance of varieties and strains of red beet (*Beta vulgaris* L.) to boron deficiency. Proc. Amer. Soc. Hort. Sci. 76, 409–415.

Kirk G J and Loneragan J F 1988 Functional boron requirement for leaf expansion and its use as a critical value for diagnosis of boron deficiency in soybean. Agron. J. 80, 758–762.

Lambert M J and Turner J 1977 Dieback in high site quality *Pinus radiata* stands - the role of sulphur and boron deficiencies. New Zealand J. For. Sci. 8, 351–368.

LeNoble M E, Blevins D G and Miles R J 1996a Prevention of aluminium toxicity with supplemental boron II. Stimulation of root growth in an acidic, high-aluminium subsoil. Plant Cell Environ. 19, 1143–1148.

LeNoble M E, Blevins D G, Sharp R E and Cumbie B G 1996b Prevention of aluminium toxicity with supplemental boron. I. Maintenance of root elongation and cellular structure. Plant Cell Environ. 19, 1132–1142.

Li B H, Li W H, Kui M C, Chao W S, Jern H P, Li C R, Chu W J and Wang C L 1978 Studies on cause of sterility of wheat. J. Northeastern Agric. Coll. 3, 1–19. *(In Chinese).*

Lõhnis M P 1940 Histology of boron deficiency in plants. Meded Landbouwhogesch 44, 3–36.

Lordkaew S 1995 Boron contamination in fertilisers. *In* Causes of Sterility in Wheat. Eds. R W Bell and B Rerkasem. pp 150–153. Env. Rep. 94/6. Murdoch University, Western Australia.

Lukaszewski K L and Blevins D G 1996 Root growth inhibition in boron-deficient or aluminium stressed squash may be a result of impaired ascorbate metabolism. Plant Physiol. 112, 1135–1140.

Marschner H 1995 Mineral Nutrition of Higher Plants, 2nd ed. Academic Press, London. 889 p.

Martens D C and Westermann D T 1991 Fertiliser applications for correcting micronutrient deficiencies. *In* Micronutrients in Agriculture. 2nd ed. Eds. J J Mordtvedt, F R Cox, L M Shuman and R M Welch. pp 549–592. SSSA Book Series no. 4. SSSA, Madison, WI.

Martinez H E P, Huag H P and Morales de M L T 1989 Micronutrientes em *Pinus caribaea* Morelet. I Efeitos da omissâo sobre o desenvolvimento e sintomas de carência. Anais da Escola Superior de Agricultura Luiz de Queiroz 46, 225–255.

Matoh T, Ishigaki K, Kaori O and Azuma J 1993 Isolation and characterisation of a boron-polysaccharide complex from radish root. Plant Cell Physiol. 34, 636–642.

Matoh T, Kawagochi S and Kobayashi M 1996 Ubiquity of a borate-rhamnogalacturonan II complex in the cell walls of higher plants. Plant Cell Physiol. 37, 636–640.

Misra R, Munankarmi R C, Panday S P and Hobbs P R 1992 Sterility work in wheat at Tarahara in the eastern Terai of Nepal. *In* Boron Deficiency in Wheat. Eds. C E Mann and B Rerkasem. pp 65–71. CIMMYT Wheat Spec Rep 11. CIMMYT, Mexico.

Mozafar A 1993 Role of boron in seed production. *In* Boron and its Role in Crop Production. Ed. U C Gupta. pp 186–206. CRC Press: Boca Raton, FL, USA.

Nable R O, Lance R C M and Cartwright B 1990. Uptake of boron and silicon by barley genotypes with differing susceptibility to boron toxicity. Ann. Bot. 66, 83–89.

Noppakoonwong R N 1993 Diagnosis of boron deficiency in black gram. Ph.D. Thesis, Murdoch University, Western Australia.

Nyomora A M S, Brown P H and Freeman M 1997 Fall foliar applied boron increases tissue boron concentration and nut set of almond. J. Amer. Soc. Hort. Sci. (*In press*).

Pant J 1993 Genotypic response of wheat to boron application under irrigated and water stress conditions. M.Sc. (Agric.) Thesis, Chiang Mai University.

Paull J G, Moody D B and Rathjen A J 1992 *In* Boron Deficiency in Wheat. Eds. C E Mann and B Rerkasem. pp 90–97. CIMMYT Wheat Special Report No. 11. CIMMYT, Mexico.

Pintasen S 1997 Responses to boron and waterlogging in wheat. M.Sc. (Agronomy) Thesis. Chiang Mai University, Thailand.

Pope D T and Munger H M 1953 The inheritance of susceptibility to boron deficiency in celery. Proc. Amer Soc. Hort. Sci. 61, 163–172

Rerkasem B 1986 Boron deficiency in sunflower and green gram at Chiang Mai. J. Agric. (Chiang Mai University) 2, 163–172. *(In Thai, with English Abstract).*

Rerkasem B 1990 Comparison of green gram (*Vigna radiata*) and black gram (*Vigna mungo*) in boron deficiency. *In* Proceedings of the Mungbean Meeting 90. Eds. C Thavarasook, P Srinives, N Bookerd, H Imai, A Pookpakdi, P Laosuwan and U Pupipat. pp 167–174. Bangkok Office of Tropical Agricultural Research Centre, Japan.

Rerkasem B 1993 Screening Wheat for Tolerance to Boron Deficiency. Agronomy Department, Chiang Mai University.

Rerkasem B and Jamjod S 1989 Correcting boron deficiency induced ear sterility in wheat and barley. Thai J. Soils Fert. 11, 200–209. *(In Thai with English abstract, tables and figure).*

Rerkasem B and Jamjod S 1997 Boron deficiency induced male sterility in wheat (*Triticum aestivum* L.) and implications for plant breeding. Euphytica in press (ref. MS No EUPH 4247).

Rerkasem B and Loneragan J F 1994 Boron deficiency in two wheat genotypes in a warm, subtropical region. Agron. J. 86, 887–890.

Rerkasem B and Lordkaew S 1992 Predicting grain set failure with tissue boron analysis. *In* Boron Deficiency in Wheat. Eds. C E Mann and B Rerkasem. pp 9–14. CIMMYT Wheat Special Report No. 11. CIMMYT, Mexico.

Rerkasem B, Lordkaew S and B Dell 1997 Boron requirement for reproductive development in wheat. 13th International Plant Nutrition Colloquium Abstract no. 05003–287.

Rerkasem B, Saunders D A and Dell B 1989. Grain set failure and boron deficiency in wheat in Thailand. J. Agric. (Chiang Mai University) 5, 1–10.

Rerkasem B, Bell R W, Lordkaew S and Loneragan J F 1993a Boron deficiency in soybean [*Glycine max* (L.) Merr.], peanut *(Arachis hypogaea* L.) and black gram [*Vigna mungo* (L.) Hepper]: Symptoms in seeds and differences among soybean cultivars in susceptibility to boron deficiency. Plant and Soil 150, 289–294.

Rerkasem B, Netsangtip R, Lordkaew S and Cheng C 1993b Grain set failure in boron deficient wheat. Plant and Soil 155/156, 309–312.

Rerkasem B, Netsangtip R, Bell R W, Loneragan J F and Hiranburana N 1988 Comparative species responses to boron on a Typic Tropaqualf in Northern Thailand. Plant and Soil 106, 15–21.

Robbertse P J, Coetzer L A, Bezuidenhout J J, Vorster L and Swart N G N 1990 The influence of boron on fruit set in avocado. Acta Hort. 275, 587–594.

Scott L E 1944 B Nutrition of the grapes. Soil Sci. 57, 55–65.

Seresinhe P S J W and Oertli J J 1991 Effects of boron on growth of tomato cell suspensions. Physiol. Plant. 81, 31–36.

Shelp B J 1988 Boron nutrition and mobility in broccoli (*Brassica oleracea* var. *italica*). Ann. Bot. 61, 83–91.

Shelp B J, Marentes E, Kitheka A M and Vevekanandan P 1995 Boron mobility in plants. Physiol. Plant. 94, 356–361.

Shelp B J, Penner R and Zhu Z 1992 Broccoli (*Brassica oleracea* var *italica)* cultivar response to boron deficiency. Can. J. Plant Sci. 72, 883–888.

Sherrell C G 1983 Effects of boron application on seed production of New Zealand herbage legumes. New Zealand J. Exp. Agric. 11, 113–117.

Shrestha G K, Thompson M M and Rigetti T L 1987 Foliar-applied boron increases fruit set in 'Barcelona' hazelnut. J. Amer. Soc. Hort. Sci. 112, 412–416.

Sillanpaa M 1982 Micronutrients and nutrient status of soils. FAO Soil Bulletin 48.

Simojoki P 1972 Tuloksia ohran boorilannoituskokeista (Boron deficiency, pollen sterility and ergot disease of barley) Ann. Agric. Fenniae 11, 333–341. *(In Finnish, with English Abstract)*

Singh R P Sharma Y P Awasti R P 1990 Influence of different cultural practices on premature cracking of pomegranate. Progressive Hort. 22, 92–96.

Singh H M Sinha S D and Prasad R B 1976 Effect of boron on seed setting in wheat under North Bihar conditions. Indian J. Agron. 21, 100–101.

Smith P E 1966 Citrus nutrition. *In* Nutrition of Fruit Crops. Ed. N F Childers. pp 174–207. Horticultural Publications. Rutgers University, New Brunswick, NJ, USA.

Smith T E, Hetherington S E, Asher C J and Stephenson R A 1997a Boron deficiency of avocado. 1. Effects on pollen viability and fruit set. *In* Proc. In Soils and Plants. Eds. R W Bell and B Rerkasem. pp 131–134. Kluwer Academic Publishers, The Netherlands.

Smith T E, Hetherington S E, Asher C J and Stephenson R A 1997b Boron deficiency of avocado. 2. Effects on fruit size and ripening. *In* Proc. In Soils and Plants. Eds. R W Bell and B Rerkasem. pp 135–138. Kluwer Academic Publishers, Dordrecht, the Netherlands.

Smithson J B and Heathcote R G 1974 Boron deficiency on cotton in northern Nigeria. II. The effect of variety. Expl. Agric. 10, 209–218.

Subedi K D, Budhathoki C B, Subedi M and Tuladhar J K 1993 Survey and Research Report on Wheat Sterility Problem 1992/93. Lumle Agricultural Research Centre Working Paper No. 93/49. Lumle, Nepal.

Tagliavini M, Scudellari D, Marangoni B, Bastianel A, Franzin F and Zamborlini M 1992 Leaf mineral compositionof apple trees: sampling date and effects of cultivar and rootstock. J. Plant Nutr. 15, 605–619.

Tandon J P and Naqvi S M A 1992 Wheat varietal screening for boron deficiency in India. *In* Boron Deficiency in Wheat. Eds. C E Mann and B Rerkasem. pp 76–78. CIMMYT Wheat Special Report No. 11. CIMMYT, Mexico.

Tehrani G, Munger H M, Robinson R W and Shannon S 1971 Inheritance and physiology of response to low boron in red beet (*Beta vulgaris* L.). J. Amer. Soc. Hort. Sci. 96, 226–230.

Tollenaar D 1966 Boron deficiency in cacao, bananas and other crops on volcanic soils of Ecuador. Neth. J. Agric. Sci. 14, 138–151.

Vaughan A K F 1977 The relation between the concentration of boron in the reproductive and vegetative organs of maize plants and their development. Rhod. J. Agric. Res. 15, 163–170.

Wall J R and Andrus C F 1962 The inheritance and physiology of boron response in tomato. Am. J. Bot. 49, 758–762.

Willis L G 1945 Defective strawberry fruit corrected by borax. Better Crops with Plant Food 29, 39–40.

Wright J W, Lemmien W A and Bright J N 1969 Early growth of ponderosa pine ecotypes in Michigan. For. Sci. 15, 121–129.

Yang Y H 1992. Wheat boron deficiency in Yunnan, China. *In* Boron Deficiency in Wheat. Eds. C E Mann and B Rerkasem. pp 72–75. CIMMYT Wheat Special Report No. 11. CIMMYT, Mexico.

Yang Y, Xue J, Ye Z and Wang K 1993 Responses of rape genotypes to boron application. Plant and Soil 155/156, 321–324.

Plant and Soil **198:** 181–198, 1997.

Chapter 12

Boron toxicity

Ross O. Nable[1], Gary S. Bañuelos[2] and Jeffrey G. Paull[3]
[1]*CSIRO Land and Water, P.M.B., P.O. Aitkenvale, QLD 4814, Australia**, [2]*Water Management Research Laboratory, USDA-ARS, 2021 S. Peach Av., CA 93727, USA and* [3]*Department of Plant Science, Waite Campus, University of Adelaide, Glen Osmond, SA 5064, Australia*

Abstract

Whilst of lesser prevalence than B deficient soils, B-rich soils are important, causing B toxicity in the field and decreased crop yields in different regions of the world. The highest naturally occurring concentrations of soil B are in soils derived from marine evaporites and marine argillaceous sediment. In addition, various anthropogenic sources of excess B may increase soil B to levels toxic for plants. The most important source is irrigation water, but others include wastes from surface mining, fly ash, and industrial chemicals. Ameliorating high-B soils is extremely difficult. A commonly used method of reclaiming high B soils is to extensively leach with low B water. Though used successfully, leaching may not be a permanent solution and causes difficulties with the disposal of the leachates. Other amelioration methods include the use of soil amendments (e.g. lime, gypsum) and the planting of plant genotypes that are tolerant of high external B concentrations. Although there are various methods available to determine the levels of B in soils, soil analysis can provide little more than a general risk assessment for B toxicity. Similarly, diagnosing B toxicity in plants, either by visible symptoms or tissue analysis has limited applicability. Thus at present, neither soil nor plant analysis can be recommended to precisely predict the growth of plants on high soil B. Recent physiological and genetic studies have provided some understanding of genetic variation in the response of plants to high concentrations of B. Moreover, these studies have facilitated the breeding of tolerant genotypes for cultivation on high B soils. Considerable genetic variation in response to high B has been identified in a wide range of plant species, most of which share a similar tolerance mechanism – reduced uptake of B in both shoots and roots. The tolerance mechanism appears to be under the control of several major additive genes, and specific chromosomal locations have been identified for the genes in some species. Considerable success has been achieved in breeding for tolerance to B toxicity, a process that is greatly aided by the ease with which genotypic variation for this characteristic can be assessed and the range of methods available to screen breeding populations.

Introduction

Boron toxicity is an important disorder that can limit plant growth on soils of arid and semi arid environments throughout the world. High concentrations of B may occur naturally in the soil or in groundwater, or be added to the soil from mining, fertilisers, or irrigation water. Although of considerable agronomic importance, our understanding of B toxicity is rather fragmented and limited. The present paper attempts to review the available literature on a broad range of topics, from the occurrence and detection of B toxic soils, to the effects of high B on plants, through to the management of B toxicity. It differs from earlier reviews on B toxicity (e.g. Gupta et al. 1985; Leyshon and Jame, 1993) by exploring in detail the physiology and genetypic variation for tolerance to B toxicity and how this variation can be utilised to maximise plant growth on high B soils.

Formation and distribution of boron laden soils

Sources of boron

Boron, the only non-metal among the elements of Group III in the periodic table, is not uniformly distributed in the earth's crust. The primary sources of

* E-mail: Ross.Nable@tvl.clw.csiro.au

B in most soils are tourmaline and the volatile emanations of volcanoes (Chesworth, 1991). Tourmaline from high temperature rocks is very resistant to chemical breakdown in the weathering zone and thus accumulates in the clastic fraction of sediments and sedimentary rocks. In igneous, metamorphic, sedimentary rocks, B occurs as borosilicates, which are resistant to weathering and not readily available to plants. Mobilisation of immobile forms of rock B occurs by weathering in the pedosphere, which includes soil reactions of acid-base, oxidation-reduction, and dissolution-precipitation. The dominant species in the soil when B from primary silicates goes into solution is $B(OH)_3$ (boric acid). This form of B is mobile and easily lost by leaching. In soils, this form of B can be taken up by vegetation, held by organic matter, or temporarily adsorbed on fine mineral fractions.

Distribution of naturally occurring high B soils

Soil is generally the primary source of trace elements for plants. However, there are exceptions in which toxic concentrations of trace elements in plants, e.g. B, can be traced directly to water from certain wells, or indirectly to land application of drainage water, and in soils with high B availability (Kubata, 1980). However, the adsorbed and solution phases of B in the soil influence potential B toxicity effects observed in the field (Cartwright et al., 1984; Shani and Hanks, 1993); and sometimes lead to decreases in crop yields grown in different regions of the world (Cartwright et al., 1986). There are few B rich soils compared to B deficient soils. Areas where high soil B are found include dry lands of South Australia (Cartwright et al., 1984, 1986), the Middle East (Ravikovitch et al., 1961), the west coast of Malaysia (Shorrocks, 1964), valleys along the southern coast of Peru (Masson, 1967), the Andes foothills in northern Chile (Caceres et al., 1992), solonchaks and solonetz soils of USSR (Zyrin and Zborishchuk, 1975), ferralsols of India (Takkar, 1982), rendzinas in Israel (Ravikovitch et al., 1961), and major $B_2O_2^{-7}$ deposits at Searles, Lake California (Chesworth, 1991). The highest concentrations of soil B are often concentrated in marine evaporites and in marine argillaceous sediment (Erd, 1980).

Activities contributing to boron laden soils

Irrigation water

Of all the potential sources, irrigation water is the most important contributor to high levels of soil B (Chauhan and Power, 1978; Table 1). Boron is often found in high concentrations in association with saline soils and saline well water (Dhankhar and Dahiya, 1980). In assessing the potential toxicity of B laden irrigation water, the physical and chemical characteristics of the soil must be considered (Goldberg, 1993). Sorption capacity of a given soil is crucial in determining the amount of B in solution. A soil that has high adsorption capacity would be expected to maintain lower soil solution B over a longer period of time than a soil with low adsorption capacity when both soils are irrigated with the same B laden water.

Soil adsorption sites may act as a pool from which B is supplied to solution or where B is adsorbed, depending on the changes in solution B concentrations and the affinity of soil for B. Thus, adsorbed B may buffer B concentration in soil solution (Keren and Bingham, 1985). The complex B sorption characteristics of a soil explain frequent observations that plant injury occurs more quickly on coarse textured soils than on fine textured soils when B laden water is used for irrigation. As the key to the assessment of B toxicity is the plant response to B in soil, it is the B concentration of the soil solution under field conditions that must be evaluated in relating soil B to plant response (Ryan et al., 1977). Moreover, water with high B levels can be used to irrigate B sensitive crops on soils that show a high affinity for B adsorption because large amounts of B in irrigation water will be adsorbed by the soil. Referring to Keren and Bingham (1985), safe concentrations of B in irrigation water range from 0.3 mg B L^{-1} for sensitive plants [i.e. avocado (*Persea americana*), apple (*Malus domestica*) and bean (*Phaseolus vulgaris*)], 1–2 mg B L^{-1} for semi tolerant plants [oat (*Avena sativa*), maize (*Zea mays*), potato (*Solanum tuberosum*)], and 2–4 mg B L^{-1} for tolerant plants [i.e. carrot (*Daucus carota*), alfalfa (*Medicago sativum*) and sugar beet (*Beta vulgaris*)]. Eventually, continued irrigation with B laden water will exceed the adsorption capacity of the soil and cause a possible reduction in crop yield (Jame et al., 1982). The level of B in solution that can be tolerated by the crop is the key information in determining criteria for allowable B concentration in irrigation water. The B concentration determined in a soil saturation extract (Bingham, 1973) is still used for

Table 1. Mean levels of tissue boron in different plant species irrigated with B laden water or grown in high B soil under field conditions

Species	B concentration in irrigation water (mg L^{-1})	Soil total extract (mg kg^{-1})	B concentration in shoots (mg B kg^{-1} dry wt)	References
Fourwing saltbush (*Atriplex canescens*)	5–10	N/A	126[1]	Watson et al. (1994)
Wavy leaf saltbush (*Atriplex undulata*)	5–10	N/A	131[1]	Watson et al. (1994)
Desert saltbush (*Atriplex deserticola*)	5–10	N/A	121[1]	Watson et al. (1994)
Old man saltbush (*Atriplex nummularia*)	5–10	N/A	142[1]	Watson et al. (1994)
Cattle saltbush (*Atriplex polycarpa*)	5–10	N/A	135[1]	Watson et al. (1994)
Alfalfa (*Medicago sativa*)	N/A	14	158	Nicholaichuk et al. (1988)
Tall fescue (*Festuca arundinacea*)	N/A	405	120[1]	Bañuelos et al. (1995)
Indian mustard (*Brassica juncea*)	N/A	535	224[2]	Bañuelos et al. (1993a)
Hibiscus cannibinus (*Hibiscus cannibinus*)	N/A	465	685	Bañuelos et al. (1993a)
Cotton (*Gossypium hirsutum*)	5	N/A	1110[3]	Ayars et al. (1990, 1994)
Wheat (*Triticum aestivum*)	5	N/A	701[3]	Ayars et al. (1990, 1994)
Silver beet (*Beta vulgaris*)	5	N/A	1043[3]	Ayars et al. (1990, 1994)
Tomato (*Lycopersicon esculentum*)	8	N/A	277[4]	Shennan et al. (1995)

[1]Mean value from multiple harvests in three years.
[2]Mean value from two plantings in two years.
[3]Mean leaf values from multiple years.
[4]Mean leaf values from two years of application.
N/A – not applicable.

defining B guidelines for plants exposed to B. If the leaching fraction or B adsorption capacity of the soil is not taking into consideration, there is doubt over the use of soil extractable B as the sole parameter to evaluate B concentrations in irrigation water.

Generally, plant uptake of B from irrigation water containing B is relatively small. Thus B will accumulate in soil irrigated with B laden water, as seen in the following example. Fifty kilograms of B would be deposited to soil after applying 100 cm of B laden water with a concentration of 5 mg B L^{-1}. Assuming a dry matter yield of 10,000 kg ha^{-1} (i.e. alfalfa) that had a tissue B concentration of 200 mg B kg^{-1} dry wt, a total of 2 kg B would be removed by plant accumulation. Irrigation water containing this level of B may not be immediately toxic to plants, but after prolonged irrigation with such water, soluble soil B levels will equal or exceed those of the irrigation water, especially in regions of low rainfall or where water used for leaching is unavailable. When such accumulation has occurred, plants sensitive to B (Francois and Clark, 1979; Maas, 1987;) will perish or plants more tolerant to B will have to be grown, or the soil must be reclaimed (Jame et al., 1982).

Surface mining

Surface mining often produces waste carbonaceous materials that can be a B source. Carbonaceous shales, coal lenses, leonardite, and coal stack are dominated by inert organic matter that often contains high B (Barth et al., 1987). Oxidation of these materials through mining processes can result in gradual release of substantial quantities of water soluble B. Total B concentrations are reported to be as high as 96 mg kg^{-1} soil in topsoil of western cool regions (Severson and Tidball, 1979), while hot water soluble B concentrations are as high as 26 mg L^{-1} in western mine soil spoil (Severson and Gough, 1983). Proposed maximum permissible levels of hot water soluble B is 8 mg kg^{-1} in mine soils in Montana, USA (Severson and Gough, 1983). Levels greater than 5 mg kg^{-1} may be considered toxic for plants (Ponnamperuma et al., 1979).

Fly ash

Fly ash is generated by electric power plants in increasing quantities because of increased use of coal for electricity production. Its use as an ameliorant on agricultural land to improve physical and chemical properties as well as its potential toxic effects have been studied (Carlson and Adriano, 1993; Pichtel et al., 1994). Contamination of land by residues of fossil fuel combustion (i.e. fly ash) can lead to high soil B concentrations (Adriano, 1986), especially if the B in fly ash is soluble and applied at a high rate to increase soil pH, i.e. 5–10% by wt (Eary et al., 1990; Kukier and Sumner, 1996). In one field study, fly ash was applied at a dry wt rate of 10% on a typical site in northern Britain with dry bulk density of 1.66 g cm^{-3} and a pH of 3.5 (Perkins, 1996). The concentrations of B in fly ash is of concern because large fractions of B in fly ash may be readily available to plants and may prevent the establishment of vegetation on soils containing fly ash (Elseewi et al., 1980b; James et al., 1982; Kukier et al., 1994; Piha et al., 1995), especially in the first growing season (Wong et al., 1996). Any general conclusions, however, have to be made with caution due to the enormous variability in fly ash chemical composition and differences in the physicochemical properties of amended soils (Eary et al., 1990; Mattigod et al., 1990). Boron concentrations of 700 mg $kg^{-1}$are often reported (James et al., 1982). The predominant forms of B in fly ash are probably soluble borates and less soluble borosilicates (James et al., 1982). Alkaline conditions decrease B solubility in fly ash, while neutral and acid reactions promote its solubility and release into solution (Elseewi et al., 1980a).

Industrial application

Boric acid and borate minerals are widely utilised in a number of industrial applications such as glass and porcelain manufacture, leather production, carpets, photographic chemicals, and fertilisers. The main application of B is, however, the use of sodium perborate as an oxidation bleaching agent in domestic and industrial cleaning products. The discharge of sodium perborate into the environment during production and end use of detergents has resulted in the accumulation of B in waste effluent and consequently in groundwater and natural aquatic systems (Vengosh et al., 1994).

Amelioration of boron laden soils

Leaching

When the amount of B added to the soil in irrigation water is greater than the amount of B removed by plant uptake and leaching, there will be a build up of B in the root zone (Ayars et al., 1994). Before soils containing high levels of B can be successfully used for agriculture, either their soluble B contents must be reduced to non-phytotoxic levels for the proposed crop or a tolerant genotype used. A commonly used method of reclaiming high B soils is to extensively leach with water (Prather, 1977; Leyshon and Jame, 1993). Applying water in excess of a plants water requirement is generally referred to as the leaching fraction (Hoffman, 1990). For efficient and continued crop production, the leaching fraction must be high enough to remove excess B but low enough to prevent loss of essential plant nutrients from the soil, especially from sandy acidic soils. The quantity of water necessary to leach B to a particular depth varies widely. If a low leaching fraction is used, the resultant soil B concentration near the soil surface will be close to those of the irrigation water and those near the bottom of the root zone will presumably be much higher. Much depends upon the original soil B concentration, the desired final B concentration, and the physical and chemical characteristics of the soil. Applying excessive water to leach B under field conditions is an old practice that is still suggested for reclaiming high B soils to satisfactory levels (Shennan et al., 1995; U.S. Salinity Lab. Staff, 1954). Recent field studies with irrigation water high in B have concluded that longer intervals of good quality water and/or intermittent ponding are necessary to remove B from the root zone (Ayars et al., 1994; Shennan et al., 1995). If an attempt is made to reduce the B level in the soil solution by leaching with low B water, a portion of the absorbed B will go into solution. Therefore, it is expected that the rate of removal by leaching will be much slower for B than for other non-absorbed elements in the soil. In this regard, studies on the kinetics of B release would be valuable in elucidating mechanisms for B desorption from soil surfaces. It is important to note that native soil B may be less soluble and more difficult to leach than B accumulated from previous irrigations (Peryea et al., 1985). However, the reduction in soluble B following leaching may not be permanent. Boron can be regenerated through the mineralisation of B from the soil organic matter, or through weathering processes of soil minerals (Peryea et al., 1985). The ability of the soil to regenerate B diminishes with each leaching, indicating a finite concentration of reclaimable B.

Soil amendments

Besides leaching with water, amendments have been used to detoxify B laden soils. As B adsorption by soils is pH dependent (Keren and Bingham, 1985), liming some soils to increase the pH and thus promote adsorption of B from soil solution may provide a short term solution (Bartlett and Picarelli, 1973). In sodic soils, the B hazard can be ameliorated by the addition of gypsum, which improves water infiltration and converts readily soluble Na-metaborate to less soluble Ca-metaborate (Bhumbla and Ckhabra, 1982). Heavy applications of $Ca(H_2PO_4)_2$ also lower plant available B, especially in acid soils. Prather (1977) reported that sulphuric acid may effectively aid in reclaiming B laden soils. Although decreasing soil pH by the addition of sulphuric acid increases the concentration of water soluble B in the soils, readsorption of B occurs with time.

The placement of soil containing low concentrations of B with materials containing excessive B, such as carbonaceous spoil, may limit both upward and downward migration of B. Such placement depends upon the use of the site, depth of water penetration, patterns of ground water movement, topography, physical, and chemical traits of surrounding soil, and organic materials. Cover depths with soil low in B for material containing soluble and mobile B have yet to be established with any precision. Some field research has indicated that 50–70 cm of soil over most spoil types is probably adequate (Barth, 1984; Barth et al., 1987). Unpublished results by Ajwa and Bañuelos have provided some insight with the planting of B tolerant plant species, e.g. saltbush (*Atriplex*), and covering high B containing soil (> 400 mg B kg^{-1} soil) with 47 cm of soil containing low B. They observed that B migrates towards the soil surface with water evaporation. Overburden only delayed B toxicity symptoms from appearing in the planted saltbush.

Under some circumstances it may be possible to alleviate B toxicity in plants by applying Zn to soils or as a foliar spray to affected plants. Graham et al. (1986) and Swietlik (1995) showed in controlled solution culture experiments with barley (*Hordeum vulgare*) and orange (*Citrus aurantium*), respectively, under conditions of Zn deficiency, excess B can be accumulated by plants and B toxicity develop, even though the levels of B in the medium do not result in B toxicity when ample Zn is supplied. To what extent this interaction occurs under field conditions remains to be shown. But as soils with low Zn and high B are encountered in some alkaline soils of semiarid regions, the possibility that B toxicity results from underlying Zn deficiency should be borne in mind.

Vegetation management

The planting of B tolerant species may be a means to revegetate soils containing elevated levels of soluble B and manage the movement of soluble B by plant uptake. However, limited information exists on how to utilise the ability of plants to take up and accumulate B as part of the revegetation strategy. Using both vegetation and water management as critical components for managing soil B has been suggested as an alternative to physically taking high B soils out of crop production (Bañuelos et al., 1993b). In a multiple year field study, Bañuelos et al. (1995) revegetated high B soil in central California and maintained the level of soluble soil B to a non-toxic level for the planted species. Some potential plant species that can be grown on high B soils are: saltbush (Watson et al., 1994), milkvetch (*Astragalus*) (Parker et al., 1991), barley (Nable, 1990a), wheat (Paull et al., 1988a), Indian mustard (*Brassica juncea*) (Bañuelos et al., 1993a), tall fescue (*Festuca arundinacea*) (Bañuelos et al., 1995), and some tree species (El-Motaium et al., 1994; Francois and Clark, 1979). Recent advances in genetic engineering are increasing the potential for breeding B tolerant plant species. Until more information is available, however, selection of B tolerant species will be dominated by trial and error experimentation. An alternative approach may be to

Table 2. Extraction methods used to assess available boron in soils

Method	Comments of referenced authors	Reference
Reflux 1:2 soil–water	Good for assessing plant available B.	Berger and Truog (1940)
Hot water soluble	Widely used; good prediction of deficient levels; site specific and no indication of B released from soil.	Bingham (1982)
Saturation extract	Comparable with solution B; B tolerance for plants based on this test, only a portion of absorbed B goes into solution; must be related to field moisture conditions.	U.S. Salinity Lab. Staff (1954), Gupta et al. (1985)
$CaCl_2$	Good for non-specifically absorbed B on soil surface at a variety of pH; indicates plant available B; similar to hot water extraction; estimates quantity factor, but not B released from soil.	Aitken and McCallum (1988); Iyenger et al. (1981); Spouncer et al. (1992)
Mannitol exchangeable B	Estimates leachable B, includes non-specifically and specifically absorbed forms; assesses regenerative power of soil for B.	Rhoades et al. (1970)
Acidified NH_2 HCl	Good for separating Mn minerals from Fe minerals; poor correlation with plant levels for calcareous soils.	Chao (1972); Cox and Kamprath (1977)
NH_4oxalate (pH 3.75)	Solubilises both crystalline and non-crystalline oxyhydroxides from soils; poor correlation to plant uptake of B.	DeEndredy (1963); Schwertmann (1964)
NH_4/DPTA	Used to determine availability of many nutrient elements in one extract.	Handreck (1990)

identify plants naturally occurring in soils high in B and utilise these for managing other high B soils.

Diagnosis of boron toxicity in soils and plants

Soil analysis

Soil sampling and analysis should identify soil horizons with excessive B. For a number of years, researchers believed that soils containing more than 5 mg L^{-1} of hot water soluble B were unsuitable for growing crops based on the response of agronomic crops to B. However, this value is not an accurate reflection of the plant species and growing conditions in arid and semi arid reclamation situations. Many native species are well adapted to B levels in excess of 5 mg L^{-1}. The plant available fraction is the best indicator for evaluating a soil's potential for inducing B toxicity in plant species. However, the form of B in soils will greatly affect its availability to plants (Fleming, 1980; Keren and Bingham, 1985; Leyshon and Jame, 1993). Generally, total B concentration is a poor indicator of plant available B. Plant available B in a specific soil is controlled by the soil's physical and chemical properties, such as pH, soil texture, clay mineralogy, organic matter, etc. (Goldberg, 1993). As these properties vary between soils, specific extraction techniques are not always reliable in predicting plant available B (Parker and Gardner, 1981; Su et al., 1994). Certain extraction reagents can specifically attack specific soil components and release different fractions of B (Table 2). In one study, 7 to 10 mg L^{-1} of B extracted with ammonium acetate reduced yields in pines (*Pinus*) by only 10% (Smidt and Whitton, 1975), while 35 mg B L^{-1} in a saturated paste extract did not significantly reduce yields (Schuman, 1969). Basin wild rye (*Leymus cinereus*) tolerates 37 mg B L^{-1} extracted from a water saturated fraction before yields are reduced by 50% (Roundy, 1985). While present knowledge is insufficient to define precisely the acceptable B levels in a growth medium, soil containing more than 5 to 8 mg B L^{-1} of hot water soluble B may require special revegetation considerations. Extraction methods that evaluate plant available B at one point in time will be different from methods that evaluate the capacity of a soil to supply B. It is also difficult to predict the B concentration of the soil solution within the root zone before a complete equilibrium is attained due to the complex B adsorption/desorption and precipitation/dissolution reactions in soil (Goldberg, 1993). Equilibrium is especially important in potentially B toxic soils that have been created by the use of high B water for irrigation. In this regard, Maas (1987) has provided a comprehensive summary of B tolerance based on plant response to B in soil solution. Although there are various methods available to determine the levels of B in soils, it seems that soil analysis can provide little more than a general risk assessment for B toxicity. It is very difficult to utilise soil analysis to precisely predict the growth of plants on high B soils.

Visible symptoms of boron toxicity in plants

Amongst a wide variety of plant species, the typical visible symptom of B toxicity is leaf burn – chlorotic and/or necrotic patches, often at the margins and tips of older leaves (Bennett, 1993; Bergmann, 1992; Eaton, 1944). These symptoms reflect the distribution of B in most species, with B accumulating at the end of the transpiration stream (Chapter 7). The chlorotic/necrotic patches have greatly elevated B concentrations compared with the surrounding leaf tissues (Oertli and Roth, 1969) and some species (e.g. barley) show characteristic patterns for different genotypes.

Contrary to the general perception, leaf burn is not the visible symptom of B toxicity in all species. In species in which B is phloem mobile (e.g. *Prunus, Malus, Pyrus*), in which B accumulates in developing sinks rather than at the end of the transpiration stream (Chapter 7), the symptoms of toxicity are fruit disorders (gummy nuts, internal necrosis), bark necrosis which appears to be due to death of the cambial tissues, and stem die back (Brown and Hu, 1996).

Visible symptoms of B toxicity do not appear to develop in roots. As B concentrations in the roots remain relatively low compared to those in leaves, even at very high levels of B supply (Nable, 1988; Oertli and Roth, 1969), perhaps toxic concentrations do not occur in root tissues. In young squash plants (*Cucurbita pepo*), an early effect of developing B toxicity is decreased chlorophyll concentrations, followed by reduced growth, loss of leaf area and decreased CO_2 fixation; all of these effects occurring well before the development of visible toxicity symptoms (Lovatt and Bates, 1984). There may also be an inhibition of ureide metabolism in the leaves of nodulated soybean (*Glycine max*) (Lukaszewski et al., 1992), and complexing of ribonucleotides which causes metabolic disturbances (Loomis and Durst, 1992). Leaf cupping, a specific visible symptom of B toxicity in some species, has been suggested to result from inhibition of cell wall expansion, through disturbance of cell wall crosslinks (Loomis and Durst, 1992).

Plant analysis

Whilst critical values for B toxicity have been established in many crop and tree species (Gupta, 1993; Table 1), there are serious problems with the use of foliar analysis for diagnosing B toxicity. In species that accumulate B in their leaves, these tissues normally contain about 40 mg to 100 mg of B kg^{-1} dry wt. However, the leaves can contain 250 mg kg^{-1} dry wt when B in the soil approaches toxic levels. Leaf concentrations of B may exceed 700 to 1000 mg kg^{-1} dry wt in extreme conditions of B toxicity. What is of particular concern is that there is an unacceptably wide range of critical values for B toxicity, sometimes even for the same species. In wheat and barley, for example, critical values range from 10 to 130 mg B kg^{-1} dry wt (Gupta, 1977; Gupta et al., 1976; Kluge and Podlesak, 1985; Paull et al., 1988a; Riley et al., 1994). The cause of the problem stems from the steep gradient of B within leaf blades, with B accumulating in tips and margins (Nable et al., 1990c; Oertli, 1994), and how this gradient is affected by environmental conditions. Depending upon the levels of transpiration, for example, leaf blades have been shown to accumulate vastly different amounts of B in their leaves, though mostly concentrated at the tips and in the margins, with the majority of the leaf blades having similar concentrations. Under these different transpiration conditions, the overall leaf B concentrations are very different, though the effect on growth can be the same (Nable et al., 1990c). Accordingly, experimental conditions lead to different critical levels for B toxicity being established, with those derived from glasshouse experiments being considerably higher, in general, than those from field experiments. Furthermore, as B is readily leached from leaves by rain (Nable and Moody, 1992), there is another reason why foliar analysis of field grown plants should be interpreted with caution.

As suggested by Shorrocks (1995), "it is perhaps preferable to think of B analysis as a relatively crude tool, and accept its limitations", rather than trying to use leaf analysis as a tool for the precise prediction of B toxicity. There seems considerable merit in his proposal to compare B concentrations in healthy regions of leaves with those in necrotic/chlorotic regions. If there is a large gradient between these regions then it is very likely that B toxicity is the causal agent, not other factors such as climate or pathogens. Such an approach could be very informative in species that accumulate B in their leaves.

As it has only recently been demonstrated that B is phloem mobile in some species (Chapter 7), and that B does not accumulate in leaves under conditions of B toxicity, there is not much information available yet on diagnosing B toxicity in these species. From what is available it appears that in species such as peach (*Prunus persica*), *Prunus*, pear, and almond, that fruit, buds, hulls and young stems are much better indicators of B toxicity than leaves (P H Brown, pers. comm.). Boron concentrations greater than 300 mg kg^{-1} dry wt indicate that B toxicity may be present.

Plant tolerance to boron toxicity

Genetic variation in response to high levels of boron

Genetic variation in response to high concentrations of B occurs at both the inter- and intra-specific levels. There have been many investigations on interspecific variation in response to B, with each species or genus represented by a single variety, for example, Eaton (1935, 1944), Francois and Clark (1979), Haas (1929), Oertli and Kohl (1961) and Purvis and Hanna (1938). All of these investigations identified a wide range in response to B, either on the basis of plant growth, the development of symptoms of B toxicity, or both. The range in variation was often such that the most tolerant entries were unaffected at treatments that killed the most sensitive entries.

More recent investigations have demonstrated a wide range of intra specific variation in response to B occurs in a number of crops, including bread wheat (*Triticum aestivum*) (Chatterjee et al., 1980; Mehrotra et al., 1980; Paull et al., 1988a), durum wheat (*Triticum turgidum* var. *durum*) (Brooks, 1991; Jamjod, 1996; Yau et al., 1995), barley (Christensen, 1934; Nable, 1988), rice (*Oryza sativa*) (Cayton, 1985, Paliwal and Mehta, 1973; Ponnamperuma et al., 1979), peas (*Pisum sativum*) (Bagheri et al., 1992; Materne, 1989), annual medics (Paull et al., 1992b), citrus (*Citrus* spp.) (Chapman and Vanselow, 1955; Haas, 1945), pecan (*Carya illinoensis*) (Picchioni and Miyamoto, 1991) and strawberry (*Fragaria ananassa*) (Blatt, 1976). Thus, the classification of each species in many of the earlier studies will be dependent on the variety tested.

Tolerance to B toxicity not only operates at the level of whole plants, it also operates at the organ and cellular level. Differences amongst wheat genotypes in susceptibility to B toxicity, expressed in whole plants, is also expressed at the organ and cellular level (Huang and Graham, 1990). In root cultures, genotypes classified as susceptible from field and glasshouse experiments produce shorter root axes and less lateral roots than tolerant genotypes in media with high B concentrations. Similarly, susceptible genotypes produce less callus than tolerant genotypes.

Following the identification of B toxicity in southern Australia (Cartwright et al., 1984, 1986), several germplasm collections were screened for response to B under a single B treatment, following the method of Moody et al. (1988), to determine the extent of variation within Australian varieties and to identify accessions that could be used in breeding more tolerant varieties (Table 3). In all cases a large variation in symptom expression and plant vigour was identified, including accessions more tolerant than all Australian varieties. Further studies, including a range of B treatments, confirmed the variation for all crops to be due to differential response to B (Bagheri et al., 1992; Jamjod et al., 1997; Nable, 1988; Paull et al., 1988a, 1992a, 1992b).

Tolerant genotypes of a number of crops originated from similar regions. In general, these correspond to where high concentrations have been reported in the soil or ground water, or symptoms of B toxicity have been observed on crops. This indicates that B has exerted a selective pressure on the crops and it is probable that tolerance of B would occur in other crops with centres of diversity in these regions. A high frequency of B tolerant bread wheat (Moody et al., 1988) and durum wheat (Jamjod, 1996) originate from the eastern Mediterranean, western Asia/north Africa (WANA), the Indian subcontinent, China, Japan and South America. While B tolerant medics (Paull et al., 1992a) and peas (Bagheri et al., 1994) originate from the eastern Mediterranean, WANA and the Indian subcontinent.

Genetics of differential tolerance to boron toxicity

The genetic control of tolerance to high concentrations of B has been investigated for several species to enable efficient strategies for breeding of B tolerant varieties

Table 3. Crops in which germplasm collections have been screened for response to B. The% tolerant accessions refers to the frequency of lines more tolerant than all Australian varieties of that crop

Crop	Species	Genotypes tested	% tolerant	References
Wheat	*Triticum aestivum*	1576	6	Moody et al. (1988)
Durum wheat	*Triticum turgidum* var. *durum*	300	2	Jamjod (1996)
Barley	*Hordeum vulgare*	>1500		Boyd, pers. comm.
		350		Jenkin (1993)
Pea	*Pisum sativum*	135	2	Paull et al. (1992b)
		617	4	Bagheri et al. (1994)
Medics	*Medicago* spp.	681	2	Paull et al. (1992b)

to be adopted. Boron tolerance of bread wheat (Paull et al., 1991a), durum wheat (Jamjod, 1996), barley (Jenkin, 1993) and field pea (*Pisum sativum*) (Bagheri et al., 1996) is controlled by partially dominant nuclear genes. However, the degree of dominance expressed by an F1 hybrid varies according to the B treatment. At low B treatments (i.e. near the threshold treatment for the tolerant parent) F1 hybrids respond in a similar manner to the tolerant parent, while at high B treatments they tend towards the sensitive parent (Jamjod, 1996; Jenkin, 1993; Paull et al., 1991a). Thus the most appropriate boron treatment for screening segregating populations will depend on the genetic composition of the test population. For example, a higher treatment would be appropriate for identifying homozygous tolerant plants in an F2 population than for identifying heterozygous plants during backcrossing where an excessively high treatment would result in the heterozygous plants being phenotypically similar to the more sensitive parent.

Studies of F2 and F3 populations of bread (Paull et al., 1991a) and durum (Jamjod, 1996) wheat, barley (Jenkin, 1993) and peas (Bagheri et al., 1996) have all identified major genes conferring tolerance to B. Several independent but additive genes occur in each species. Despite being similar phenotypically, tolerant lines of both bread and durum wheat might differ genetically with respect to tolerance to B (Chantachume, 1995; Jamjod, 1996; Paull et al., 1991a). Transgressive segregation occurred among the progeny derived from crosses between these tolerant genotypes and included lines more tolerant and more sensitive than both parents. Thus, by understanding the genetic relationships among moderately tolerant and tolerant accessions it is possible to develop lines with a greater level of B tolerance than that available in a germplasm collection.

Genes conferring tolerance to B have been located to chromosomes of cereals by several methods, including intervarietal substitution lines, interspecific substitution and addition lines and monosomic analysis, and linkage has been established between some genes conferring tolerance to B and other loci. In particular, chromosome 4A and chromosomes of homologous group 7 have been implicated in response to B. Chromosome 4A of the Chinese Spring/Kenya Farmer and Chinese Spring/Sapporo intervarietal substitution line series has a major effect on response to B, with both substitution lines being more sensitive than Chinese Spring (Paull et al., 1988b, 1995). Evaluation of the progeny of CS×CS(KF4A) indicated that the bo3 gene is linked to the stem rust resistance gene Sr7a of Kenya Farmer, located on the long arm of chromosome 4A (Paull et al., 1995). Further evidence of the significant effect of chromosome 4A on response to B was provided by backcross reciprocal monosomic analysis (Snape and Law, 1980), with the tolerant lines G61450 (from Greece), Benventuto Inca (Argentina) and India 126 (India) crossed to the Condor (moderately sensitive) monosomic series (Chantachume, 1995; Chantachume et al., 1994). Progeny containing chromosome 4A derived from the tolerant lines were significantly more tolerant than progeny with chromosome 4A of Condor. In addition, an RFLP marker, XksuG10 (Gill et al., 1991) located on chromosome 4A, was identified as linked to a B response gene by QTL analysis of F7 lines of G61450×Kenya Farmer (very sensitive) (Paull et al., 1995). Monosomic analysis has demonstrated the Bo1 gene of bread wheat (Chantachume, 1995; Chantachume et al., 1994) and BoT1 and BoT2 genes of durum wheat (Jamjod, 1996) are located on chromosome 7B.

Wild species are a potential source of genes for related cultivated crops, and a number of members of

Triticeae and aneuploid stocks of wheat including chromosomes of these related species have been evaluated for response to B. Tall wheatgrass (*Agropyron elongatum*) is tolerant to B (Schuman, 1969) and its tolerance is expressed in the amphiploid with bread wheat. However, the amphiploid is not significantly more tolerant than the most tolerant bread wheats (Paull et al., 1991b, 1992b). The chromosome 7E addition line is more tolerant than Chinese Spring (Paull et al., 1992a) and this is further evidence of the significance of the chromosomes of homologous group 7 in controlling response to B. Chromosomes of homologous group 5 have also been shown to have a significant effect on response of cereals to B (Manyowa and Miller, 1991). The Chinese Spring/"alien" addition lines including chromosomes 5R of Imperial rye and 5S1 of Aegilops sharonensis were significantly more tolerant then Chinese Spring.

Physiology of tolerance to boron toxicity

Exclusion mechanisms

An examination of tolerance to B toxicity amongst diverse plants species reveals that the B concentration in leaves and shoots are usually not closely related to B tolerance. However, in more closely related species, genotypes susceptible to B toxicity generally have higher concentrations of B in leaves and shoots than do tolerant genotypes (Eaton, 1944; Francois and Clark, 1979). For example: *Helianthus tuberosus* (Jerusalem artichoke) has higher concentrations of B in its leaves than the more B tolerant *Helianthus annuus* (sunflower); *Citrus limonin* (lemon), accumulates more B in its leaves than its B tolerant relative *Severina buxifolia* (Chinese box orange) (Eaton and Blair, 1935); *Lycopersicon esculentum* (cultivated tomato) is more susceptible to B toxicity and accumulates more B in its shoots than its wild relative *L. cheesmanii* (Toledo and Spurr, 1984); and in several cereal and legume species, genotypes susceptibility to B toxicity have shoot B concentrations much higher than in those of tolerant genotypes (Bagheri et al., 1992; Chhipa and Lal, 1990; Paull et al., 1988a, 1992a, 1992b). Similarly, stem B concentrations are closely related to the development of B toxicity in *Prunus* and can be used to rank rootstocks for B tolerance (El-Motaium et al., 1994).

Thus, many B tolerant genotypes are able to maintain lower B concentrations in their shoots than related, susceptible genotypes. Whether or not these B tolerant genotypes operate exclusion mechanisms that result in decreased B accumulation in the entire plant, cannot for the most part be determined, for there is little data on B concentrations in roots. However, where both roots and shoots have been examined in solution culture experiments, exclusion mechanisms have been shown to operate in a wide range of species. For example, B tolerant genotypes of wheat (Figure 1) and barley (Nable, 1988) and peas and medics (Paull et al., 1992b) maintain lower concentrations of B in both root and shoots than do susceptible genotypes. That the ranking of these genotypes for B tolerance (and leaf B concentrations) is the same for plants grown at high B concentrations in solution culture or in soil in pots and the field, indicates that the exclusion mechanism operates under all conditions. To understand how

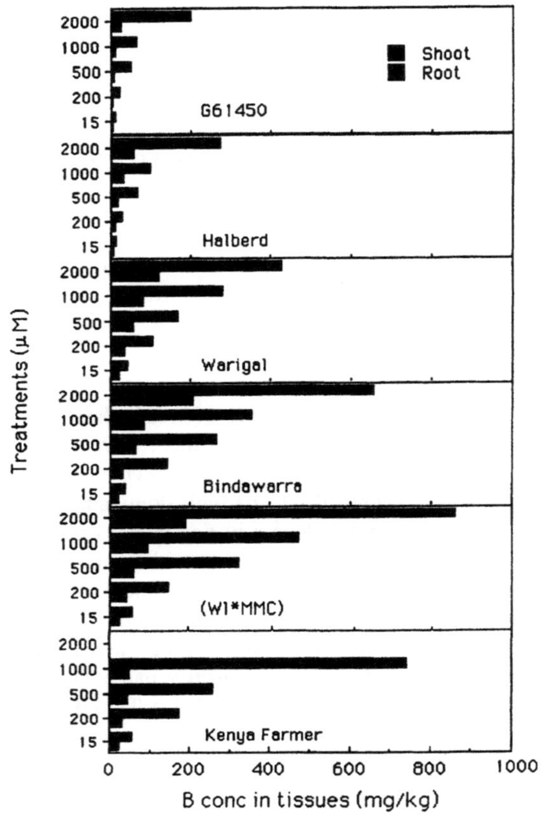

Figure 1. Boron concentration in the roots and shoots of six wheat (*Triticum aestivum*) genotypes grown in solution culture at five levels of B. The relative responses of the genotypes to B toxicity were: G61450 – tolerant; Halberd – moderately tolerant; Warigal – moderately sensitive; Bindawarra – moderately sensitive; and (WI*MMC) – sensitive. Adapted from Nable (1988).

B accumulation may be regulated under conditions of potential B toxicity, it is necessary to re-examine some aspects of B absorption.

It was noted in Chapter 4 that B is absorbed from solutions as the small, neutral molecule, $B(OH)_3$, and the transmembrane movement of boric acid is primarily determined by its ability to partition into the lipid bilayer. For most molecules of MW > 50, there is a direct relationship between the transmembrane permeability and a measure of its ability to partition into the lipid bilayer, such as an oil/water partition coefficient. Using a partition coefficient for $B(OH)_3$ in water/ether of 0.035 (Bachelet et al., 1947), and based upon the relationship between partition coefficient and membrane permeability (Walter and Gutknecht, 1986), $B(OH)_3$ is predicted to have a very high transmembrane permeability of approximately 4.6×10^{-4} cm s^{-1}. Based upon this value, assuming an external $B(OH)_3$ concentration of 1m*M* (i.e. 1 mol m^{-3}; which is in the range to induce B toxicity) and an (initial) internal concentration of zero, the flux per unit area of membrane would be 4.6×10^{-6} mol m^{-2} s^{-1}. Such a flux is several orders of magnitude greater than the observed protein mediated transmembrane flux of most solutes across membranes (e.g. see Tables 6.1 and 6.2 in Hope and Walker, 1975), suggesting that most of the flux is not protein mediated. Similarly, the structure of the boric acid molecule indicates high membrane permeability that would effectively short circuit any active regulatory mechanism for B absorption at high external concentrations (Raven, 1980). Thus, under conditions of potential B toxicity, B absorption into roots appears to be governed by passive transport processes.

Experimental evidence supporting passive transport processes for B absorption (at least at normal to high external B concentrations) can be found in several uptake studies. In barley (Nable et al., 1990b), sunflower, squash and cultured tobacco (*Nicotiana tobaccum*) cells (Brown and Hu, 1994), B accumulation increases linearly with increasing external B concentrations, shows no multiphasic kinetics, and is not saturable over a wide range of external B concentrations (Figure 2). Furthermore, the inclusion of metabolic inhibitors in the uptake solution (Brown and Hu, 1994) or exposure of plants to temperatures from 2–42 °C (Brown and Hu, 1994; Nable et al., 1990b) does not inhibit B accumulation in any species.

Different abilities to passively transport B appears to be associated with differential tolerance to B toxicity. For in barley, the comparative tolerance of genotypes to B toxicity is reflected in the relative differences in passive B accumulation rates over a wide range of B supply (Nable et al., 1990b; Figure 2). Each genotype has a characteristically different ability to passively accumulate B, a difference that is apparently constitutive and not induced by high B. Furthermore, whilst root temperatures from 5–25 °C markedly affect the growth of barley genotypes, temperature has no effect on the relative susceptibilities of genotypes to B toxicity. Similar results from less detailed experiments with other species (medics, wheat, oats; Nable, unpublished data), indicate a general phenomenon in which the relative tolerance of genotypes to B toxicity is associated with genetic differences in passive B accumulation.

Interestingly, Si accumulation, which also differs greatly amongst barley genotypes, closely reflects B accumulation and the relative tolerance of barley genotypes to B toxicity (Nable et al., 1990b; Walker and Lance, 1991). Thus, the mechanism that governs differences in passive B accumulation also appears to govern Si accumulation in barley. A number of similar properties of $B(OH)_3$ and $Si(OH)_4$ may account for the observed association. Both molecules are weak acids in aqueous solutions, both are mainly undissociated at physiological pH values (pKa1 values of 9.25 and 9.82, respectively at 25 °C), indicating that both are absorbed as neutral molecules, both are thought to move passively across biological membranes and the accumulation of both by plants is greatly influenced by transpiration rates (Raven, 1980, 1983). Yet, addition of Si to plants does not reduce B accumulation, suggesting there is no competition for an active site on a transport protein, despite the apparent similarity of their absorption pathway. Moreover, there is no close relationship between B and Si accumulation in medics. Medic genotypes accumulate very little Si, irrespective of their vastly different levels of B accumulation (Jones and Handreck, 1969; Paull et al., 1992b). Thus, it appears that in the case of medics, and perhaps other dicotyledonous species, a different additional mechanism to that which regulates both B and Si absorption in cereals, controls Si absorption independently of B absorption.

Internal tolerance mechanisms

Except for certain halophytes in which solutes such as sorbitol might inactivate excess B (Rozema et al., 1992), we are unaware of data showing genotypic variation in the an ability of tissues to tolerate high B concentrations.

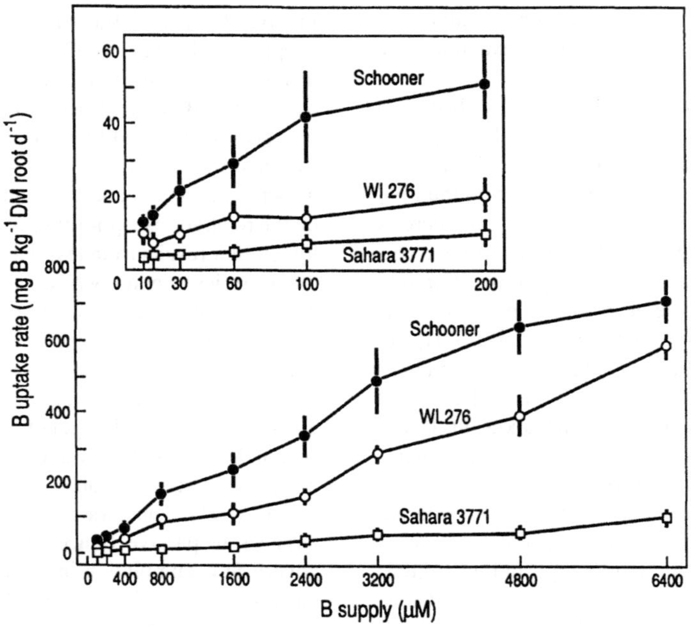

Figure 2. The relationship between external boron concentration and B accumulation by three barley (*Hordeum vulgare*) genotypes differing in response to B toxicity. The relative responses of the genotypes to B toxicity were: Sahara 3771 – tolerant; WI 276 – moderately tolerant; and Schooner – sensitive. From Nable et al. (1990b).

There is also the possibility that B tolerance is related to altered distribution of accumulated B. That is, preferential distribution at the cellular, tissue, or organ level. Again, in cereals, the pattern of B distribution amongst individual leaves and roots is similar in genotypes having greatly different levels of tolerance to B toxicity, despite great differences in total amounts of B accumulated and the actual B concentrations of individual parts (Nable, 1989).

It was noted earlier that there are differences in the phloem mobility of B amongst species which results in B being accumulated in either leaf margins (phloem immobile) or in fruits, cambial tissue, and stems (phloem mobile) (Brown and Hu, 1996). As most species that transport B in the phloem are susceptible to B toxicity, phloem immobility could be viewed as an internal tolerance mechanism. As such, phloem immobility keeps B away from key metabolic sites, retaining it instead in leaf margins, where despite suffering leaf burn, plants are still able to maintain considerable photosynthetic area.

It thus appears that in some species tolerance to B toxicity is governed by an exclusion mechanism – the ability of plants to restrict B accumulation. However, given that differences in B accumulation can be so large, and given that B absorption is not controlled by a protein mediated process, how might plants regulate B accumulation? Several possibilities exist, and these will be discussed briefly.

Due to the readiness by which B binds to polysaccharides and organic acids of various kinds, particularly those with alcohol groups, the secretion of B chelating compounds into rhizosphere, perhaps associated with mucilage, would be likely to reduce the amount of free B around the root, and thus of B entering the plant. Yet, in numerous solution culture experiments with a wide range of species, usually with a mixture of genotypes in the same container, differential tolerance to B toxicity continues to be expressed (Nable, 1988; Nable et al., 1990a, 1990b). Whilst under such conditions, rhizosphere complexation seems unlikely to be the tolerance mechanism, there remains the possibility, particularly under non-sterile conditions, that exudates are important, as they may act immediately adjacent to the root surface, but be metabolised before they have diffused to the bulk solution. However, as seen with pollen tube and tobacco cell cultures, in the presence of high B concentrations very high sugar concentrations are necessary to alleviate B toxicity (Yokota and Konishi, 1990). Furthermore, the exudate would be required over the entire surface of the root to bind the high amounts of $B(OH)_3$ which would be diffusing to the root surface under conditions of potential B toxicity.

Similarly, it seems unlikely that inactivation in the cell wall or cytoplasm of root cells could be the mechanism controlling B exclusion. For although such complexation has been shown to occur (Brown and Hu, 1994), the amounts of $B(OH)_3$ which would need to be bound are enormous, particularly for plants grown in solution culture, where there is a continuously replenished, well mixed supply of B. Also, some threshold pattern of accumulation might be expected, with plants being able to resist increasing B accumulation until a level is reached where the ability to complex is exceeded and B floods into the plant. No such threshold has been reported. If B is being complexed in root cell walls or cells, high B concentrations would be expected in roots, with tolerant genotypes having higher concentrations than susceptible genotypes. In fact, B concentrations in roots are generally quite low, with tolerant genotypes of several species having lower concentrations than susceptible genotypes (Nable, 1988; Nable et al., 1990b; Paull et al., 1992b).

As the transmembrane flux of $B(OH)_3$ is not mediated by proteins (at least at moderate to high external B concentrations), would it possible for an exclusion mechanism to operate through differences in the ability of $B(OH)_3$ to cross the lipid bilayer? That is, by changing the ability of the molecule to partition into the bilayer. For other molecules (including water, glucose and ions), it is known that both major lipid components of plant membranes, phospholipids and sterols, can affect the transport across the lipid bilayer. There is no theoretical reason why changes in the composition of the lipid bilayer could not have similar effects on the transport of $B(OH)_3$. Whether or not such changes form the basis of the observed exclusion mechanism is unknown.

Finally, as B is known to be greatly influenced by transpiration rates (Raven, 1980), there is the possibility that genotypic differences in B accumulation are governed by differences in transpiration rates and the transport of B in the xylem. However, to achieve the differences in B accumulation that have been observed, several fold differences in water use efficiency (WUE) are required (Nable, 1988). In fact, observed differences in WUE between B tolerant and susceptible barley genotypes are much lower, with a range from 2.5-4.0 g dry wt kg^{-1} water.

Breeding for tolerance to boron toxicity

High concentrations of B in the soil result in a significant yield reduction to barley crops (e.g. 17% reduction in yield, Cartwright et al., 1984), a problem that is widespread throughout the cereal growing districts of southern Australia (Cartwright et al., 1986). Additionally, two other lines of evidence indicate the full agronomic and economic significance of B toxicity, and the potential to overcome this problem through breeding. Namely: (1) the association between high levels of B and the cultivation of tolerant wheat varieties (Paull, 1990; Rathjen and Pederson, 1986); and (2) the performance of near isogenic lines under normal and high levels of B (Campbell et al., 1994, 1995; Moody et al., 1993).

There is a wide range in genetic variation in response to B among Australian wheat varieties released during the twentieth century ranging from very sensitive to moderately tolerant in comparison to the total range in variation in wheat (Moody et al., 1988). Distinct differences occur in the response of varieties selected in different regions and virtually all moderately tolerant varieties were selected in South Australia or Victoria, while varieties from New South Wales, Queensland and Western Australia are generally sensitive (Campbell et al., 1994; Paull et al., 1990). All moderately tolerant Australian varieties are related and can be traced to Federation and Currawa. Varieties in this family, especially Federation, Ghurka, Quadrat, Insignia, Heron, Olympic, Halberd and Spear, have been the most widely cultivated varieties in South Australia and Victoria throughout the twentieth century and have regularly accounted for greater than 50% of the total wheat production in South Australia, as was the case in Victoria prior to the introduction of stripe rust (*Puccinia graminis tritici*) in 1979. Furthermore, the regions of maximum cultivation of tolerant varieties correspond to those where high levels of soil B occur, such as the Eyre Peninsula of South Australia and the Mallee and Wimmera of Victoria, and tolerant varieties often account for 90 percent of wheat production in high B regions (Rathjen and Pederson, 1986). In recent years Halberd and Spear have been cultivated extensively on the sodic soils of Western Australia where high concentrations of B would be expected. High concentrations of B therefore appear to have had a major effect on both selection during breeding and subsequent adoption of varieties by farmers.

The concentrated distribution of B tolerant wheat varieties in regions where high levels of soil B predominate might be considered circumstantial evidence of the impact of B status on Australian agriculture. More compelling data have been obtained from variety trials and the comparative yields of tolerant and sensitive near isogenic lines (NILs). The association between grain yield and the B response of wheat varieties and breeding lines, as measured by the concentration of B in plant tissues, was demonstrated soon after the recognition of B toxicity in South Australia. Cartwright et al. (1987) analysed the grain of 150 lines grown at a high B site and found the B concentration in grain of the high yielding lines (130% of site mean) was 5.9 mg kg^{-1}, compared with 8.5 mg kg^{-1} for the low yielding lines (69% of site mean). Similarly, Rathjen et al. (1987) demonstrated that three years of selection at five sites from a broadly based population of breeding lines resulted in disruptive selection that could be attributed to the B status of the site of selection. The highest yielding lines selected at high B sites yielded significantly more, and had lower concentrations of B in grain than the highest yielding lines selected at low B sites, when compared at a high B site.

As response to B is conferred by major additive genes, backcrossing is an appropriate breeding method to transfer tolerance into otherwise well adapted varieties. Several sets of NILs of wheat have been developed through backcrossing, mostly transferring the Bo1 allele from Halberd to moderately sensitive varieties. Evaluation of these NILs over a range of sites in southern Australia has consistently demonstrated a yield advantage in the order of 5–10% for the tolerant selections when grown at high B sites (Campbell et al., 1994; Moody et al., 1993).

Breeding for tolerance to B toxicity is greatly aided by the ease with which genotypic variation for this characteristic can be assessed and the resulting range of methods that can be utilised to screen breeding populations. For example, as tolerance to B toxicity is consistently related to B concentrations in various plant tissues (e.g. leaf, grain), whether plants are grown under controlled conditions in the glasshouse or in the field, the measurement of tissue B concentrations has been an approach that is successfully utilised as a screening method in several breeding programs (e.g. Bagheri et al., 1996; Jamjod, 1996; Paull et al., 1990). The effect of high B concentrations on the early plant growth, either shoot growth of plants grown in soil (e.g. Paull et al., 1990) or root growth of plants grown on filter paper (Chantachume et al., 1995), is an alternative approach that has also been successfully utilised as screening strategies. Both of these approaches give the same ranking of genotypes.

Acknowledgment

Thanks to Dr M Tester, University of Cambridge for helpful input and discussions.

References

Adriano D C 1986 Trace Elements in the Terrestrial Environment. pp 73–105. Springer-Verlag, New York.

Aitken R L and McCallum L E 1988 Boron toxicity in soil solution. Aust. J. Soil Res. 26, 605–610.

Ayars J E, Hoffman G J and Schrale G 1990 Strategies to reduce salt load in drainage water. Proceedings of the 1990 National Conference IrDiv/ASCE. Durango, CO. pp 331–338.

Ayars J E, Hutmacher R B, Schoneman R A, Vail S S and Pflaum T 1994 Long term use of saline water for irrigation. Irrig. Sci. 14, 27–34.

Bachelet M, Cheylan E and le Bris J 1947 Solubilite des principaux acides fixes des minerals d'uranium dan l'ether ethylique et coefficient de partage entre l'eau et l'ether ethyliques. J. Chim. Phys. Phys.-Chim. Biol. 44, 302–305.

Bagheri A, Paull J G and Rathjen A J 1994 The response of *Pisum sativum* L. germplasm to high concentrations of soil boron. Euphytica 75, 9–17.

Bagheri A, Paull J G and Rathjen A J 1996 Genetics of tolerance to high concentrations of soil boron in peas (*Pisum sativum* L.). Euphytica 87, 69–75.

Bagheri A, Paull J G, Rathjen A J, Ali S M and Moody D B 1992 Genetic variation in the response of pea (*Pisum sativum* L.) to high soil concentrations of boron. Plant and Soil 146, 261–269.

Bañuelos G S, Akohoue S, Zambrzuski S and Mead R 1993a Trace element composition of different plant species used for remediation of boron-laden soils. *In* Plant Nutrition From Genetic Engineering to Field Practice. Ed. N J Barrow. pp 425–428. Kluwer Academic Publishers, Dordrecht, The Netherlands.

Bañuelos G S, Mackey B, Wu L, Zambrzuski S and Akohoue S 1995 Bioextraction of soil boron by tall fescue. Ecotoxicology Environ. Safety 31, 110–116.

Bañuelos G, Cardon G, Mackey B, Ben-Asher J, Wu L, Beuselinck P, Akohoue S and Zambrzuski S 1993b Boron and selenium removal in boron laden soils by four sprinkler irrigated plant species. J. Environ. Qual. 22, 786–792.

Barth R C 1984 Soil depth requirements to re-establish perennial grasses on surfaced-mined areas in the Northern Great Plains. Mineral Energy Resources 27, 1–20.

Barth R C, Severson R C and Weiler G 1987 Boron. *In* Reclaiming Mine Soils and Overburden in the Western US: Analytical Parameters and Procedures. Eds. R D Williams and G E Schuman. pp 135–153. Soil Conservation Soc. Am., Ankey, IA, USA.

Bartlett R J and Picarelli C J 1973 Availability of boron and phosphorus as affected by liming an acid potato soil. Soil Sci. 116, 77–80.

Bennett, W F 1993 Nutrient Deficiencies and Toxicities in Crop Plants. APS Press, St Paul, MN, USA.

Berger K C and Truog E 1940 Boron deficiencies as revealed by plant and soil tests. J. Am. Soc. Agron. 32, 297–301.

Bergmann W 1992 Colour Atlas: Nutritional Disorders of Plants. pp 204–239. Gustav Fischer, New York.

Bhumbla D R and Ckhabra R 1982 Chemistry of sodic soils in review of soil research in India. *In* Trans. 12th Int. Cong. Soil Sci. New Delhi, India. 169 p.

Bingham F T 1973 Boron in cultivated soils and irrigation waters. *In* Trace Elements in the Environment. Advances in Chemistry, Ser. 123. Ed. E L Kothnz. pp 130–138. Am. Chem. Soc., Washington, DC, USA.

Bingham F T 1982 Boron. *In* Methods of Soil Analysis. Ed. A L Page. pp 431–447. Am. Soc. Agron., Madison, WI, USA.

Blatt C R 1976 Phosphorus and boron interactions on growth of strawberries. HortScience 11, 597–599.

Brooks B J 1991 The adaptation of *Triticum turgidum* L. var. durum (durum wheat) to South Australia. Honours Thesis, The University of Adelaide.

Brown P H and Hu H 1994 Boron uptake by sunflower, squash and cultured tobacco cells. Physiol. Plant. 91, 435–441.

Brown P H and Hu H 1996 Phloem mobility of boron is species dependent: evidence for phloem mobility in sorbitol-rich species. Ann. Bot. 77, 497–505.

Caceres L, Gruttner D. and Contreras N 1992 Water recycling in arid regions: Chilean case. Ambio 21, 138–144.

Campbell T A, Rathjen A J and Jefferies S P 1994 Breeding wheat (*Triticum aestivum* L.) for tolerance to boron toxicity. *In* Proc. 7th Assembly Wheat Breeding Society of Australia. Eds. J G Paull, I S Dundass, K W Shepherd and G J Hollamby. pp 111–114. Wheat Breeding Society of Australia.

Campbell T A, Moody D B, Jefferies S P, Cartwright B and Rathjen A J 1995 Grain yield evaluation of near isogenic lines for boron tolerance. *In* Proc. 8th Int. Wheat Genet. Symp., Beijing. Eds. Z S Li and Z Y Xin. pp 1021–1027.

Carlson C L and Adriano D C 1993 Environmental impact of coal combustion residues. J. Environ. Qual. 22, 227–247.

Cartwright B, Zarcinas B A and Mayfield A H 1984 Toxic concentrations of B in a red-brown earth at Gladstone, South Australia. Aust. J. Soil Res. 22, 261–272.

Cartwright B, Zarcinas B A and Spouncer L A 1986 Boron toxicity in South Australian barley crops. Aust. J. Agric. Res. 37, 351–359.

Cartwright B, Rathjen A J, Sparrow D H B, Paull J G and Zarcinas B A 1987 Boron tolerance in Australian varieties of wheat and barley. *In* Genetic Aspects of Plant Mineral Nutrition. Eds. H W Gabelman and B C Loughman. pp 139–151. Martinus Nijhoff, Dordrecht, The Netherlands.

Cayton M T C 1985 Boron toxicity in rice. IRRI Research Paper Series 113.

Chantachume Y 1995 Genetic studies on the tolerance of wheat to high concentrations of boron. Ph.D. Thesis, The University of Adelaide, South Australia.

Chantachume Y, Shepherd K W, Paull J G and Rathjen A J 1994 Chromosomal location of genes in wheat controlling tolerance to high concentrations of boron. *In* Proc. 7th Assembly Wheat Breeding Society of Australia. Eds. J G Paull, I S Dundas, K W Shepherd and G J Hollamby. pp 27–30. Adelaide, South Australia.

Chantachume Y, Smith D, Hollamby G J, Paull J G and Rathjen A J 1995 Screening for boron tolerance in wheat (*T. aestivum*) by solution culture in filter paper. Plant and Soil 177, 249–254.

Chao T T 1972 Selective dissolution of manganese oxides from soils and sediments with acidified NH_2OH-HCl. Soil Sci. Soc. Am. Proc. 36, 764–768.

Chapman H D and Vanselow A P 1955 Boron deficiency and excess. Calif. Citrograph 41, 31–34.

Chatterjee B W, Chatterjee m and Das N R 1980 Note on the differences in the response of wheat varieties to boron. Indian J. Agric. Sci. 50, 796.

Chauhan R P S and Powar S L 1978 Tolerance of wheat and pea to boron in irrigation water. Plant and Soil 50, 145–149.

Chesworth W 1991 Geochemistry of micronutrients. *In* Micronutrients in Agriculture. Eds. J J Mortvedt, F R Cox, L M Shuman and R M Welch. pp 1–30. Soil Sci. Soc. Am., Madison, WI, USA.

Chhipa B R and Lal P 1990 A comparative study on the effect of soil B on yield, yield attributes and nutrient uptake by susceptible and tolerant varieties of wheat. An. Edafol. Agrobiol. 48, 489–498.

Christensen J J 1934 Nonparasitic leaf spots of barley. Phytopath. 24, 726–742.

Cox R R and Kamprath E J 1977 Essential micronutrient soil tests. *In* Micronutrients in Agriculture. Eds. J J Mortvedt, P M Giordano and W L Lindsay. pp 289–317. Soil Science Soc. Amer., Madison, WI, USA.

DeEndredy A S 1963 Estimation of free iron oxides in soils and clays by a photolytic method. Clay Miner Bull. 9, 209–217.

Dhankhar D P and Dahiya S S 1980 The effect of different levels of boron and soil salinity on the yield of dry matter and its mineral composition in Ber (*Zizyphus rotundifola*). Int. Symp. on Salt Affected Soils. Karnal, India, pp 396–403.

Eary L E, Dhanpot R, Mattigod S V and Ainsworth C C 1990 Geochemical factors controlling the mobilisation of inorganic constituents from fossil fuel combustion residues. II Review of the minor elements. J. Environ. Qual. 19, 202–210.

Eaton F M 1935 Boron in soil and irrigation waters and its effect on plants. USDA Tech. Bull. 448.

Eaton F M 1944 Deficiency, toxicity and accumulation of boron in plants. J. Agric. Res. 69, 237–277.

Eaton F M and Blair G Y 1935 Accumulation of boron by reciprocally grafted plants. Plant Physiol. 10, 411–424.

El-Motaium R, Hu H and Brown P H 1994 the relative tolerance of six *Prunus* rootstocks to boron and salinity. J. Am. Soc. Hort. Sci. 119, 1169–1175.

Elseewi A A, Page A L and Grimm S R 1980a Chemical characterisation of fly ash aqueous systems. J. Environ. Qual. 9, 424–428.

Elseewi A A, Straughan I R and Page A L 1980b Sequential cropping of fly ash-amended soils: Effect on soil chemical properties and yield, and elemental composition of plants. Sci. Total Environ. 15, 247–259.

Erd R C 1980 The minerals of boron. *In* Mellor's Comprehensive Treatise on Inorganic and Theoretical Chemistry. Suppl. to Vol. V. Ed. R Thompson. pp 7–71. Longman, New York.

Fleming G A 1980 Essential micronutrients: boron and molybdenum. *In* Applied Soil Trace Elements. Ed. B E Davies. pp 155–176. John Wiley, New York.

Francois L E and Clark R A 1979 Boron tolerance of twenty-five ornamental shrub species. J. Amer. Soc. Hort. Sci. 104, 319–322.

Gill K S, Lubbers E L, Gill B S, Raupp W J and Cox T S 1991 A genetic linkage map of *Triticum tauschii* (DD) and its relationship to the D genome of bread wheat (AABBDD). Genome 34, 362–374.

Goldberg S 1993 Chemistry and mineralogy of boron in soils. *In* Boron and Its Role in Crop Production. Ed. U C Gupta. pp 344. CRC Press, Boca Raton, FL, USA.

Graham R D, Welch R M, Grunes D L, Cary E E and Norvell W A 1986 Effect of zinc deficiency on the accumulation of boron and other mineral nutrients in barley. Soil Sci. Soc. Am. J. 51, 652–657.

Gupta U C 1977 Effects of boron and limestone on cereal yields and on B and N concentrations of plant tissue. Plant and Soil 47, 283–287.

Gupta U C 1993 Deficiency, sufficiency, and toxicity levels of boron in crops. *In* Boron and Its Role in Crop Production. Ed. U C Gupta. pp 137–145. CRC Press, Boca Raton, FL, USA.

Gupta U C, MacLeod J A and Sterling J D E 1976 Effects of boron and nitrogen on grain yield and boron and nitrogen concentrations of barley and wheat. Soil Sci. Soc. Am. J. 40, 723–726.

Gupta U C, Jame Y W, Campbell C A, Leyshon A J and Nicholaichuk W 1985 Boron toxicity and deficiency: a review. Can. J. Soil Sci. 65, 381–409.

Haas A R C 1929 Toxic effect of boron on fruit trees. Bot. Gaz. 88, 113–131.

Handreck K A 1990 Methods of assessing boron availability in potting media with special reference to toxicity. Commun. Soil Sci. Plant Anal. 21, 2265–2280.

Hoffman G J 1990 Leaching fraction and root zone salinity control. *In* Agricultural Salinity Assessment and Management. Ed. K K Tanji. pp 238–261. Amer. Soc. of Civil Eng., New York.

Hope A B and Walker N A 1975 The Physiology of Giant Algal Cells. pp 77. Cambridge University Press, UK

Huang C and Graham R D 1990 Resistance of wheat genotypes to boron toxicity is expressed at the cellular level. Plant and Soil 126, 295–300.

Iyenger S S, Martens D C and Miller W P 1981 Distribution and plant availability of soil zinc fractions. Soil Sci. Soc. Am. J. 45, 735–739.

Jame Y W, Nicholaichuk W, Leyshon A J and Cambell C A 1982 Boron concentration in the soil solution under irrigation: a theoretical analysis. Can. J. Soil. Sci. 62, 461–470.

James W D, Graham C C, Glascock M D and Hanna A G 1982 Water-leachable B from coal ashes. Environ. Sci. Tech. 16, 195–199.

Jamjod S 1996 Genetics of boron tolerance in durum wheat. Ph.D. Thesis, The University of Adelaide, South Australia.

Jamjod S, Paull J G, Brooks B J and Rathjen A J 1997 Genetic variation in the tolerance of durum wheat (*Triticum turgidum* L. var. *durum*) to high concentrations of boron. *In* Proceedings of Boron in Soils and Plants. Eds. R W Bell and B Rerkasem. Kluwer Academic Publishers, Dordrecht, The Netherlands (in press).

Jenkin M J 1993 The genetics of boron tolerance in barley. Ph D Thesis, The University of Adelaide, South Australia.

Jones L H P and Handreck K A 1969 Uptake of silica by *Trifolium incartanum* in relation to the concentration in the external solution and to transpiration. Plant and Soil 30, 71–80.

Keren R and Bingham F T 1985 Boron in water, soils, and plants. Adv Soil Sci. 1, 230–276.

Kluge R and Podlesak W 1985 Plant critical levels for the evaluation of boron toxicity in spring barley (*Hordeum vulgare* L.). Plant and Soil 83, 381–388.

Kubata J 1980 Regional distribution of trace element problems in North America. *In* Applied Soil Trace Elements. Ed. B. Davies. pp 443–466. John Wiley, London.

Kukier U and Sumner M E 1996 Boron availability to plants from coal combustion by-products. Water, Air, and Soil Pollution. 87, 93–110.

Kukier U, Sumner M E and Miller W P 1994 Boron released from fly ash and its uptake by corn. J. Environ. Qual. 23, 596–603.

Leyshon A J and Jame Y M 1993 Boron toxicity and irrigation management. *In* Boron and It's Role in Crop Production. Ed. U C Gupta. pp 207–226. CRC Press, Boca Raton, FL, USA.

Loomis W D and Durst R W 1992 Chemistry and biology of boron. BioFactors 3, 229–239.

Lovatt C J and Bates L M 1984 Early effects of excess boron on photosynthesis and growth of *Cucurbita pepo*. J. Exp. Bot. 35, 297–305.

Lukaszewski K M, Blevins D G and Randall D D 1992 Asparagine and boric acid cause allantoate accumulation in soybean leaves by inhibiting manganese-dependent allantoate amidohydrolase. Plant Physiol. 99, 1670–1676.

Maas E V 1987 Salt tolerance of plants. *In* Handbook of Plant Science in Agriculture, vol. 2. Ed. B R Christie. pp 57. CRC Press, Boca Raton, FL, USA.

Manyowa N M and Miller T E 1991 The genetics of tolerance to high mineral concentrations in the tribe Triticeae – a review and update. Euphytica 57, 175–185.

Masson L 1967 Alaunos problemas relacionados con la calinidad in los vales de la costa sur. Oficeing Nacional de Evaluacion di Recursos Nationals. Lima, Peru.

Materne A M 1979 Genetic variability in the response of field pea varieties to soil boron. Honours Thesis, The University of Adelaide.

Mattigod S V 1983 A method of estimating the standard free energy of information of borate minerals. Soil Sci. Soc. Amer. J. 47, 654–655.

Mehrotra O N, Srivastava R D L and Hishra P H 1980 Some observations on the relative tolerance of wheat genotypes to boron. Indian Agric. 24, 223–238.

Moody D B, Rathjen A J and Cartwright B 1993 Yield evaluation of a gene for boron tolerance using backcross-derived lines. *In* Genetic Aspects of Plant Mineral Nutrition. Eds. P J Randall, E Delhaize, R A Richards and R Munns. pp 363–366. Kluwer Academic Publishers, Dordrecht, The Netherlands.

Moody D B, Rathjen A J, Cartwright B, Paull J G and Lewis J 1988 Genetic diversity and geographical distribution of tolerance to high levels of soil boron. *In* Proc. 7th Int. Wheat Genet. Symp., Cambridge. Eds. T E Miller and R M D Koebner. pp 859–865.

Nable R O 1988 Resistance to boron toxicity amongst several barley and wheat cultivars: a preliminary examination of the resistance mechanism. Plant and Soil 112, 45–57.

Nable R O 1989 Effects of boron toxicity upon the mineral nutrient composition of barley and wheat cultivars. CSIRO Div. Soils Report No. 104.

Nable R O and Moody D B 1992 Effects of rainfall on the use of foliar analysis for diagnosing boron toxicity in field-grown wheat. Plant Soil 140, 311–314.

Nable R O, Cartwright B and Lance R C M 1990a Genotypic differences in boron accumulation in barley: relative susceptibilities to boron deficiency and toxicity. *In* Genetic Aspects of Plant Mineral Nutrition. Ed. N El Bassam. pp 243–251. Kluwer Academic Publishers, Dordrecht, The Netherlands.

Nable R O, Lance R C M and Cartwright B 1990b Uptake of boron and silicon by barley genotypes with differing susceptibilities to boron toxicity. Ann. Bot. 66, 83–90.

Nable R O, Paull G P and Cartwright B 1990c Problems associated with the use of foliar analysis for diagnosing boron toxicity in barley. Plant and Soil 128, 225–232.

Nicholaichuk W A, Leyshon J, Jame Y W and Campbell C A 1988 Boron and salinity survey of irrigation projects and the boron adsorption characteristics of some Saskatchewan soils. Can. J. Soil Sci. 68, 77–90.

Oertli J J 1994 Non-homogeneity of boron distribution in plants and consequences for foliar diagnosis. Comm. Soil Sci. Plant Anal. 25, 1133–1147.

Oertli J J and Kohl H C 1961 Some considerations about the tolerance of various plant species to excessive supplies of boron. Soil Sci. 92, 243–247.

Oertli J J and Roth J A 1969 Boron nutrition of sugar beet, cotton, and soybean. Agron. J. 61, 191–95.

Paliwal K V and Mehta K K 1973 Interactive effect of salinity, SAR and boron on the germination and growth of seedlings of some paddy (*Oryza sativa*) varieties. Plant and Soil 39, 603–609.

Parker P R and Gardner E H 1981 The determination of hot water soluble boron in some acid Oregon soils using a modified azomethine-H procedure. Commun. Soil Sci. Plant Anal. 12, 1311–1377.

Parker D R, Page A L and Thomas D N 1991 Salinity and boron tolerances of candidate plants for the removal of selenium from soils. J. Environ. Qual. 20, 157–164.

Paull J G 1990 Genetic studies on the tolerance of wheat to high concentrations of boron. Ph.D. Thesis, The University of Adelaide, South Australia.

Paull J G, Cartwright B and Rathjen A J 1988a Responses of wheat and barley genotypes to toxic concentrations of soil boron. Euphytica 39, 137–144.

Paull J G, Nable R O and Rathjen A J 1992a Physiological and genetic control of the tolerance of wheat to high concentrations of boron and implications for plant breeding. Plant and Soil 146, 251–260.

Paull J G, Rathjen A J and Cartwright B 1988b Genetic control of tolerance to high concentrations of soil boron in wheat. *In* Proc. 7th Int. Wheat Genet. Symp., Cambridge. Eds. T E Miller and R M D Koebner pp 871–877.

Paull J G, Rathjen A J and Cartwright B 1991a Major gene control of tolerance of bread wheat (*Triticum aestivum* L.) to high concentrations of soil boron. Euphytica 55, 217–288.

Paull J G, Rathjen A J and Cartwright B 1991b Tolerance to high concentrations of boron for the amphiploid of *Triticum aestivum*×*Agropyron elongatum*. Plant and Soil 133, 297–299.

Paull J G, Rathjen A J, Cartwright B and Nable R O 1990 Selection parameters for assessing the tolerance of wheat to high concentrations of boron. *In* Genetic Aspects of Plant Mineral Nutrition. Ed. N El Bassam. pp 361–369. Kluwer Academic Publishers, Dordrecht, The Netherlands.

Paull J G, Rathjen A J, Langridge P and McIntosh R A 1995 Location of genes controlling boron tolerance of wheat. *In* Proc. 8th Int. Wheat Genet. Symp, Beijing. Eds. Z S Li and Z Y Xin pp 1065–1069.

Paull J G, Nable R O, Lake A W H, Materne M A and Rathjen A J 1992b Response of annual medics (*Medicago* spp.) and field peas (*Pisum sativum*) to high concentrations of boron: genetic variation and the mechanism of tolerance. Aust. J. Agric. Res. 43, 203–213.

Perkins P V 1996 A theoretical approach to the assessment of boron phytotoxicity resulting from the amelioration of minespoil with pulverised fuel ash. Land Degradation and Development. 7, 239–256.

Peryea F J, Bingham F T and Rhoades J D 1985 Mechanisms for boron regeneration. Soil Sci. Soc. Am. J. 49, 840–843.

Picchioni G A and Miyamoto S 1991 Growth and boron uptake of five pecan cultivar seedlings. HortScience 26, 386–388.

Pichtel J R, Dick W A and Sutton P 1994 Comparison of amendments and management practices for long-term reclamation of abandoned mine lands. J. Environ. Qual. 23, 766–777.

Piha M J; Vallack H W, Michael N and Reeler B M 1995 A low impact approach to vegetation establishment on mine and coal ash wastes in semi-arid regions. II. Lagooned pulverized fuel ash in Zimbabwe. J. Applied Ecology 32, 382–390.

Ponnamperuma F N, Lantin R S and Cayton M T C 1979 Boron toxicity in rice soils. Int. Rice Res. Newsletter 4, 8.

Prather R J 1977 Sulphuric acid as an amendment for reclaiming soils high in boron. Soil Sci. Soc. Am. Proc. 41, 1098–1101.

Purvis E R and Hanna J 1938 Boron studies: I The susceptibility of various plants to boron toxicity as influenced by soil types. Soil Sci. Soc. Am. Proc. 3, 205–209.

Rathjen A J and Pederson D G 1986 Selecting for improved grain yields in variable environments. *In* Plant Breeding Symposium DSIR. Eds. T A Williams and G S Wratt. pp 104–115. Agronomy Society of New Zealand, Special Publication No. 5.

Rathjen A J, Cartwright B, Paull J G, Moody D B and Lewis J 1987 Breeding for tolerance of mineral toxicities in Australian cereals with special reference to boron. *In* Priorities in Soil/Plant Relations Research for Plant Production. Eds. P G E Searle and B G Davey. pp 111–130. School of Crop Sciences, The University of Sydney.

Raven J A 1980 Short- and long-distance transport of boric acid in plants. New Phytol. 84, 231–249.

Raven J A 1983 The transport and function of silicon in plants. Biol. Rev. 58, 179–207.

Ravikovitch S, Margolin m and Navroth J 1961 Microelements in soils of Israel. Soil Sci. 92, 85–89.

Rhoades J D, Ingvalson R D and Hatcher J T 1970 Laboratory determinations of leachable soil boron. Soil Sci. Soc. Amer. Proc. 34, 871–875.

Riley M M, Robson A D, Dellar G A and Gartrell J W 1994 Critical toxic concentrations of boron are variable in barley. J Plant Nutr. 17, 1701–1719.

Roundy B A 1985 Germination and seedling growth of tall wheatgrass and basin wildrye in relation to boron. J. Range Management 38, 270–272.

Rozema J, De Bruin J and Broekman R A 1992 Effect of boron on growth and mineral economy of some halophytes and nonhalophytes. New Phytol. 121, 249–256.

Ryan J, Mirjamoto S and Stroehlein J L 1977 Relation of solute and sorbed B to the B hazard in irrigation water. Plant and Soil. 47, 253–256.

Schuman G E 1969 Boron tolerance of tall wheatgrass. Agron. J. 61, 445–447.

Schwertmann U 1964 The differentiation of iron oxides in soil by extraction with NH4-oxalate solution. Z. Pflanzenernahr Bodenkd. 105, 194–202.

Severson R C and Gough Z P 1983 Boron in mine soils and rehabilitation plant species at selected cool mines in western U.S. J. Environ. Qual. 12, 142–146.

Severson R C and Tidball R R 1979 Spatial variation in total element concentration in soil within the northern great plains cool region. U.S. Geological Survey Paper 1134-A. Washington, DC, USA.

Shani Y and Hanks R J 1993 Model of integrated effects of boron, inert salt, and water flow on crop yield. Agron. J. 85, 713–717.

Shennan C, Grattan S R, May D M, Hillhouse C J, Schachtman D P, Wander M, Roberts B, Tafoya S, Burau R G, McNeish C and Zelinski L 1995 Feasibility of cyclic reuse of saline drainage in a tomato cotton rotation. J. Environ. Qual. 24, 476–486.

Shorrocks V M 1964 Boron toxicity in *Hevea brasiliensis*. Nature 204, 599–600.

Shorrocks V M 1995 Micronutrient News and Information, 15(1), 3–4.

Smidt R E and Whitton J S 1975 Note on boron toxicity in a stand of radiata pine in Hawkes Bay, New Zealand. J. Science 18, 104–113.

Snape J W and Law C N 1980 The detection of homologous chromosome in wheat using backcross reciprocal monosomic lines. Heredity 45, 187–200.

Spouncer L R, Nable R O and Cartwright B 1992 A procedure for the determination of soluble boron in soils ranging widely in Boron concentrations, sodicity, and pH. Commun. Soil Sci. Plant Anal. 23, 441–453.

Su C, Evans L J, Bates T E and Spiers G A 1994 Extractable soil boron and alfalfa uptake: calcium carbonate effects on acid soil. Soil Sci. Soc. Am. J. 58, 1445–1450.

Swietlik D 1995 Interaction between zinc deficiency and boron toxicity on growth and mineral nutrition of sour orange seedlings. J. Plant Nutrition 18, 1191–1207.

Takkar P N 1982 Micronutrients: forms, content, distribution in profile, indices of availability and soil test methods. *In* 12th Int. Soil Sci. Cong., Part 1. New Delhi. 136p.

Toledo J and Spurr J 1984 plant growth and boron uptake by *Lycopersicon esculentum* and*L. cheesmanii* f. minor. Turrialba 34, 111–115.

United States Salinity Lab. Staff 1954 Diagnosis and improvement of saline and alkali soils. USDA Handbook. No. 60.

Vengosh A, Heumann K G, Juraski S and Kasher R 1994 Boron isotope application for tracing sources of contamination in groundwater. Environ. Sci. Tech. 28, 1968–1974.

Walker C D and Lance R C M 1991 Silicon accumulation and ^{13}C composition as indices of water-use efficiency in barley cultivars. Aust. J. Plant Physiol. 18, 427–434.

Walter A and Gutknecht J 1986 Permeability of small non-electrolytes through lipid bilayer membranes. J. Membrane Biol. 90, 207–217.

Watson M C, Bañuelos G S, O'Leary J W and Riley J J 1994 Trace element composition of *Atriplex* grown with saline drainage water. Agri. Ecosys. Environ. 48, 157–167.

Wong J W C, Jiang R F and Su D C 1996 Boron availability in ash-sludge mixture and its uptake by corn seedling (*Zea mays* L.). Soil Sci. 161, 182–187.

Yau S K, Nachit, M M, Ryan J and Hamblin J 1995 Phenotypic variation in boron-toxicity tolerance at seedling stage in durum wheat (*Triticum durum*). Euphytica 83: 185–191.

Yokota H and Konishi S 1990 Effect of the formation of a sugar-borate complex on the growth inhibition of pollen tubes of *Camellia sinensis* and cultured cells of *Nicotiana tabacum* by toxic levels of borate. Soil Sci. Plant Nutr. 36, 275–281.

Zyrin N G and Zborishchuk J N 1975 Boron in the ploughed layer of soils of the European part of the USSR. Pochvovedenise 5, 44.

Plant and Soil **193:** 199–208, 1997.

Chapter 13

Boron in human and animal nutrition

Forrest H. Nielsen
*USDA, ARS, Grand Forks Human Nutrition Research Centre, P.O. Box 9034, University Station, Grand Forks, ND 58202–9034, USA**

Abstract

This review describes the findings from human and animal studies indicating that B is a dynamic trace element which, in physiological amounts, can affect the metabolism or utilisation of numerous other substances involved in life processes including macrominerals, energy substrates such as triglycerides and glucose, nitrogen containing substances such as amino acids and proteins, reactive oxygen species, and estrogen. Through these effects, B can affect the function or composition of several body systems, including the brain, skeleton and immune system, generally in a beneficial fashion. Moreover, homeostatic mechanisms apparently exist for B because it is rapidly excreted in the urine, does not accumulate in tissues, and is maintained in a relatively narrow range of concentrations in blood of healthy individuals. Thus, even though B has not been conclusively established as essential because a biochemical function for it has not been identified, its beneficial actions suggest that an intake of over 1 mg day^{-1} (but probably not more than 13 mg day^{-1}) is desirable; diets low in fruits, vegetables, legumes and nuts may not provide this amount of B. Boron may be of more practical nutritional importance than currently acknowledged.

Brief introduction and early history of boron in food and nutrition

Since 1857, B has been known to be present in plants (Ploquin, 1967). Thus, B is a constant constituent of foods. The recognition that there are benefits and detriments to having B present in foods apparently began in the 1870s. At that time it was discovered that borax (sodium borate) and boric acid could be used to preserve foods. For about the next 50 years, borate addition was considered one of the best methods of preserving or extending the palatability of foods such as fish, shellfish, meat, sausages, bacon, ham, cream, butter and margarine. According to a brief historical review (Gordon, 1987) of B as a food preservative, an English Royal Commission appointed in 1899 to investigate preservatives and colourings in foods recognised in 1901 that borates were used to preserve all foodstuffs except milk. Also in England, an act in 1907 named borax and boric acid as the only permitted preservatives in butter and margarine. Boron had a vital role as a preservative in preventing food crises during World War I. In other words, for the last 30 years of the past century and the first part of this century, B was considered a beneficial element only, and medical opinion was that B was rather innocuous because no corpses resulted from the use of B as a preservative.

As early as 1902, however, German and American scientists began to question the orthodox view that large amounts of borates in foods were innocuous. Foremost among the works that changed perceptions about B was a report by Wiley (1904) in which it was stated that consumption of boric acid in doses greater than 0.5 g day^{-1} for 50 days resulted in disturbances in appetite, digestion and health in human volunteers. Wiley (1904) concluded that 0.5 g day^{-1} of boric acid was too much for a normal man to receive regularly, and 4.0 g day^{-1} of boric acid was the limit beyond which a normal man cannot go without harm. Subsequent to his report, the opinion that B posed a risk to health gained momentum; by the mid-1920s, many countries of the world began legislating against the addition of borates to food. Only during World War II were the restrictions involving B in foods eased; food shortages were making food preservation a major concern in many countries (Gordon, 1987). After the war, restrictions were gradually reimposed; by the middle 1950s, B as a food preservative was essentially for-

* E-mail: fnielsen@gfhnrc.ars.usda.gov

bidden throughout the world. This was also the time when B began to receive attention from toxicologists. Because of some accidental poisonings and inappropriate uses in the medical profession, boric acid and borates became important in human health only from the toxicological point of view between 1950 and 1980.

About 15 years after Warington (1923) and Sommer and Lipman (1926) showed that B was an essential nutrient for plants, attempts were made to show that B is essential for higher animals. However, these attempts were apparently unsuccessful (Hove et al., 1939; Orent-Keiles, 1941; Teresi et al., 1944). It was reported in 1945 that high dietary B, 100 to 1000 $\mu g\ g^{-1}$ diet, enhanced survival, and increased body fat and liver glycogen in potassium deficient rats (Skinner and McHargue, 1945). However, attempts to confirm these findings using a different diet with an unknown B content and with different amounts of B supplementation were unsuccessful (Follis, 1947). Thus, the inability to produce a B deficiency in animals in these early studies apparently resulted in generations of students of biochemistry and nutrition being taught that B was a unique element in that it was essential for plants but not for higher animals including humans.

In the early 1980s, the dogma about B in animal and human nutrition began to change; this is when studies of the nutritional importance of B in my laboratory began. These were stimulated by a rather serendipitous finding. In an arsenic study, chicks grew about 50% slower than expected and exhibited leg abnormalities. Examination of the diet which had been recently reformulated revealed that the routine additions of B, fluoride and nickel to the diet had been eliminated and a new source of cholecalciferol (vitamin D_3) was used. Thus, a preliminary study with chicks was performed to ascertain whether supplemental B, fluoride or nickel would improve the performance of the chicks fed the basal diet used in the arsenic studies. As the result of the surprising finding, that B seemed to stimulate growth of these chicks and partially prevented the occurrence of leg abnormalities, further studies of the importance of B in nutrition were, and continue to be, performed.

Further study of the diet used in the first B experiment revealed that it contained inadequate cholecalciferol because the indicated potency of the supplement used apparently was not correct. This led to the study of the possible interaction between B and cholecalciferol; in 1981, the first results of this study appeared (Hunt and Nielsen, 1981). This report showed that B deprivation depressed the growth of chicks with the effect seemingly more marked when dietary cholecalciferol was deficient. Morphological examination of the tibias of the chicks indicated that an interaction between B and cholecalciferol affected bone formation. When dietary cholecalciferol was low, rachitic long bones were found in 17 of 21 B deprived chicks, but only 9 of 22 B supplemented chicks exhibited rachitic long bones; moreover the lack of calcification generally was more severe in the B deprived chicks. In the 15 years since this report, circumstantial evidence has been accumulating which strongly suggests that B is an essential nutrient for higher animals and humans.

Nutritional importance of boron for animals

Findings from numerous experiments indicate that chicks and rats fed low dietary B ($<0.3\ \mu g\ g^{-1}$) exhibit altered bone development, brain function, macromineral metabolism, energy substrate utilisation, immune function and insulin secretion. In these experiments, the responses to low dietary B were most marked when the experimental animals were exposed to a stressor that adversely altered hormonal or macromineral metabolism, possibly at the cell membrane level, such as calcium, cholecalciferol or magnesium deprivation. Some of these findings are described here.

Bone development

Boron deprivation was found to exacerbate gross bone abnormalities in chicks caused by a diet deficient, but not completely lacking, in cholecalciferol (Hunt and Nielsen, 1981; Hunt et al., 1994). At the microscopic level, B deprivation exacerbated the distortion of marrow sprouts caused by cholecalciferol deprivation and delayed the initiation of cartilage calcification (Hunt, 1989). Boron deprivation also decreased chondrocyte density in the zone of proliferation of the growth plate in cholecalciferol deficient chicks (Hunt et al., 1994). It also has been found that *in ovo* injections of B reduced the abnormal height of the growth plate of chicks hatched from cholecalciferol deficient eggs (King et al., 1991). Based on these findings, Hunt et al. (1994) have suggested that B enhances the maturation of the growth plate in the long bones.

Brain function

Boron deprivation was found to systematically influence brain electrical activity assessed by an electrocorticogram in mature rats; the principal effect was on the frequency distribution of electrical activity (Penland and Eberhardt, 1993). This finding suggests that B may play an important role in the maintenance of brain activation. Brain mineral composition also is affected by B deprivation. Calcium concentrations in total brain and in brain cortex, as well as the phosphorus concentration in the cerebellum, were found to be higher in B deprived than in B supplemented rats fed a cholecalciferol deficient diet (Hegsted et al., 1991). Penland (1990) found that B deprivation increased the copper concentration in brain.

Macromineral metabolism

In the cholecalciferol deficient rat, B deprivation was found to decrease the apparent absorption and balance of calcium, magnesium, and phosphorus (Hegsted et al., 1991). Boron deprivation also has been found to alter the plasma concentrations of substances involved in macromineral metabolism. Boron deprivation exacerbated the abnormally elevated plasma concentration of total alkaline phosphatase activity in chicks caused by cholecalciferol deficiency (Hunt and Nielsen, 1981; Hunt et al., 1994). Also in cholecalciferol deficient chicks fed low dietary B (about 0.3 $\mu g\ g^{-1}$), B supplementation (about 1.4 $\mu g\ g^{-1}$) markedly increased plasma 25-hydroxycholecalciferol (Bai and Hunt, 1995b) and 1,25-dihydroxycholecalciferol (Bakken and Hunt, 1995) concentrations. In the cholecalciferol adequate chick, low dietary B increased plasma 1,25-dihydroxycholecalciferol concentrations (Bakken and Hunt, 1995). Regardless of cholecalciferol status, Hunt et al. (1994) found that B supplementation of B deficient chicks significantly increased femur calcium, phosphorus and magnesium concentrations.

Energy substrate utilisation

Hunt (1997) reported that B supplementation (about 1.3 mg kg^{-1} diet) of chicks fed a low B diet (about 0.2 mg kg^{-1}) increased the concentration of 2-phosphoglycerate and decreased the concentration of dihydroxyacetone in liver. These findings suggest that B intake has an influence on the hepatic glycolytic pathway and thus affects energy substrate metabolism. Further evidence for B affecting energy metabolism is the findings that B deprivation exacerbates the elevation in plasma glucose caused by cholecalciferol deficiency in chicks (Hunt and Nielsen, 1987), and decreases plasma triglycerides in both chicks (Bai and Hunt, 1995b; Hunt and Herbel, 1993) and rats (Aasen and Hunt, 1993; Herbel and Hunt, 1992).

Immune function

Evidence is emerging which indicates that B nutriture can affect the immune response or the inflammatory process. Bai and Hunt (1996b) reported that B deprivation depresses the antibody response in rats immunised with bacterial antigens. Moreover, Bai and Hunt (1995a) found that luxuriant amounts of dietary B (20 $\mu g\ g^{-1}$ diet) delayed the onset and severity of adjuvant induced arthritis in rats. Boron as borax also has been reported to have an anti-arthritic effect on formaldehyde induced arthritis in rats (Shah and Vohora, 1990).

Insulin secretion

Boron deprivation has been found to increase plasma insulin concentrations in the cholecalciferol deficient rat (Hunt and Herbel, 1991–1992). Additionally, it has been found that peak insulin release was markedly higher from isolated, perfused pancreata from chicks fed a low B diet (about 0.3 mg kg^{-1}) compared to chicks fed about 1.4 mg B kg^{-1} diet (Bakken and Hunt, 1995).

Nutritional importance of boron for humans

Findings indicating that B is of nutritional importance to humans have come mainly from two studies in which men over the age of 45, postmenopausal women, and postmenopausal women on estrogen therapy were fed a low B diet (about 0.25 mg/2000 kcal) for 63 days and then fed the same diet supplemented with 3.0 mg B day^{-1} for 49 days (Nielsen, 1994). The major differences between the two experiments were the intakes of copper and magnesium; in one experiment they were marginal (1.6 mg Cu/2000 kcal) or inadequate (115 mg Mg/2000 kcal), in the other, they were adequate (2.4 mg Cu and 300 mg Mg/2000 kcal). Findings from these experiments, some described in the following, showed that B affects the metabolism of macrominerals, energy, nitrogen, and reactive oxygen in humans; they also showed that B affects brain function, psychomotor performance and the response to estrogen ingestion.

Table 1. Examples of findings indicating that boron affects macromineral, nitrogen, reactive oxygen species and estrogen metabolism in humans[1]

	Dietary B[2] (mg day^{-1})	Serum calcitonin[3,4] (pg mL^{-1})	Blood urea nitrogen[5] (ng mL^{-1})	Erythrocyte superoxide dismutase[6] (μg haemoglobin)	Serum 17β-estradiol (pg mL^{-1})
Men over the age of 45	0.25	71	14.6	3091	20
	3.25	60	12.9	3231	17
	p value	0.16	0.01	0.71	0.12
Postmenopausal women	0.25	78	13.8	2666	11
	3.25	52	12.6	3169	11
	p value	0.02	0.08	0.04	0.86
Postmenopausal women on estrogen therapy	0.25	61	13.4	2520	99
	3.25	55	11.5	3327	157
	p value	0.02	0.03	0.03	0.02
Above groups combined plus one premenopausal woman	0.25	74	13.8	2735	48
	3.25	59	12.2	3243	69
	p value	0.0008	0.0001	0.04	0.06

[1]Four to five individuals in each group.
[2]After an equilibration period of 14 days when dietary B was about 3.25 mg day^{-1}, there was a depletion period of 63 days when dietary B was about 0.25 mg 2000 kcal^{-1} followed by a repletion period of 49 days when the basal diet was supplemented with 3 mg B day^{-1} as sodium borate.
[3]From experiment where magnesium was inadequate (115 mg 2000 kcal^{-1}) and copper was marginal (1.6 mg 2000 kcal^{-1}) throughout the study (Nielsen et al., 1990).
[4]In experiment where dietary magnesium was adequate (about 300 mg 2000 kcal^{-1}) and copper was luxuriant (about 2.4 mg 2000 kcal^{-1}) calcitonin values were close to 37 pg mL^{-1}, an expected normal-type value, and were not significantly affected by dietary B.
[5]From experiment where magnesium was adequate (about 300 mg 2000 kcal^{-1}) and copper was luxuriant (about 2.4 mg 2000 kcal^{-1}) (Nielsen, 1994; Nielsen et al., 1991, 1992).
[6]Does not include premenopausal woman.

Macromineral metabolism

Evidence that B affects macromineral metabolism includes changes in hormones that are involved in this metabolism. In both experiments, the serum 25-hydroxycholecalciferol concentration was lower during B depletion than B repletion (Nielsen et al., 1990, 1992). In the experiment where dietary copper was marginal and magnesium was inadequate, calcitonin values were much higher than in the experiment where these two elements were adequate. This finding suggests that, because the calcitonin values obtained with adequate copper and magnesium were close to those reported by others as being normal, the combined magnesium-low, copper-marginal diet caused elevated serum calcitonin indicative of an abnormal calcium metabolism. As shown in Table 1, B depletion exacerbated this abnormality (Nielsen et al., 1990). Similar findings were obtained with serum osteocalcin. Further evidence that B affects macromineral metabolism is that plasma ionised calcium and serum magnesium concentrations were lower during B depletion than repletion in the experiment where dietary copper and magnesium were marginal or inadequate.

Energy substrate metabolism

Similar to the findings with animals, B deprivation affects circulating glucose and triglyceride concentrations in humans. In the experiment where dietary copper and magnesium were low, serum glucose concentrations were significantly higher during B depletion than B repletion (Nielsen, 1989). Unfortunately, serum glucose was not determined in the other experiment. However, serum triglycerides were determined and were significantly lower during B depletion than repletion (Nielsen, 1992).

Nitrogen containing biosubstances metabolism

In both experiments with volunteers, blood urea nitrogen (BUN), serum creatinine and urinary urea were significantly higher during B depletion than B repletion (Nielsen, 1989; Nielsen et al., 1991). In the experiment where dietary copper and magnesium were adequate, urinary hydroxyproline excretion was significantly lower during B depletion than B repletion (Nielsen, 1994). The changes in these nitrogen containing metabolites suggest an alteration in amino acid or protein metabolism. In other words, the utilisation of some amino acids or proteins are affected by B such that the incorporation of amino acids into, or the breakdown of proteins, is changed and results in altered concentrations in nitrogen metabolites in blood and urine.

Reactive oxygen species metabolism

Both superoxide dismutase and ceruloplasmin are enzymes involved in the protection against damage caused by reactive oxygen species. In both experiments, erythrocyte superoxide dismutase was significantly lower (Table 2) during B depletion than B repletion (Nielsen, 1989, 1994). The ceruloplasmin findings apparently were modified by dietary copper and magnesium; when these were low, enzymatic ceruloplasmin was significantly lower during B depletion than B repletion. When dietary copper and magnesium were adequate, dietary B did not affect enzymatic ceruloplasmin, but immunoreactive ceruloplasmin was significantly lower during B depletion than repletion. Most likely, B does not directly participate in the conversion of reactive oxygen species into harmless metabolites, but instead affects their formation during normal metabolism of energy (discussed above), which apparently affects the need for enzymes involved in reactive oxygen species metabolism.

Response to estrogen ingestion

In both experiments, estrogen ingestion elevated both serum 17β-estradiol (Table 1) and plasma copper; these elevations were higher during B repletion than B depletion (Nielsen et al., 1992). Dietary B did not affect plasma copper nor serum 17β-estradiol in men or postmenopausal women not on estrogen therapy. These findings indicate that B can enhance the effects of estrogen ingestion.

Brain function and cognitive and psychomotor performance

Penland (1994) has reported that a low B intake results in electroencephalogram (EEG) changes suggestive of reduced behavioural activation (e.g. drowsiness) and mental alertness. In addition, Penland (1994) found that the EEG changes seemed to be in concert with the finding that a low B intake results in poorer performance in tasks that involve psychomotor skills and the cognitive processes of attention and memory. Penland (1994) has stated that the changes induced by low dietary B are similar, but not as severe, as those found with malnutrition or some metal toxicities.

Physiological (biochemical) function

About the only item lacking for the unequivocal acceptance of B as an essential nutrient for higher animals and humans is a defined biochemical function. This lack should not be surprising because a biochemical function for B in plants is only now emerging, which is over 70 years after it was found essential for plants to complete their life cycle (Sommer and Lipman, 1926; Warington, 1923). A consensus seems to be forming that the specific biochemical role of B in plants is structural and/or functional in the cell wall and probably in the cell membrane. This role supports a hypothesis advanced for the biochemical function of B in higher animals. This hypothesis is that B has a role in cell membrane function or stability such that it influences the response to hormone action, transmembrane signalling, or transmembrane movement of regulatory cations or anions (Nielsen, 1991). This hypothesis is supported by the finding that B influences the transport of extracellular calcium and the release of intracellular

Table 2. Changes in rat[1] platelet ionised calcium, $[Ca^{2+}]_i$, in response to thrombin suggesting that boron can influence cell membrane function or transmembrane signalling

Dietary treatment B ($\mu g\ g^{-1}$)	K (%)	$\Delta\ [Ca^{2+}]_i$ without external Ca^{2+}	$\Delta [Ca^{2+}]_i$ in presence of 1.0 mM Ca^{2+}
0	0.36	74	627
3	0.36	64	538
0	1.00	71	648
3	1.00	54	582
		P Values	
B		0.04	0.05
K		0.31	0.41
B×K		0.59	0.76
Root MSE		16	127

[1]Data from groups of 12 rats fed their respective diets for 12 weeks. After platelets were obtained they were loaded with the fluorescent marker FURA-2 so that cellular $[Ca^{2+}]_i$ could be measured in the resting stage and after activation with 0.25 units mL^{-1} thrombin.

calcium in rat platelets activated by thrombin (Table 2), and that B influences redox actions involved in cellular membrane transport in plants (Blevins and Lukaszewski, 1994). Another hypothesis which accommodates a large and varied response to B deprivation and the known biochemistry of B is that B acts as a metabolic regulator through complexing with a variety of substrate or reactant compounds in which there are hydroxyl groups in favourable positions (Hunt, 1994). Because this complexing usually results in a competitive inhibition of two classes of enzymes *in vitro*, the regulation by B is hypothesised to be mainly negative.

Metabolism of boron

One of the criteria often stated for an element to be considered essential is that homeostatic control mechanisms must exist for it. Evidence that B is homeostatically controlled includes the rapid urinary excretion of absorbed B, the lack of accumulation of B in tissues, and the relatively narrow range of B concentrations in blood of apparently healthy individuals.

Because there is no useable radioisotope of B, the study of its metabolism has been made difficult. However, sodium borate, boric acid and most likely food B are rapidly absorbed, and excreted largely in the urine. Most ingested B probably is converted into $B(OH)_3$, the normal hydrolysis end product of most B compounds and the dominant inorganic species at the pH of the gastrointestinal tract. Thus, B probably is absorbed, transported throughout the body and excreted mainly as undisassociated $B(OH)_3$. During transport in the body, the $B(OH)_3$ most likely is weakly attached to biosubstances containing cis-hydroxyl groups.

An inductively coupled plasma mass spectrometry method using the ratio of the two staple isotopes, $^{11}B/^{10}B$, has been developed to study B metabolism (Vanderpool et al., 1994). This method was used to show that B in broccoli, intrinsically enriched with ^{10}B, was absorbed as well as extrinsic ^{10}B in boric acid from a test meal in rats. When 20 μg of ^{10}B isotope were fed to rats, 95% of this isotope was detected in the urine and 4% in the faeces after 3 days. This agrees with other urinary findings indicating that >90% of ingested B is usually absorbed (Jansen et al., 1984). The high urinary excretion indicates that this is the major homeostatic mechanism for controlling body content of B. Moreover, the mechanism probably is something more than just the movement of B down a concentration gradient because the urinary concentration of B apparently can be markedly different from the blood concentration. Hunt et al. (1997) found that in 11 postmenopausal women, an increase in dietary B from 0.36 (probably deficient) to 3.3 mg day^{-1} (luxuriant) increased the mean fasting plasma B concentration only from 64 to 95 ng mL^{-1} whereas the mean daily excretion of B increased from 0.37 to 2.87 mg. Nielsen (1996) found that a B supplement of 2.5 mg day^{-1} given to 43 perimenopausal women consuming a mean of about 1.2 mg B day^{-1} (based upon urinary excretion) increased the mean fasting plasma B concentration from 34 to 53 ng mL^{-1} while mean urinary B excretion increased from 1.19 to 3.29 mg day^{-1}. Urine volume was not apparently affected by the B supplementation in either study.

Boron is distributed throughout soft tissue and fluids of animals and humans at concentrations mostly between 0.015 and 0.6 $\mu g\ g^{-1}$ fresh tissue (Bai and Hunt, 1996a; Ward, 1993; Shuler et al., 1990). Bone, fingernails, hair and teeth usually contain several times these concentrations. Spleen also apparently contains relatively high amounts of B (Bai and Hunt, 1996a). Because the concentration of B in bone increases with increased B intakes, and the increased concentrations are maintained for a period of time after B intake is decreased, bone may be a storage site for B. As with other mineral elements, overcoming homeostatic mechanisms by high B intakes will elevate tissue B concentrations.

Boron toxicity

Boron has low toxicity when administered orally. Toxicity signs in animals generally occur only after dietary B exceeds 100 $\mu g\ g^{-1}$. Boron toxicity was a focus of a recent symposium (Health Effects of Boron, 1994). In this symposium, it was stated that boric acid has a low, acute oral toxicity of about 4000 mg kg^{-1} body wt in rats. In mice and rats, the threshold toxicity effect of B (about 4500 mg boric acid kg^{-1} diet for mice) was found to be testicular cell damage and atrophy in males. Studies of the effect of high B on development revealed that this was not affected in rabbits fed 125 mg boric acid kg^{-1} body wt, or in mice fed 450 mg boric acid kg^{-1} body wt.

In humans, the signs of acute toxicity include nausea, vomiting, diarrhoea, dermatitis and lethargy (Linden et al., 1986). The signs of chronic B toxicity have been described as including poor appetite, nausea, weight loss, and decreased sexual activity, seminal volume, and sperm count and motility (Hunt, 1993). Two infants who had their pacifiers dipped into a preparation of borax and honey for a period of several weeks exhibited scanty hair, patchy dry erythema, anaemia and seizures (Gordon et al., 1973). The seizures stopped and the other abnormalities were alleviated when the use of the borax and honey preparation was discontinued. It should be noted that high B intake induces riboflavinuria (Pinto et al., 1978); thus high B intakes could possibly exacerbate the consequences of low dietary riboflavin.

The safe daily intake of B is in the process of being determined through the International Programme on Chemical Safety of the World Health Organisation. However, another group analysed both human and animal data for the World Health Organisation (WHO/FAO/IAEA, 1996) and suggested that an acceptable safe B intake could well be 13 mg day^{-1}.

Dietary considerations of boron

In the human depletion-repletion experiments discussed above, the subjects responded to a B supplement after consuming a diet supplying only about 0.25 mg B 2000 $kcal^{-1}$ for 63 or more days. Thus, humans apparently have a dietary requirement higher than this. An analysis of both human and animal data resulted in the suggestion that an acceptable safe range of population mean intakes of B for adults could well be 1 to 13 mg day^{-1} (WHO/FAO/IAEA, 1996).

For most people the major source of B is food. Most of the reported values for the concentrations of B in foods reported before 1985 are of questionable validity because of inadequate analytical methods. Two recent reports (Anderson et al., 1994; Hunt et al., 1991) provide an adequate indication of the amounts of B in various foods (Table 3). The richest sources of B are fruits, vegetables, pulses, legumes and nuts. Dairy products, fish, meats and most grains are poor sources of B. Based on the recent analyses of foods and food products, estimations of daily intakes of various age/sex groups have been made. Rainey et al. (1996) estimated that the median, mean and 95th percentile daily intakes of B for 27 age/sex groups in the United States ranged from 0.29, 0.49 and 1.53 mg day^{-1}, respectively, for infants aged 0 to 5 months to 1.02, 1.25 and 3.15 mg day^{-1}, respectively, for males aged 60 to 65 years. The estimated median, mean and 95th percentile daily intakes of B were 0.75, 0.93 and 2.19 mg day^{-1}, respectively, for all groups, and 0.79, 0.98 and 2.33 mg day^{-1}, respectively for adults aged 17 and older. Using foods included in the United States Food and Drug Administration Total Diet Studies, Iyengar et al. (1990) determined the mean adult male daily intake of B to be 1.52 mg day^{-1}, whereas Anderson et al. (1994) determined the intake to be 1.21 mg day^{-1}. Based on the United Kingdom National Food Survey (UK Ministry of Agriculture, Fisheries and Food, 1991), the dietary intake of B in the United Kingdom ranges from 0.8 to 1.9 mg day^{-1}. It should be noted that the increased consumption of specific foods with high B content will increase B intake significantly; for example, one serving of wine or avocado provides 0.42 or

Table 3. Boron content of selected foods

Food	B fresh wt ($\mu g\ g^{-1}$) ICP[1]	B fresh wt ($\mu g\ g^{-1}$) PGAA[2]	Food	B fresh wt ($\mu g\ g^{-1}$) ICP[1]	B fresh wt ($\mu g\ g^{-1}$) PGAA[2]
Fruits, fruit juices			*Meats*		
Apple juice	1.88	2.38	Beef, ground	<0.015	<0.05
Apple sauce	2.83	–	Chicken	<0.015	0.34
Apple, red w/peel	–	2.73	Cod/haddock	–	0.24
Banana	–	1.04	Ham	–	0.20
Cherries	1.47	7.00	Lamb	–	0.14
Grapes, purple/green	–	4.60	Liver (beef)	–	<0.07
Grape juice	2.02	3.72	Pork roast	-	0.06
Orange (naval)	-	2.17	Turkey breast	-	0.09
Orange juice	0.41	0.92	Milk, eggs		
Peach, canned	1.87	2.06	Cheese, cream	<0.015	–
Pear, canned	1.22	1.59	Cheese, cottage	–	0.19
Pineapple juice	0.27	0.57	Milk, 2%	<0.015	0.23
Wine	-	3.52	Eggs	<0.015	0.12
Fruits, dried			*Cereal grain products*		
Prunes	–	21.5	Bread, white, enriched	0.20	0.48
Raisins	-	19.0	Corn flakes	0.31	0.92
Vegetables			Flour, wheat, white	0.28	–
Avocado	-	11.1	Oatmeal	–	0.10
Beans, green	0.46	1.56	Rice	<0.015	0.32
Broccoli, flowers	1.85	2.47	Spaghetti/ macaroni	<0.015	0.14
Carrots	0.75	2.59	*Miscellaneous*		
Corn	–	0.49	Catsup	0.85	1.39
Peas	–	1.28	Chocolate powder	–	4.25
Potato	0.17	1.25	Honey	–	6.07
Squash, winter	–	2.65	Sugar	<0.015	0.29
Sweet potato	–	1.08	Vegetable oil, corn	–	<0.04
Tomato	–	0.75			
Pulses, nuts					
Cow peas (blackeyed)	–	4.76			
Lima beans	–	3.43			
Red beans	–	3.14			
Peanuts	–	13.8			
Pecans	–	6.6			

[1]Inductively coupled plasma emission spectrometric analysis by Hunt et al. (1991).
[2]Neutron capture prompt x-ray activation analysis by Anderson et al. (1994).

1.11 mg, respectively (Anderson et al., 1994). Moreover, for the population obtaining their drinking water from the 10% of the public water systems in the United States which provide water containing 0.4 mg B L^{-1}, water used for drinking and cooking may be the major, or significant source of B.

Clinical considerations

Many people apparently consistently consume less than 1.0 mg B day^{-1}, the lower value given for the safe range of intakes above. For example, in a group of 43 perimenopausal women studied in the eastern North Dakota area of the United States, two women apparently consumed an average of less than 0.5 mg B day^{-1}, and 14 women consumed between 0.5 and 1.0 mg B day^{-1} over a 90 day period (Nielsen, 1996). Rainey et al. (1996) also reported that many people consistently consume less than 1.0 mg B day^{-1}. These findings suggest that B could be a practical nutritional or clinical concern. Based on the findings from animal and B deprivation studies, it may be of special concern to people exposed to certain nutritional stressors such as vitamin D, copper or magnesium deficiencies.

Although knowledge about B nutrition, biochemistry and metabolism is growing, more is needed before specific clinical disorders can be attributed to subnormal B nutrition. Reports such as those suggesting that low B status may enhance the susceptibility or exacerbate some forms of arthritis (Newnham, 1994; Travers et al., 1990) should be recognised as evidence that further study of the clinical importance of B is urgently needed. Because B clearly is a biologically dynamic element in higher animals and humans, its dietary lack may have a role in some disorders of unknown cause such as arthritis or osteoporosis.

References

Aasen G H and Hunt C D 1993 Biochemical responses to dietary boron over a 20-fold physiological range in the male rats. Proc. ND Acad. Sci. 47, 57.

Anderson D L, Cunningham W C and Lindstrom T R 1994 Concentrations and intakes of H, B, S, K, Na, Cl, and NaCl in foods. J. Food Comp. Anal. 7, 59–82.

Bai Y and Hunt C D 1995a Dietary boron alleviates adjuvant-induced arthritis (AIA) in rats. FASEB J. 9, A576.

Bai Y and Hunt C D 1995b Dietary boron improves indices of marginal vitamin D status but does not substitute for the vitamin. New Approaches, Endpoints, and Paradigms for RDAs of Mineral Elements Abstracts, p. 29. University of North Dakota and USDA, ARS, Grand Forks Human Nutrition Research Center, Grand Forks, ND.

Bai Y and Hunt C D 1996a Absorption and distribution of boron in rats following a single oral administration of boron. Proc. ND Acad. Sci. 50, 53.

Bai Y and Hunt C D 1996b Dietary boron (B) increases serum antibody concentrations in rats immunised with heat-killed mycobacterium tuberculosis (MT). FASEB J. 10, A819.

Bakken N A and Hunt C D 1995 Dietary boron affects plasma 1, 25-dihdroxyvitamin D (1, 25 Vit D) concentrations and peak pancreatic insulin secretion in the chick. New Approaches, Endpoints, and Paradigms for RDAs of Mineral Elements Abstracts, p. 30. University of North Dakota and USDA, ARS, Grand Forks Human Nutrition Research Center, Grand Forks, ND.

Blevins D G and Lukaszewski K M 1994 Proposed physiological functions of boron in plants pertinent to animal and human metabolism. Environ. Health Perspect. 102 (Suppl 7) , 31–33.

Follis R H Jr 1947 The effect of adding boron to a potassium-deficient diet in the rat. Am. J. Physiol. 150, 520–522.

Gordon A S, Prichard J S and Freedman M H 1973 Seizure disorders and anaemia associated with chronic borax intoxication. Can. Med. Assoc J. 108, 719–721.

Gordon V (Ed) 1987 The case of the toxic life-preserver. Borax Review No. 2, 10–12.

Health Effects of Boron 1994 Environ. Health Perspect. 102 (Suppl 7), 150.

Hegsted M, Keenan M J, Siver F and Wozniak P 1991 Effect of boron on vitamin D deficient rats. Biol. Trace Elem. Res. 26, 243–255.

Herbel J L and Hunt C D 1992 Dietary boron modifies the effects of thiamine nutriture in the male rat. Proc. ND Acad. Sci. 46, 71.

Hove E, Elvehjem C A and Hart E B 1939 Boron in animal nutrition. Am. J. Physiol. 127, 689–701.

Hunt C D 1989 Dietary boron modified the effects of magnesium and molybdenum on mineral metabolism in the cholecalciferol-deficient chick. Biol. Trace Elem. Res. 22, 201–220.

Hunt C D 1993 Boron. *In* Encyclopedia of Food Science, Food Technology and Nutrition, vol 1. Eds. R Macrae, R K Robinson and M J Sadler. pp 440–447. Academic Press, London.

Hunt C D 1994 The biochemical effects of physiological amounts of dietary boron in animal nutrition models. Environ. Health Perspect. 102 (Suppl 7), 35–43.

Hunt C D 1997 Boron and vitamin D deficiencies affect hepatic glycolytic metabolite concentrations in the chick. *In* Ninth International Symposium on Trace Elements in Man and Animals, TEMA-9. Eds. P W F Fischer, M R L'Abbe, K A Cockell and R S Gibson. pp 599–601. NRC Research Press, Ottawa, Ontario, Canada.

Hunt C D and Herbel J F 1991-1992 Boron affects energy metabolism in the streptozotocin-injected, vitamin D_3-deprived rat. Magnesium Trace Elem. 10, 374–386.

Hunt C D and Herbel J L 1993 Physiological amounts of dietary boron improve growth and indicators of physiological status over a 20-fold range in the vitamin D_3-deficient chick. *In:* Trace Element Metabolism in Man and Animals, TEMA-8. Eds. M Anke, D Meissner and C F Mills. pp 714–718. Verlag Media Touristik, Gersdorf.

Hunt C D and Nielsen F H 1981 Interaction between boron and cholecalciferol in the chick. *In* Trace Element Metabolism in Man and Animals (TEMA-4). Eds. J McC Howell, J M Gawthorne and C L White. pp 597–600. Australian Academy of Science, Canberra, Australia.

Hunt C D and Nielsen F H 1987 Interactions among dietary boron, magnesium, and cholecalciferol in the chick. Proc. ND Acad. Sci. 41, 50.

Hunt C D, Herbel J L and Idso J P 1994 Dietary boron modifies the effects of vitamin D_3 nutrition on indices of energy substrate utilisation and mineral metabolism in the chick. J. Bone Min. Res. 9, 171–182.

Hunt C D, Herbel J L and Nielsen F H 1997 Metabolic responses of postmenopausal women to supplemental dietary boron and aluminium during usual and low magnesium intake: Boron, calcium and magnesium absorption and retention and blood mineral concentrations. Am. J. Clin. Nutr. 65, 803–813.

Hunt C D, Shuler T R and Mullen L M 1991 Concentration on boron and other elements in human foods and personal-care products. J. Am. Diet Assoc. 91, 558–568.

Iyengar G V, Clarke W B and Downing R G 1990 Determination of boron and lithium in diverse biological matrices using neutron activation-mass spectrometry (NA-MS). Fres. J. Anal. Chem. 338, 562–566.

Jansen J A, Schou J S and Aggerbeck B 1984 Gastro-intestinal absorption and in vitro release of boric acid from water-emulsifying ointments. Food Chem. Toxicol. 22, 49–53.

King N, Odom T W, Sampson H W and Yersin A G 1991 The effect of *in vivo* boron supplementation on bone mineralisation of the vitamin D-deficient chicken embryo. Biol. Trace Elem. Res. 31, 223–233.

Linden C H, Hall A H, Kulig K W and Rumack B H 1986 Acute ingestions of boric acid. Clin. Toxicol. 24, 269–279.

Newnham R E 1994 Essentiality of boron for healthy bones and joints. Environ. Health Perspect. 102 (Suppl 7), 83–85.

Nielsen F H 1989 Dietary boron affects variables associated with copper metabolism in humans. *In* 6th International Trace Element Symposium, vol. 4. Eds. M Anke, W Baumann, H Braunlich, C Bruckner, B Groppel and M Grun. pp 1106–1111. Friedrich-Schiller-Universitat, Jena.

Nielsen F H 1991 Nutritional requirements for boron, silicon, vanadium, nickel and arsenic: current knowledge and speculation. FASEB J. 5, 2661–2667.

Nielsen F H 1994 Biochemical and physiologic consequences of boron deprivation in humans. Environ. Health Perspect. 102 (Suppl 7), 59–63.

Nielsen F H 1996 Dietary supplementation of physiological amounts of boron increases plasma and urinary boron of perimenopausal women. Proc. ND Acad. Sci. 50, 52.

Nielsen F H, Mullen L M and Gallagher S K 1990 Effect of boron depletion and repletion on blood indicators of calcium status in humans fed a magnesium-low diet. J. Trace Elem. Exp. Med. 3, 45–54.

Nielsen F H, Mullen L M and Nielsen E J 1991 Dietary boron affects blood cell counts and haemoglobin concentrations in humans. J. Trace Elem. Exp. Med. 4, 211–223.

Nielsen F H, Gallagher S K, Johnson L K and Nielsen E J 1992 Boron enhances and mimics some effects of estrogen therapy in postmenopausal women. J. Trace Elem. Exp. Med. 5, 237–246.

Orent-Keiles E 1941 The role of boron in the diet of the rat. Proc. Soc. Exp. Biol. Med. 44, 199–202.

Penland J G 1990 Dietary boron affects brain in mature Long-Evans rats. Proc. ND Acad. Sci. 44, 78.

Penland J G 1994 Dietary boron, brain function, and cognitive performance. Environ. Health Perspect. 102 (Suppl 7), 65–72.

Penland J G and Eberhardt M J 1993 Effects of dietary boron and magnesium on brain function of mature male and female Long-Evans rats. J. Trace Elem. Exp. Med. 6, 53–64.

Pinto J, Huang Y P, McConnell R J and Rivlin R S 1978 Increased urinary riboflavin excretion resulting from boric acid ingestion. J. Lab. Clin. Med. 92, 126–134.

Ploquin J 1967 Le bore dans l'alimentation. Bull. Soc. Sci. Hyg. Aliment. 55, 70–113.

Rainey C J, Christensen R E, Nyquist L A, Strong P L and Coughlin J R 1996 Boron daily intake from the American diet. FASEB J. 10, A785.

Shah S A and Vohora S B 1990 Boron enhances anti-arthritic effects of garlic oil. Fitoterapia 61, 121–126.

Shuler T R, Pootrakul P, Yamsukon P and Nielsen F H 1990 Effect of thalassemia/haemoglobin E disease on macro, trace, and ultratrace element concentrations in humans tissues. J. Trace Elem. Med. 3, 31–43.

Skinner J T and McHargue J S 1945 Response of rats to boron supplements when fed rations low in potassium. Am. J. Physiol. 143, 385–390.

Sommer A L and Lipman C B 1926 Evidence of the indispensable nature of zinc and boron for higher green plants. Plant Physiol. 1, 231–249.

Teresi J D, Hove E, Elvehjem C A and Hart E B 1944 Further study of boron in the nutrition of rats. Am. J. Physiol. 140, 513–518.

Travers R L, Rennie G C and Newnham R E 1990 Boron and arthritis: the results of a double-blind pilot study. J. Nutr. Med. 1, 127–132.

UK Ministry of Agriculture, Fisheries and Food 1991 Household food consumption and expenditure 1991. Annual Report of the National Food Survey Committee, HMSO, Table B1.

Vanderpool R A, Hoff D and Johnson P E 1994 Use of inductively coupled plasma-mass spectrometry in boron-10 stable isotope experiments with plants, rats and humans. Environ. Health Perspect. 102 (Suppl 7), 13–20.

Ward N I 1993 Boron levels in human tissues and fluids. *In* Trace Elements in Man and Animals, TEMA-8. Eds. M Anke, D Meissner and C F Mills. pp 724–728. Verlag Media Touristik, Gersdorf.

Warington K 1923 The effect of boric acid and borax on the broad bean and certain other plants. Ann. Bot. 37, 629–672.

WHO/FAO/IAEA: 1996 Trace Elements in Human Nutrition and Health. World Health Organisation, Geneva. pp. 175–179.

Wiley H W 1904 Influence of food preservatives and artificial colours on digestion and health. I. Boric acid and borax. US Department of Agriculture Bulletin No. 84, Pt, 1, Government Printing Office, Washington, DC.

Author index

Bañuelos, G.S., 181
Bell, R.W., 149
Brown, P.H., 15, 49, 85
Cakmak, I., 71
Dell, B., 103
Goldberg, S., 35
Hu, H., 49
Huang, L., 103
Jamjod, S., 169
Loneragan, J.F., vii
Matoh, T., 59
Nable, R.O., 181
Nielsen, F.H., 199
Paull, J.G., 181
Power, P.P., 1
Römheld, V., 71
Rerkasem, B., 169
Sah, R.N., 15
Shelp, B.J., 85
Shorrocks, V.M., 121
Woods, W.G., 1

SUBJECT INDEX

6-phosphogluconate, 8
6-phosphogluconate dehydrogenase, 72
6-phosphogluconic acid, 72

Absorption, 49, 204
Absorption active, 51
Absorption, multiphasic, 52
Abundance, 2
Acrisols, 134
Active transport, 52
Adenosine, 10
Adsorption, 8, 36, 38, 137
Adsorption capacity, 40
Adsorption isotherm, 45
Adsorption maxima, 36, 39, 40
Adsorption, rate, 39
Adsorption, temperature, 39
Africa, 133
Air temperature, 54
Alazarin Red S, 17
Alcohol dehydrogenase, 8
Alkali fusion, 23
Allantoate amidohydrolase, 9
Allophane, 134
Aluminium oxides, 37, 38, 41–44
Amberlite IRA-743, 24
Americas, 133
Andosols, 134
Animal nutrition, 199
Animal tissue, 2
Apiose, 4, 6, 50, 62, 65
Arenosols, 134
Argon ICP, 18
Arsenazo, 17
Arthritis, 201
Ascharite, 142
Ascorbic acid, 74–77, 80
Asia, 132
Assimilate, 108
Atomic and physical properties, 2
Atomic spectrometric methods, 18
Auxin, 7, 73, 78
Azomethin-H, 17
Azomethine-HR, 17

B efficiency, 169, 170
B exclusion mechanism, 55
B-mannitol complex, 50
Biochemical function, 203
Bone development, 200
Boranes, 1
Borate complexes, 3
Borate esters, 3, 4
Borate-diol complex, 41, 50, 62
Borates, 3, 5
Borax, 2, 141, 142
Borax-rich potash, 126
Boric acid, 2, 49, 141, 142
Boron deficient areas, 130, 135
Boron determination techniques, 16
Boron-polyol complexes, 94
Borosilicate, glassy frits, 141
Brain function, 200, 203
Breeding for B efficiency, 97
Breeding for toxicity tolerance, 193
Brown pigments, 74, 75
Buds, 92

Caffeic acid, 72, 73
Calcite, 41, 42
Calcium borate ion pair, 40
Calcium carbonate, 40
Calcium clay, 39
Callose, 110, 113, 115
Carbohydrate, 114
Carmine, 17
Cartilage calcification, 200
Catechin, 75
Cell division, 104
Cell enlargement, 106
Cell wall, 5, 51, 56, 59, 72, 153
Cell wall binding capacity, 56
Cell wall bound boron, 62
Cell wall elasticity, 104
Cell wall thickening, 65, 104
Cellular B functions, 71
Chalcone synthase, 113, 114
Chemistry of B, 1
Chlorogenic acid, 72, 73
Chromotropic acid, 17
Cinnamic acid 4-hydrolase, 113
Cis-diol complexes, 4, 50, 72
Clay, 36, 37, 39, 41–43, 54, 136, 162

Clay, edge sites, 40
Climate, 139
Colemanite, 2, 3, 141–143, 145
Colorimetric methods, 25
Colorimetric techniques, 17
Compartmentation, 53
Compound fertilisers, 143
Concentration of B in animal/human tissue, 205
Concentration of B in plant tissue, 151–165
Concentration of B in soil solution, 35
Correction of boron deficiency, 141
Coumarins, 72
Critical concentrations in plants, 150
Crop removal of boron, 138
Cross pollination, 114
Crystal violet, 17
Crystallinity, 38
Curcumin, 17

Datolite, 142
Desorption, 44, 53
Desorption isotherm, 45
Desorption period, 52
Diagnosis of B deficiency, 98, 149
Diagnosis of B toxicity, 186
Dibenzoylmethane-isobutyl methyl ketone, 17
Dietary intake, 205
Diffusible B, 37
Diffusion of B, 45
Diffusivity of B, 37
Direct current plasma, 18
Distance from sea, 138
Dry ashing, 22
Dulcitol, 90, 94

Electrothermal vaporisation (ET)-AAS method, 18
Elovich rate equation, 45
Endothecium, 110
Energy substrate metabolism, 201, 203
Environmental conditions, deficiency diagnosis, 157
Environmental factors, B absorption, 54
Environmental factors, deficiency prognosis, 159
Enzymes, 8
Estrogen response, 203
Ether-water partition coefficient, 50
Europe, 133
Exchange separation, 24
Excretion, 204
Exudation, 55

Ferralsols, 182
Fertilisation strategies, 98
Fertiliser, aerial application, 145
Fertiliser, application to seed, 145
Fertiliser, rate of application, 145
Fertiliser, soil application, 143
Fertiliser, timing soil application, 143
Ferulic acid, 72
Field response to boron fertiliser, 126, 151
First order rate equation, 45
Fixation of B, 37, 40
Flame atomic absorption spectrometry (AAS), 18
Flame atomic emission spectrometry (AES), 18
Flame atomic spectrometric methods, 18
Flowering response, 109
Fluorimetric method, 17, 25
Fly ash, 184
Foliar applied boron, 92
Food boron content, 206
Food preservative, 199
Formation and distribution of boron laden soils, 181
Free radical, 75
Freundlich adsorption isotherm equation, 41, 42
Freundlich parameters, 45
Frost damage, 140
Fructose, 9, 10, 50
Functional nutrient requirements, 150
Fused magnesium phosphate, 143

Genetic control of tolerance, 188
Genetic engineering, 97
Genetic variation, 169
Genotype differences, 55, 169
Germanic acid, 63
Germination, 116
Glauconite, 136
Global consumption, 135
Glow discharge plasma, 18
Glucose-6-phosphate dehydrogenase, 8, 72
Glutathione, 75
Glutathione reductase, 75, 80
Glycerol, 50
Glycolipid, 7, 8, 79
Glycoproteins, 79
Gneiss, 136
Granite, 135
Gypsum, 185

H^+ extrusion, 77, 78
Heterocyst, 76
High-performance liquid chromatography, 20
Hot water extraction, 21, 40, 160
Human nutrition, 201
Humic acid, 41
Humus, 40
Hydrides, 1

Hydroboracite, 142
Hydrogen peroxide, 75
Hydrous oxides of iron and aluminium, 137
Hydroxy aluminium, 42
Hydroxyferulic acid, 72
Hysteresis, 45

Illite, 39, 42, 136
Immune function, 201
Inactivation, 193
Indoleacetic acid, 106
Inductively coupled plasma-mass spectrometry, 53
Industrial uses, 1
Inefficiency, 170
Inflorescence development, 87
Inner-sphere adsorption mechanism, 44
Inner-sphere B surface complexes, 43
Insulin secretion, 201
Intraveinal exchange, 86
Ion chromatography, 20
Ion fluxes, 77, 79
Ionic strength, 43
Iron oxide, 37, 41, 44
Irrigation water, 182
Isotope analysis, 52, 87, 92, 93

Kaolinite, 39, 42, 43
Kernite, 2

Langmuir adsorption isotherm equation, 41–43, 45
Layer silicate clay minerals, 39
Leaching, 138, 184
Leaf growth, 107
Leaf senescence, 86
Leakage of K^+, 7, 78
Lewis acid, 35
Ligand exchange, 38, 41, 43, 45
Light intensity, B deficiency, 54, 72, 74, 78, 80, 140, 157
Lignification, 113
Lignin, 5, 72, 106
Lime, 36
Liming, 137, 185
Lithosols, 134
Luvisols, 134

Magnesium borate, 142
Magnesium hydroxide, 38
Male sterility, 109, 111, 112, 153, 177
Mannitol, 5, 8, 10, 50, 90, 91, 94, 96, 144
Mannose, 8
Matrix assisted laser desorption Fourier transform mass spectrometry, 94
Measurement of B, 17–28
Meiosis, 111
Membrane, 68, 79
Membrane conformation, 53
Membrane effects, 7
Membrane function, 115, 154
Membrane permeability, 49–51, 56, 191
Membrane-bound ATPases, 77
Metabolism of B, 204
Methods of diagnosis and prognosis, 151
Methylene blue, 17
Microwave induced plasma, 18
Mineral deposits, 2
Mitosis, 106
Modelling of boron adsorption, 41
  adsorption isotherm, equation, 41
  chemical model, 43
  constant capacitance model, 43, 44
  empirical models, 41, 44
  Keren model, 42
  Stern variable surface charge variable surface potential model, 44
  triple layer model, 43
Modelling of boron desorption, 45
Montmorillonite, 39, 42
Mucilage, 55

Neutron activation analysis, 20
Neutron capture radiography, 20, 27
Nicotinamide-adenine-dinucleotide, 50
Nitrogen fixation, 76
Nitrogen metabolism, 203
Nitrogenase, 76, 77
NMR spectroscopy, 4, 41
Nutrition, human and animal, 199

Organic matter, 40, 42, 45, 138
Organic soil, 41
Outer-sphere adsorption mechanism, 43
Outer-sphere surface complexation, 44
Outer-sphere surface complexes, 43
Oxides, 37

Pectic polysaccharides, 62
Pectin, 56, 128
Pentose phosphate pathway, 72
Permeability, cell membrane, 50
Permeability coefficient, 50
Peroxidase, 74
pH, 36, 42, 43, 54, 162
pH dependency, 51
pH of the soil, 54
Phenol, 7, 115
Phenol metabolism, 72, 75, 80

Phenolic alcohols, 72
Phenylalanine ammonia lyase, 73, 113
Phloem exudate, 86, 87
Phloem mobility, 90, 91, 93
Phloem sap, 88, 94
Phloem transport, 112
Phosphate transport, 7
Photon flux density, 107
Photooxidation, 80
Photosynthesis, 80, 107
Physical barriers, cell wall, 56
Piper-Steenbjerg curvature, 155
Pistil, 112
Pistil development, 114
Plant age, 156, 158
Plant analysis, 25, 187
Plant analysis for diagnosis, 153
Plant analysis for prognosis, 158
Plant families, 128
Plant part for sampling, 154
Plasma membrane, 78, 86
Plasma source MS, 19, 27
Plasma-source methods, 18
Plasma-source OES, 19, 25
Plasmalemma, 7
Podzols, 134
Polarographic methods, 17
Pollen, 78, 108, 140, 177
Pollen tube, 7
Pollen development, 109, 111
Pollen germination, 114, 138
Pollen sterility, 109, 113
Pollen tube, 59, 62, 67, 114, 115, 138, 177, 192
Pollen viability, 138
Pollen wall, 111, 113
Pollen-stigma interactions, 114
Polyol, 4, 5, 9, 90, 95, 98
Polyphenoloxidase, 73
Potentiometric methods, 17
Power function equation, 45
Pressure jump kinetic data, 44
Pressure jump relaxation, 38
Pressure jump relaxation procedure, 45
Prognosis, 150, 151
Prompt-γ activation analysis, 21, 27
Protein, 7, 8, 10
Pyrophyllite, 42

Quinalizarine, 17
Quinones, 72, 74–76

Rainfall, 139
Reactive oxygen, 203
Recommended intake, 205
Relative humidity, 54
Rendzinas, 182
Residual effects of soil applied boron, 143
Retranslocation, 89
Rhamnogalacturonan II, 6, 51, 59, 112
Rhizosphere, 55
Rhizosphere complexation, 192
Rhizosphere exudates, 55
Ribose, 4, 8, 10, 50
River water, B in, 2
Rocks, B in, 2
Root B-adsorption, 55
Root exudates, 55
Root growth, 104
Root temperature, 54
Rosocyanin, 17

Sample preparation, 21
Sampling techniques, 98
Sandstone, 136
Sea water, B in, 2
Secondary ion mass spectrometry (SIMS), 20
Seed and fruit development, 115
Seed B, 152
Seed viability, 116
Selection, deficiency, tolerance, 97
Serine protease, 8
Sexual reproduction, 108
SH-compounds, 75, 76
Silicate, 39
Silicate minerals, 38
Sink demand, 93
Sodium clay, 39
Soil ameloration, 184
Soil amendments, 185
Soil analysis and crop responses, 129
Soil analysis for prognosis, 25, 160
Soil B adsorption, 41–43
Soil B analysis, 22, 129, 186
Soil B availability, 36
Soil B maps, 129, 132
Soil groups, 130, 131, 134
Soil moisture, 37, 159
Soil, parent material, 135
Soil solution, 160
Soil texture, 36, 136
Soils high in B, 182
Solonchaks, 182
Solonetz, 182
Solubor, 141, 142, 144, 145
Solution ionic strength, 38, 41, 44
Solvent B extraction, 25

Sorbitol, 8, 10, 50, 90, 91, 93, 94, 97, 144, 191
Sorption capacity, 38
Sorption sites, 38
Sources of boron, 141
Speciation of boron in plants, 5
Specific adsorption, 38
Spectrophotometric techniques, 17
Stachyose, 97
Stem infusion technique, 88
Sterility, 140
Succinic dehydrogenase, 8
Superoxide radical, 75
Surface charge density, 44
Surface complexation models, 43, 44
Surface complexes, bidentate, 45
Surface functional group, 43, 44
Surface hydroxyl groups, 38
Surface mining, 183
Symptoms of B deficiency, 59, 85, 86, 126, 139, 152, 187
Symptoms of B toxicity, 86, 91, 187

Tactoids, 40
Tapetum, 110
Temperature effects, 37, 53, 139, 140, 158
Thermal ionisation mass spectrometry, 20
Thiols, 75
Tincal, 3
Tolerance to boron toxicity, 188
Tourmaline, 35, 182
Toxicity, animals/humans, 205
Toxicity, plants, 68, 129, 181
Toxixity, tolerance mechanisms, 188–193
Translocation of B, 86
Translocation of sugars, 9
Transpiration, 54
Transpiration rate, 54

Ulexite, 2, 142
Uptake, 49
Uptake, metabolic control, 55
Urease, 8

Vanillic acid, 72
Vapour pressure deficit, 112, 154
Vascular bundles, 114, 116
Vascular exudate, 94
Vascular tissue, 106, 112
Vegetation management and B toxicity, 185
Vitamin D, 200

Wall, 104
Water soluble boron, 60
Wet ashing, 23
Wetting and drying cycles, 39

Xylem sap, 86, 88
Xylem to phloem transfer, 88

COMMON NAME CITATION INDEX

Alfalfa, 35, 36, 163, 182, 183
Almond, 90–93, 97, 125, 138, 158, 159, 164, 173, 188
Alsike clover, 170
Annual medic, 188
Apartagus, 125
Apple, 6, 9, 86, 90, 91, 93, 94, 97, 122, 123, 125, 127, 128, 134, 138–140, 144, 145, 154, 170, 171, 177, 182
Apricot, 91, 95, 97, 122, 124, 126, 128
Ash, 95, 97, 140
Asian pear, 97
Asparagus, 97, 129
Aubergine, 124, 126
Avocado, 109, 111, 114–116, 124, 125, 127, 173, 177, 182

Bamboo, 6
Banana, 123, 125, 127
Barley, 50–52, 54–56, 109, 111, 122, 123, 125, 128, 145, 162, 171, 172, 177, 185, 187–193
Basin wild rye, 186
Bean, 75, 97, 159–161, 182
Bean, field, 121, 122, 124, 126, 145
Bean, French, 122, 124, 125, 128, 145
Beet, 62, 126, 133–135, 138, 139, 141, 144, 145, 165, 170, 171
Bermuda grass, 125
Birch, 122, 140
Black cherry, 97
Black gram, 80, 105, 107, 115, 116, 124, 140, 152, 155, 157, 159–161, 170–172, 176
Black mustard, 165
Blue-gum, 105
Blueberry, 126, 140
Brassica, 128
Bread wheat, 188–190
Broccoli, 85, 87–89, 98, 112, 113, 160, 161, 171, 174, 204
Brussel sprout, 63, 64, 160, 161

Cabbage, 9, 63, 64, 97, 122, 123, 125, 127, 128, 134
Cacao, 171, 176
Camellia, 115
Cape gooseberry, 124
Caraway, 95
Carnation, 122
Carrot, 51, 52, 63, 64, 66, 67, 72, 77, 78, 95, 97, 122, 125, 128, 134, 138, 139, 145, 165, 182
Cassava, 54, 123, 125, 157
Castor oil, 124
Casuarina, 123
Cattle saltbush, 183
Cauliflower, 85, 87, 97, 122, 123, 125, 127, 128, 145, 155, 160, 161
Celeriac, 122
Celery, 10, 50, 67, 87, 90–92, 94–97, 106, 122, 123, 125, 127, 128, 139, 165, 170, 171
Cherry, 91, 95, 97, 122, 124, 126, 128, 144
Chestnut, 123
Chick pea, 22, 123, 138
Chinese box orange, 190
Chinese cabbage, 123, 139, 145
Chinese elm, 87
Christmas berry, 97
Chrysanthemum, 122
Citrus, 122, 123, 125, 127, 171, 188
Clover, 61, 122, 124, 126, 138, 139
Cocoa, 22, 126, 127
Coconut, 123, 125, 127, 129, 138
Coffee, 95, 97, 123, 125, 127, 128, 135, 139
Cotton, 106, 114, 122, 123, 125, 127, 128, 133, 135, 138, 139, 144, 170, 171, 183
Cowpea, 63, 64
Crimson clover, 126
Cucumber, 63, 64, 122, 123, 125
Cultivated tomato, 190
Cyanobacteria, 7, 8, 71, 76, 77, 80, 107

Dandelion, 165
Date palm, 23
Dayflower, 77
Desert saltbush, 183
Diatoms, 8
Douglas fir, 6, 126
Duckweed, 52, 59, 63–65, 80, 140
Durum wheat, 188, 189

Elodea, 78
English hawthorn, 97
Eucalypt, 123, 125, 127, 128, 138–140

Faba bean, 7, 59, 60, 77, 106, 165
Fennel, 95, 97
Field pea, 174, 189
Fig, 93, 125, 144

Finger millet, 123
Flax, 122, 123, 128, 165
Flowering almond, 97
Fodder beet, 129
Fourwing saltbush, 183
Fragrant olive, 87, 91

Gardenia, 95
Garland chrysanthemum, 63, 64
Garlic, 123, 125, 134
Ginger, 124
Gourd, 63, 64
Grape, 87, 122, 124, 126–128, 133, 135, 138–140, 156, 170, 171
Green gram, 107, 109, 115, 124, 157, 170–172, 176
Grevillea, 123, 127
Guatyule, 127
Guava, 124

Hazelnut, 125, 173
Hop, 122, 123, 134

Imperial rye, 190
Indian mustard, 183, 185
Italian ryegrass, 63, 64

Japanese radish, 124
Jerusalem artichoke, 190

Kale, 123
Kenaf, 123
Kola, 127
komatsuna, 9

Lavender, 122
Leek, 63, 64, 122, 125, 165
Lemon, 190
Lentil, 123, 165, 174
Lettuce, 63, 64, 92, 122, 123, 125
Lily, 63, 64, 78, 115, 116
Litchi, 123
Longan, 123
Loquat, 97
Lucerne, 122, 123, 125, 127, 128, 134, 135, 137, 139, 141, 142, 145, 165, 170, 176
Lupin, 87, 88

Macadamia, 123, 159
Maize, 60, 63, 64, 77, 85, 109, 110, 112, 114–116, 121, 122, 124, 126–129, 139, 143, 152, 156, 158, 165, 171, 172, 177, 182
Mango, 27, 114, 116, 123, 127, 152, 154, 155, 173
Maple, 125
Medic, 188–191
Melon, 123, 125, 127
Milkvetch, 185
Mint, 123
Mulberry, 123
Mushroom, 75
Mustard, 123, 124

Nectarine, 93, 95, 97
Norway spruce, 122, 140

Oats, 121, 122, 182, 191
Oil palm, 123, 125, 127–129, 135, 138, 139
Oilseed rape, 108, 109, 111, 114–116, 121–123, 125, 128, 135, 138, 139, 150, 154–157, 161, 171, 174, 176
Okra, 79
Old man saltbush, 183
Olive, 91, 95, 122, 124, 125, 128, 138, 139, 144, 162
Onion, 97, 122, 123, 125, 129, 165
Orange, 128, 185

Papaya, 22, 138
Parsley, 63, 64, 95
Pea, 6–8, 62–64, 76, 97, 116, 122, 124, 125, 145, 165, 188–190
Peach, 10, 90, 91, 94, 95, 97, 109, 124, 126, 127, 140, 173, 188
Peanut, 87, 109, 115, 116, 123, 125, 127, 128, 135,152, 155, 158, 160, 161, 163, 170–172
Pear, 91, 93, 97, 122, 124, 126–128, 138, 140, 144, 170, 188
Pecan, 188
Petunia, 7, 114, 115
Pigeon pea, 123
Pine, 5, 62, 122, 124, 125, 127, 128, 133, 134, 138–140, 163, 164, 171, 186
Pineapple, 123, 125
Pinto bean, 6
Pistachio, 90, 91, 93, 144, 159
Plum, 90, 92, 95, 97, 122, 124, 126, 140, 144, 171
Pomegranate, 97, 124, 171
Poppy, 122, 124, 140, 165
Potato, 75, 122, 124, 126–128, 134, 145, 158, 160, 165, 176, 182
Prune, 91–93, 97, 140, 144, 173

Quince, 97
Quinoa, 129

Radish, 6, 9, 59, 62, 63, 66, 85, 87, 121, 122, 124, 126, 134, 145, 165
Raspberry, 124, 126, 134
Red beet, 51, 60, 122, 123, 125, 128, 145, 175

Red clover, 75, 142, 170
Rhubarb, 126
Rice, 6, 7, 63, 64, 109, 114, 122, 124, 125, 129, 172, 177, 188
Rutabaga, 85, 87, 89, 152, 154, 160, 161
Rye, 165, 173
Ryegrass, 170

Safflower, 123
Sainfoin, 122
Saltbush, 185
Sesame, 124, 126
Silver beet, 183
Sisal, 127
Snapdragon, 95
Sorghum, 124, 126
Sour cherry, 92
Soybean, 63, 64, 75–77, 87, 107, 115, 122, 123, 125, 145, 150, 152, 154, 155, 157, 165, 170–172, 187
Spinach, 63, 64, 80, 126, 129, 165
Spruce, 75
Squash, 53, 62, 74, 86, 104, 106, 187, 191
Strawberry, 122, 125, 128, 171, 188
Stylosanthes, 127
Subterranean clover, 87, 115, 124
Sugar beet, 6, 52, 122, 123, 125, 127, 128, 129, 139, 145, 152, 155, 160, 170, 182
Sugar cane, 124, 126
Sunflower, 7, 53, 62, 65, 66, 71–80, 104–107, 109, 122, 123, 125, 127, 128, 135, 137–140, 145, 152, 154, 157, 159, 160–162, 170, 171, 190, 191
Swede, 122, 123, 125, 128, 134, 139, 145
Sweet cherry, 93
Sweet pepper, 63, 64
Sweet potato, 123, 125
Swiss chard, 60, 61, 63
Sycamore, 6, 62

Table beet, 63, 64
Tall fescue, 183, 185
Tall wheatgrass, 190
Tea, 122, 123, 127, 133
Tobacco, 53, 60, 65, 66, 68, 72, 97, 98, 110, 113, 114, 122, 123, 125, 127, 128, 133, 165, 191, 192
Tomato, 6, 63–67, 79, 86, 104, 106, 122, 123, 125, 127, 139, 145, 156, 165, 171, 175, 183
Towel gourd, 63, 64
Tulip, 122, 124
Tung, 125
Turnip, 85, 87, 121–123, 127, 128, 134, 145, 165

Vegetables, 135

Walnut, 86, 90, 91, 93, 125, 128, 144, 154
Water melon, 123, 125
Wattle, 127
Wavy leaf saltbush, 183
Wax-leaf privet, 92
Welsh onion, 63, 64
Wheat, 55, 56, 73, 85, 105, 109, 111–116, 121, 122, 124, 126–129, 138–140, 145, 152–154, 156, 159, 160–162, 165, 170–173, 175–178, 183, 185, 187–191, 193
White cabbage, 165
White clover, 142, 170
White lupin, 174
White mustard, 165
Wild sea beet, 129

Yam bean, 124
Yeast, 8
Yucca, 89

Developments in Plant and Soil Sciences

50. P.J. Randall, E. Delhaze, R.A. Richards and R. Munns (eds.): *Genetic Aspects of Plant Mineral Nutrition.* 1993 ISBN 0-7923-2118-9
51. K.S. Kumarasinghe and D.L. Eskew (eds.): *Isotopic Studies of Azolla and Nitrogen Fertilization of Rice.* 1993 ISBN 0-7923-2274-6
52. F.A. Bliss and G. Hardarson (eds.): *Enhancement of Biological Nitrogen Fixation of Common Baen in Latin America.* 1993 ISBN 0-7923-2451-X
53. M.A.C. Fragoso and M.L. van Beusichem (eds.): *Optimization of Plant Nutrition.* 1993 ISBN 0-7923-2519-2
54. N.J. Barrow (ed.) *Plant Nutrition – From Genetic Engineering to Field Practice.* 1993 ISBN 0-7923-2540-0
55. A.D. Robson (ed.): *Zinc in Soils and Plants.* 1993 ISBN 0-7923-2631-8
56. A.D. Robson, L.K. Abbott and N. Malajczuk (eds.): *Management of Mycorrhizas in Agriculture, Horticulture and Forestry.* 1994 ISBN 0-7923-2700-4
57. P.H. Graham, M.J. Sadowsky and C.P. Vance (eds.): *Symbiotic Nitrogen Fixation.* 1994 ISBN 0-7923-2781-0
58. F. Baluška, M. Čiamporová, O. Gašparíkova and P.W. Barlow (eds.): *Structure and Function of Roots.* 1995 ISBN 0-7923-2832-9
59. J. Abadía (ed.): *Iron Nutrition in Soils and Plants.* 1995 ISBN 0-7923-2900-7
60. P.S. Curtis, E.G. O'Neill, J.A. Teeri, D.R. Zak and K.S. Pregitzer (eds.): *Belowground Responses to Rising Atmospheric CO_2.* Implications for Plants, Soil Biota, and Ecosystem Processes. 1994 ISBN 0-7923-2901-5
61. P.C. Struik, W.J. Vredenberg, J.A. Renkema and J.E. Parlevliet (eds.): *Plant Production on the Threshold of a New Century.* 1994 ISBN 0-7923-2903-1
62. L.O. Nilsson, R.F. Hüttl and U.T. Johansson (eds.): *Nutrient Uptake and Cycling in Forest Ecosystems* (CEC/IUFRO Symposium, Halmstad, Sweden, 1993). 1995 ISBN 0-7923-3030-7
63. H.P. Collins, G.P. Robertson and M.J. Klug (eds.): *The Significance and Regulation of Soil Biodiversity* (International Symposium, East Lansing, 1993). 1995 ISBN 0-7923-3138-9
64. R.A. Date, N.J. Grundon, G.E. Rayment and M.E. Probert (eds.): *Plant-Soil Interactions at Low pH: Principles and Management* (3rd International Symposium, Brisbane, 1993). 1995 ISBN 0-7923-3198-2
65. J.K. Ladha and M.B. Peoples (eds.): *Management of Biological Nitrogen Fixation for the Development of More Productive and Sustainable Agricultural Systems.* (15th Congress of Soil Science, Acapulco, Mexico, 1994) 1995 ISBN 0-7923-3413-2; Pb 0-7923-3414-0
66. C. Rodriguez-Barrueco (ed.): *Fertilizers and Environment.* 1996 ISBN 0-7923-3729-8
67. B. Buerkert, B.E. Allison and M. von Oppen (eds.): *Wind Erosion in Niger.* Implications and Control Measures in a Millet-based Farming System. 1996 ISBN 0-7923-3885-5
68. O. van Cleemput, G. Hofman and A. Vermoesen (eds.): *Progress in Nitrogen Cycling.* 1996 ISBN 0-7923-3962-2
69. N. Ahmad (ed.): *Nitrogen Economy in Tropical Soils.* 1996 ISBN 0-7923-4094-9

Developments in Plant and Soil Sciences

70. M. Rahman, A. Kumar Podder, C. van Hove, Z.N. Tahmida Begum, T. Heulin and A. Hartmann (eds.): *Biological Nitrogen Fixation Associated with Rice Production.* 1996 ISBN 0-7923-4197-X
71. D. Rosen, E. Tel-Or, Y. Hadar and Y. Chen (eds.): *Modern Agriculture and the Environment.* 1997 ISBN 0-7923-4295-X
72. G.H. Elkan and R.G. Upchurch (eds.): *Current Issues in Symbiotic Nitrogen Fixation.* 1997 ISBN 0-7923-4367-0
73. H.M. Anderson, P.W. Barlow, D.T. Clarkson, M.B. Jackson and P.R. Shewry (eds.): *Plant Roots – From Cells to Systems.* 1997 ISBN 0-7923-4369-7
74. Zhu Zhao-liang, Wen Qi-xiao and J.R. Freney (eds.): *Nitrogen in Soils of China.* 1997 ISBN 0-7923-4372-7
75. J.K. Ladha, F.J. de Bruijn and K.A. Malik (eds.): *Opportunities for Biological Nitrogen Fixation in Rice and Other Non-Legumes.* 1997 ISBN 0-7923-4514-2
76. R.W. Bell and B. Rerkasem (eds.): *Boron in Soils and Plants.* 1997 ISBN 0-7923-4705-6
77. B. Dell, P.H. Brown and R.W. Bell (eds.): *Boron in Soils and Plants: Reviews.* 1997 ISBN 0-7923-4729-3

Kluwer Academic Publishers – Dordrecht / Boston / London

MIX
Papier aus verantwortungsvollen Quellen
Paper from responsible sources
FSC® C105338

If you have any concerns about our products,
you can contact us on
ProductSafety@springernature.com

In case Publisher is established outside the EU,
the EU authorized representative is:
Springer Nature Customer Service Center GmbH
Europaplatz 3, 69115 Heidelberg, Germany

Printed by Libri Plureos GmbH
in Hamburg, Germany